The New Common Weeds *of the* United States

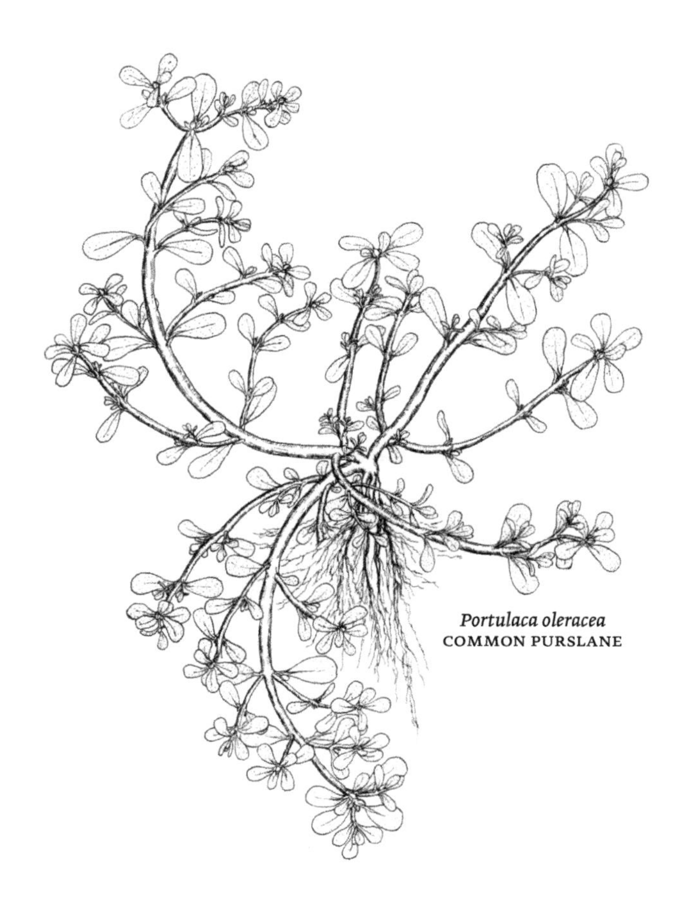

Portulaca oleracea
COMMON PURSLANE

STEVE W. CHADDE
Illustrations by REGINA OLSON HUGHES

Originally published as *Selected Weeds of the United States,* Agriculture Handbook No. 366, Agricultural Research Service, U.S. Department of Agriculture, Washington, D.C., March, 1970. Descriptions by Clyde F. Reed, research botanist and plant explorer, Crops Research Division, Agricultural Research Service; drawings by Regina O. Hughes, scientific illustrator, Crops Research Division, Agricultural Research Service. This new edition features an all-new format, updates taxonomy of all families and taxa treated, includes a biography of botanical illustrator Regina Hughes, and provides current (2018 data) county-level distribution maps for each species.

The New Common Weeds of the United States

Steve W. Chadde
Illustrations by Regina Olson Hughes

Printed in the United States of America

ISBN: 978-1-951682-14-9

Grateful acknowledgment is given to the Biota of North America Program (*www.bonap.org*) for permission to use their data to generate the distribution maps.

The author can be reached via email: *steve@chadde.net*

Ver. 1 (11/09/2019)

CONTENTS (Scientific Family Name)

CONTENTS (Common Family Name)

INTRODUCTION

PLANT SPECIES are considered weeds when they interfere with human activities or welfare. Such plants grow where they are not wanted. They reduce yield and quality of crop and forage species; they make roadsides, utility rights-of-way, and landscape plantings unsightly; they poison livestock and humans either directly or as allergens; they clog lakes, irrigation ditches, and drainageways; and they reduce the availability of fish and wildlife. Weeds thrive in close association with people and their domesticated plants and animals.

Losses caused by weeds exceed losses caused by any other class of agricultural pest. Losses in developed countries in the temperate zones, where efficient methods of weed control have been developed, range from 10 to 15 percent of the total value of agricultural and forest products. Losses from weeds are greater in tropical regions. Here, half or more of the total effort of farming may be devoted to combating invading vegetation.

The life duration of the different weeds is not uniform. Many weeds are annual or winter annual; some are biennial; and many are perennial. Also, any particular species may have both annual and winter annual forms, or annual and biennial, or biennial and perennial. We still have much to learn about life duration of weeds. Investigations of this type are complicated by the influences of climate, disease organisms, insects, and other factors.

Annuals complete their life cycles within one year. Annuals have no means of survival from one year to the next except by seeds, although some spread, as in purslane, by the rooting of prostrate stems or stem segments. Winter annuals germinate in the fall and pass through one winter as seedlings or rosettes; biennials germinate in the spring or early summer and survive the winter as rosettes. Both winter annuals and biennials complete their life cycles after producing seed in the second year. Control measures for these groups of weeds thus include the prevention of seed production by destroying the weeds before flowering and the encouragement of germination by cultural practices. This latter reduces the reservoir of seed in the soil.

Simple perennials persist for more than 2 years. Since these weeds have no means of spreading other than by seeds, they may be treated much like annuals or biennials, but deeper root cutting may be necessary to eliminate them.

More difficult weeds to control are the creeping or persistent perennials spreading by means of seed, underground lateral roots, rhizomes, aboveground stolons,

AUTHOR'S NOTE: This 2019 edition largely retains the Introduction as worded in the 1970 publication. However, the descriptions for each plant species have been updated to reflect taxonomic changes. Plants are presented in alphabetical order, first by plant family, then by genus. Distribution maps reflect verified presence or absence within counties of the contiguous United States as of 2018.

and stems that root upon contact with the soil (layering). It is apparent that in spite of the application of severe control measures, many of these weeds are able not only to survive but also to increase through lateral spread. Control depends on prevention of seed production and destruction of shoots and roots.

Since 1945, the use of herbicides in agriculture has expanded remarkably. Use of these chemicals has revolutionized weed control practices; in fact, no other innovation in the history of agriculture has been accepted so rapidly and widely as the use of chemical weed killers. Such herbicides have been devised for many control purposes and we expect that even more effective and more specific herbicides will be discovered in the future.

In spite of the successful application of herbicides to control weeds, other methods continue to be important. Very frequently, a combination of tillage and application of herbicides to weeds in a crop will be more effective than either method alone. Also, because some weeds respond only slightly, if at all, to herbicides, and some crops are more sensitive to herbicides than the weeds themselves, mechanical methods of control must be used. Herbicidal sprays are not persistently effective in reclaiming grassland or in removing weeds from lawns if soil fertility is too low or drainage too poor for vigorous growth of grasses and clover. Under such conditions, either fertilization or drainage, or both, are also required to enable the grasses and clover to compete more successfully with weed and shrub seedlings, even after herbicide has been applied. Also, grazing often has to be adjusted to prevent deterioration of the grassland.

Control methods are developing so rapidly that recommendations are soon out of date. Also, recommendations for weed control vary from one area to another. For these reasons, we omitted recommendations on control in this publication. These will be found in other publications that are more easily kept current.

Methods of weed control are determined by the specific weed species to be controlled. Therefore, the first step in deciding how to solve a weed problem is to correctly identify the weed species. It is for this purpose that this handbook has been prepared. This information will be useful to research workers, weed specialists, agricultural extension agents, and others interested in weeds and their control. This handbook will help to identify some of the more common and important weeds infesting lawns, gardens, crops, pastures, ponds, and other areas.

Handbooks such as this can only assist in identification; much effort has to be made by the person identifying the weeds. The great number of weed species, the variability within a single species, and the problems attendant on recognizing weeds in different stages of growth, all make identification difficult. In addition, this handbook does not describe all the weeds that may be found. Because few people will be able to identify all the weeds growing on their land, their only recourse is to send specimens of the unknown weeds to the county agent or to the weed specialist at the agricultural college at the state university.

However, too frequently people send specimens for identification without first taking proper care in collecting and packaging. The material arrives at its destination in such poor condition that it cannot be identified. The sender should collect the best possible material, preferably with flowers, seed-bearing parts, and part of the underground system. Lack of a complete plant, however, should not deter anyone from sending a specimen. Unless the material is to be mailed promptly, it should be dried between papers and cardboards placed under some heavy object. The packaging should be adequate to insure safe arrival. When several plants are submitted, each specimen should bear a numbered tag corresponding to duplicates kept by the sender. Notes on habit, history of infestation, and place and habitat of the collection are valuable for identification and sometimes give information useful in giving advice for control.

This handbook illustrates and describes many of the important weeds in the United States and it indicates their geographical distribution in this country. In 1848, the species of plants considered as weeds was about 10 percent of the species known in the United States; today, over 50 percent of our flora is made up of species that are considered undesirable by some segment of our society. Most of our weeds have been introduced, either accidentally or intentionally. A few of our native species have become more common weeds because of human activities and because of modern means of transportation.

The 224 species of weeds included in this handbook were orogonally selected by weed scientists largely from the composite list of 1,775 weeds published by the Weed Science Society of America in *Weeds* 14: 347–386, 1966. Those selected are some of the prevalent weeds in croplands, grazing lands, non-croplands, and aquatic sites.

Nomenclature of plant families, genera, and species generally follows that of the published volumes of *The Flora of North America* series (1993+), the *Synthesis of the North American Flora* (Kartesz 2014), and *The Plant List*, a collaboration between the Royal Botanic Gardens (Kew), and the Missouri Botanical Garden (online at *www.theplantlist.org*).

The organization of the plant families in this book is in alphabetical order. Within each family the genera and species are arranged alphabetically according to the scientific name. Occasionally synonyms are given, especially those appearing in various current flora and weed lists. No attempt is made to give complete scientific synonomy or a complete list of common names.

Each species of plant has a unique name made up of three parts. The first part is the genus to which a species belong; the second part is termed the specific epithet. The name is followed by the name of the person or persons who first named a species, and are often abbreviated. For example, "L." refers to Carolus Linnaeus, the 18th-century Swedish botanist considered the father of modern plant taxonomy. Together the two form the scientific, or botanical, name of the species.

DESCRIPTIONS AND MAPS

The descriptions in this publication have been prepared and compiled from regional, state, and local publications on the flora of the United States.

In the original 1970 edition of this work, the included maps gave only a general approximation of the area in which a plant may be considered a weed, because weeds come and go in a region and because detailed information often was not been available. In this new edition, detailed, county-level distribution maps were available through the efforts of the "Biota of North America Program" (BONAP), and were prepared by the author and incorporated into this publication (see *www.bonap.org*).

The descriptions are semi-technical, with the terminology simplified wherever possible. A glossary of technical terms will be found on page 453. Some of the terms are illustrated for clarity on pages 458–460. Where a plant characteristic is found in only one species (as in *Chara*, where many technical terms apply to algae only), that characteristic is identified in the description and not in the glossary. Most confusing is the use of the words "fruit" and "seed" by the layperson because their meaning differs from that used by botanists. For example, a corn "seed" or "grain" is really a "caryopsis." A caryopsis is a particular kind of fruit.

All measurements in this publication are in the metric system and pertain to the life size of plants.

The time of year indicates when the species is in flower, and unless otherwise indicated, is given as a fruiting range. In any particular region the dates of flowering may be earlier or later than that indicated.

DRAWINGS

The drawings of the species of weeds illustrated in this book have been prepared from herbarium specimens. Herbarium specimens for this work were loaned to us by the National Arboretum, the U.S. National Herbarium, and the Reed Herbarium.

A habit sketch and magnified areas of leaves, stems, flowers, fruits, and seeds are included for the species illustrated. Size of parts illustrated relative to life size of plants varies according to the part being shown; some are less than life size, others are magnified. Relative size is shown in the legend by means of a ruler. The structures of some species discussed are illustrated here for the first time in a manual of this type.

REGINA OLSON HUGHES

REGINA OLSON HUGHES (1895–1993) was a pre-eminent scientific illustrator, specializing in botanical illustrations, with a special interest in orchids. Born on February 1, 1895, in rural Herman, Nebraska, she contracted scarlet fever at age 10, and by age 14 was completely deaf. Hughes, however, soon became proficient in lip-reading, developed her speech skills, and continued to speak fluently throughout her life.

Hughes attended Gallaudet College (now Gallaudet University) in Washington, D.C., then the nation's only post-secondary school for the deaf. She received her bachelor's degree in art in 1918 and her master's degree in 1920. In 1923, Hughes married Frederick H. Hughes and the couple lived on the Gallaudet University campus for the next thirty years. Frederick H. Hughes was Gallaudet's long-time economics professor, theater enthusiast, and football coach. In 1956, he passed away following a heart attack.

After graduation, Hughes worked at the Veterans Administration for a short time before moving on to the State Department. In 1930, Hughes began working at the U.S. Department of Agriculture (USDA), first as a research clerk, then as a scientific illustrator and language translator (Hughes was fluent in four languages: French, Spanish, Portuguese, and Italian). In 1936, she became the scientific illustrator in botany for the Agricultural Research Service. In 1967, she was awarded an honorary Ph.D. from Gallaudet University. In 1969, she retired from the USDA, but continued working for the Smithsonian Institution.

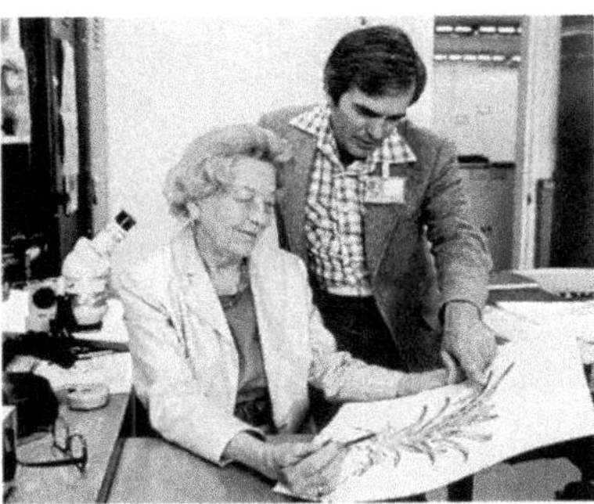

Regina Hughes discussing the finishing touches of an orchid painting with Dr. Robert R. Reed.

During her long career, Hughes worked closely with botanists and other scientists to illustrate a wide range of flowers, weeds, plants, and seeds collected from around the world. Her illustrations were done in a variety of media: watercolor, oils, gouache, and pencil. She was known for her precise yet beautiful depictions of orchids, weeds (as in this work), and exotic species. Her illustrations of small, intricate plant parts often required her to use a microscope. Today, her orchid paintings are permanently on display at the Smithsonian's National Museum of Natural History.

In recognition of her many contributions to botany, in 1979 a new species of Bromeliad, *Billbergia reginae,* was named after her by the Smithsonian. In 1981, a member of the Asteraceae was named *Hughesia reginae* in her honor. Hughes continued to work at the Smithsonian Institution until the day of her death in 1993.

Sagittaria latifolia Willd. COMMON ARROWHEAD

Perennial herb, aquatic or growing in marshes, erect, stoloniferous herb, with milky juice.

DESCRIPTION Scape erect, simple or branched, sheathed at base by the bases of the long petioles, 1-15 dm high, angled, with one or more of the lower whorls pistillate, or all unisexual flowers. **Leaf blades** ovate to linear, rarely without lobes, mostly sagittate, the basal lobes triangular-ovate to linear, from one-half as long as the body of the leaf to longer than the body of the leaf, 5-40 cm long, 2-25 cm broad. Bracts distinctly united or slightly joined, thin, dry, membranaceous, obtuse to acute, 1 cm or less long. **Flowers** in whorls of 2-15, the upper staminate or perfect on short pedicels, the lower usually pistillate on longer, terete pedicels. Sepals ovate, obtuse, 5-7 mm long. Petals showy, white, broadly ovate, 1-2 cm long. **Achene** inverted ovate, 2.3-4 mm long, 1.5-3 mm broad, with broad marginal wings on both margins, the broad-based beak usually subhorizontal to slightly incurved or erect, 1-2.5 mm long. Plants exceedingly variable, especially in leaf outline.

FLOWERING July-September.

WHERE FOUND In water and wet places; along lakes, ponds, rivers, bays, and swamps.

ORIGIN Native.

NOTE The genus includes a number of additional wetland species, separated on technical details, and many having distinct geographic distributions. *Sagittaria latifolia* is the most widely distributed of our species.

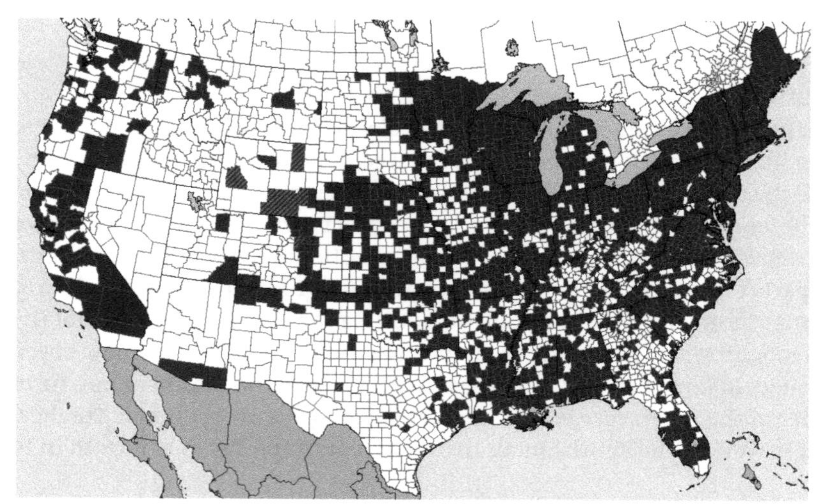

Distribution of **Sagittaria latifolia**, COMMON ARROWHEAD

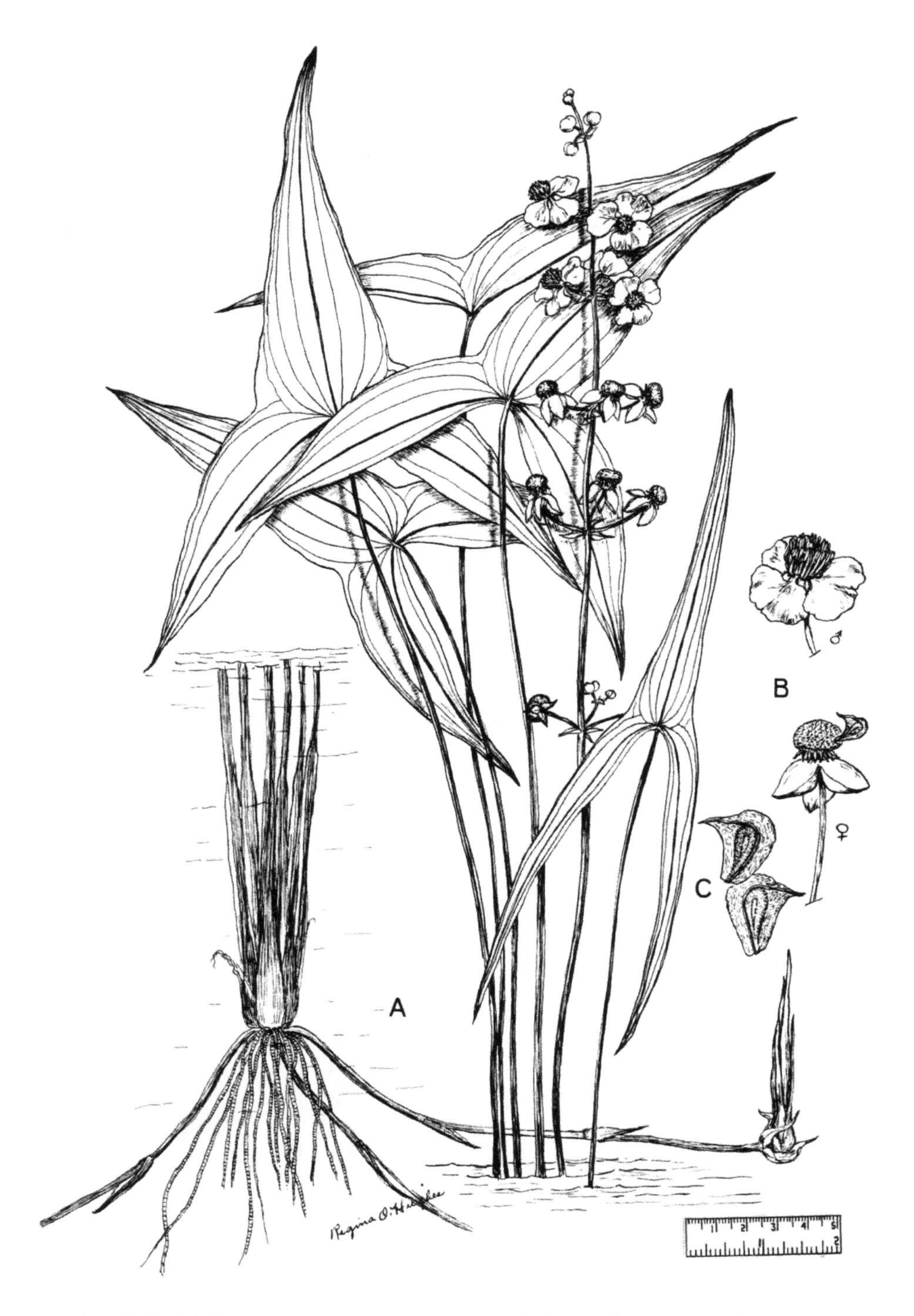

Sagittaria latifolia, COMMON ARROWHEAD. A, habit. B, flowers. C, achenes.

Allium vineale L. WILD GARLIC

Perennial bulbous herb, the outer layers of the bulb formed from the sheathing leaves of the foliage leaves.

DESCRIPTION Stem becoming stiff, erect, leafy to near the middle, 30-100 cm tall. **Leaves** 2-ranked, with sheathing bases, the leaf blades circular, hollow in cross section, striped, the younger ones easily flattened, slenderly tapering. **Spathe** usually 1, dry and thin, short, beaked, its edges united above. **Umbel** projecting through the base of the deciduous spathe, nearly head-shaped, 2-5 cm in diameter. **Two plant types**: the larger, scapigerous plants bearing scapes that produce aerial bulblets and sometimes flower; the flowers having greenish to purplish perianths with lanceolate to elliptic segments, the segments obtuse to acutish and about as long as the stamens; seeds black, flat on one side, about 3 mm long; the smaller, nonscapigerous plants bearing slender foliage and fewer leaves, and not producing a scape at the end of the growing season. **Four types of bulbs** found at the end of season in late spring: **aerial bulblets** (as many as 300 in one scape of the scapigerous plant); **hard-shell bulbs**, having a single bladeless storage leaf that contains a growing point at its base; formed underground, in the axils of the outer leaves; **central bulbs**, formed underground by nonscapigerous plants, conspicuous at the end of the season's growth, formed around the main axis of the plant, circular in cross section, varying from the size of an aerial bulblet to that of a soft offset bulb; **soft offset bulbs** formed underground in the axil of the innermost leaves of scapigerous plants, largest of the four types of bulbs, ovate in longitudinal section, with a convex abaxial face and a flat adaxial face that form two distinct ridges where the faces meet, these ridges clasping the sides of the flattened scape to which they are attached. **Seed** produced in the spring, germinate the following fall; usually spreading by bulbs rather than by seed; and most usually by the numerous aerial bulbs.

FLOWERING May-July.

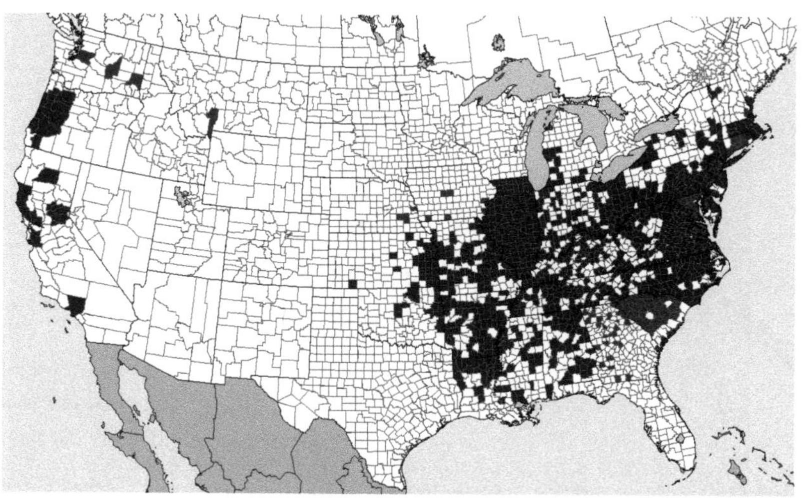

Distribution of **Allium vineale**, WILD GARLIC

WHERE FOUND A ubiquitous weed, drought-hardy, cold-hardy, tolerant to wet soils, liking heavy soils best; poorly drained land along rivers and streams, hillsides; a pest and troublesome weed in fields of small grain, pastures, hayfields used for milk cows (the allyl sulfide imparting the garlicky flavor to agricultural products, as milk, along with a disagreeable odor), gardens, roadsides, lawns, and noncrop areas. Difficult to eradicate.

ORIGIN Native of Eurasia.

Allium vineale, WILD GARLIC. A, habit. B, flower cluster. C, bulblets. D, flower. E, bulbs, hard-shell. F, bulbs, soft-shell.

Alternanthera philoxeroides (Mart.) Griseb. ALLIGATORWEED

Perennial aquatic herb, nearly glabrous, reproducing by seeds.

DESCRIPTION **Stems** or their branches prostrate or decumbent, freely creeping, the ascending portion often 1-5 dm long, the upper internodes often slightly pubescent. **Leaves** opposite, 3-11 cm long, the leaf blades somewhat fleshy, linear to linear-oblanceolate, 2-10 cm long, usually acute or small short abrupt tip, entire, narrowed to the sessile base. **Flower heads** in spikes, nearly globose or cylindric, long-peduncled, silvery-white, the peduncles 2-7 cm long, glabrous or pubescent in lines. **Flowers** perfect. Sepals 5, glabrous, 5-6 mm long, unequal, 4-nerved, lanceolate, acute, firm. Sterile stamens narrow, usually entire. **Utricle** flattened, indehiscent. **Seed** lens-shaped, smooth.

FLOWERING June-October.

WHERE FOUND Wet waste places, often choking watercourses.

ORIGIN Introduced from South America.

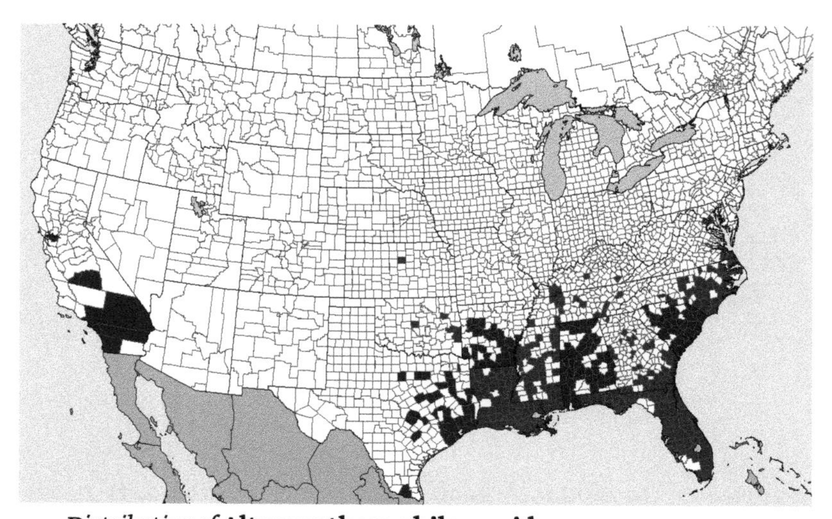

Distribution of **Alternanthera philoxeroides,** ALLIGATORWEED

Alternanthera philoxeroides, ALLIGATORWEED. A, habit. B, roots and young plant. C, part of aquatic growth, new shoot from rooting node. D, flower. E, persistent chaffy flower with the single mature achene. F, achenes. G, seeds.

Amaranthus albus L. TUMBLE PIGWEED, TUMBLEWEED

Annual herb, reproducing by seed.

DESCRIPTION Stem pale-green, slender, with whitish erect or ascending slender stems, diffusely branched, 0.1-1 m tall, plants breaking off at ground at maturity and rolling over open ground, often piled up along fence rows. **Leaves** egg-shaped to spatulate-oblong, blunt or notched at the rounded apex, with veiny blades, 1-7 cm long, short-petioled. **Flowers** monoecious in small axillary clusters, greenish. Bracts rigid, awl-shaped, pungent, about twice as long as the calyx. Sepals of pistillate flowers commonly 3, uneven, the longest about equaling the utricle, oblong. Stamens 3. **Utricle** lens-shaped, wrinkled when dry, 1.3-1.7 mm long, dehiscing by a transverse line at the middle. **Seed** to 1 mm broad, round, shiny, black, each contained in a bladder-like hull.

FLOWERING July-October.

WHERE FOUND Waste ground, fallow ground.

ORIGIN Originally probably native of only the Great Plains; now throughout United States.

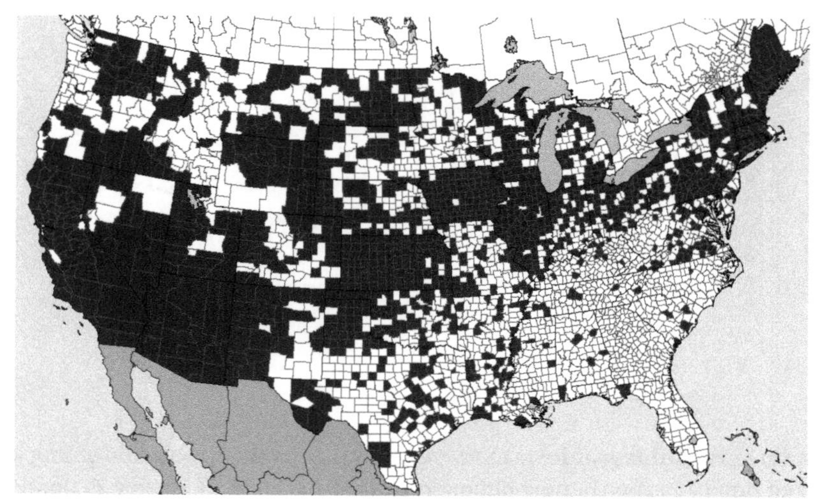

Distribution of **Amaranthus albus**, TUMBLE PIGWEED

Amaranthus albus, TUMBLE PIGWEED. A, habit. B, enlarged node showing flowers and leaf. C, flower spike. D, utricles. E, seeds.

Amaranthus blitoides S. Wats. PROSTRATE PIGWEED

Annual, reproducing by seeds.

DESCRIPTION Stems prostrate, smooth, much-branched, 2-6 dm long, reddish, spreading flat over the ground, mostly erect at the tips. **Leaves** numerous, small, slenderly tapering, simple, oblong to obovate, 1-4 cm long, blunt or rounded, slenderly tapering into a long petiole. **Flowers** inconspicuous in short dense axillary clusters, without petals. Bracts about equaling the sepals, tapering to a protracted point, nearly bristlelike. Sepals of the pistillate flowers normally 5 (occasionally 4), egg-shaped to oblong, unequal in length. **Utricle** thick, lens-shaped, 2-2.5 mm long, about equaling the longest sepal, smooth or nearly so, dehiscing by a transverse line at the middle. **Seed** flattened, shiny black, nearly circular, 1.4-1.7 mm wide.

FLOWERING July-October.

WHERE FOUND Fields, wastes, gardens, and vacant lots.

ORIGIN Native of Western United States.

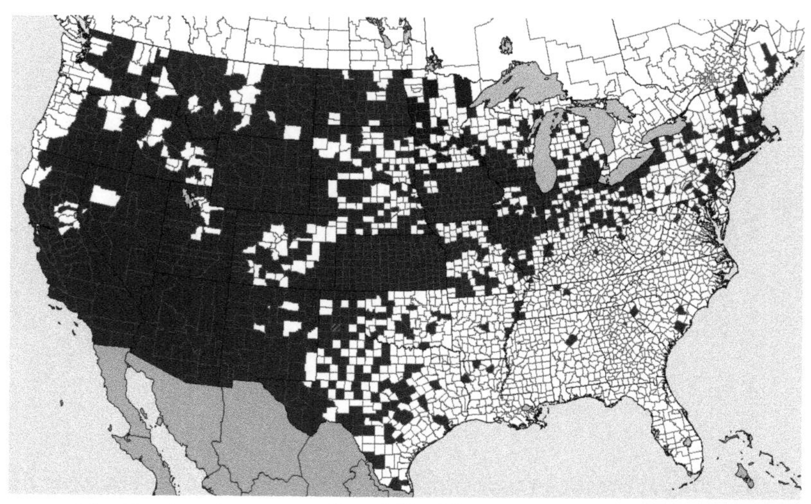

Distribution of **Amaranthus blitoides**, PROSTRATE PIGWEED

Amaranthus blitoides, PROSTRATE PIGWEED. A, habit, as seen from above. B, flowering branchlet. C, rootstock. D, utricles. E, seeds.

Amaranthus retroflexus L. REDROOT PIGWEED

Annual, reproducing by seeds. Taproot shallow, red.

DESCRIPTION Stems erect, branching freely if not crowded, up to 2 m high, finely hairy. **Leaves** dull-green, long-petioled, egg-shaped or rhombic-ovate, up to 1 dm long. **Flowers** green, small, the terminal panicle of several to many, short, densely crowded, ovoid, blunt spikes (1-5 cm long), the whole 5-20 cm long; smaller panicles produced in the upper axils; each flower surrounded by 3 spiny bracts, the bracts rigid, awl-shaped, much exceeding the calyx, 4-8 mm long. **Utricle** flattened, 1.5-2 mm long, dehiscing by a transverse line at the middle, the upper part wrinkled. **Seed** oval to egg-shaped, lens-shaped, notched at the narrow end, 1-1.2 mm long, shiny-black to dark red-brown.

FLOWERING August-October.

WHERE FOUND Cultivated fields, yards, fence rows, and waste ground. A troublesome semi-cosmopolitan weed.

ORIGIN Native of tropical America.

NOTE Plants sometimes accumulate excess nitrites and thus poisonous to cattle, causing them to bloat.

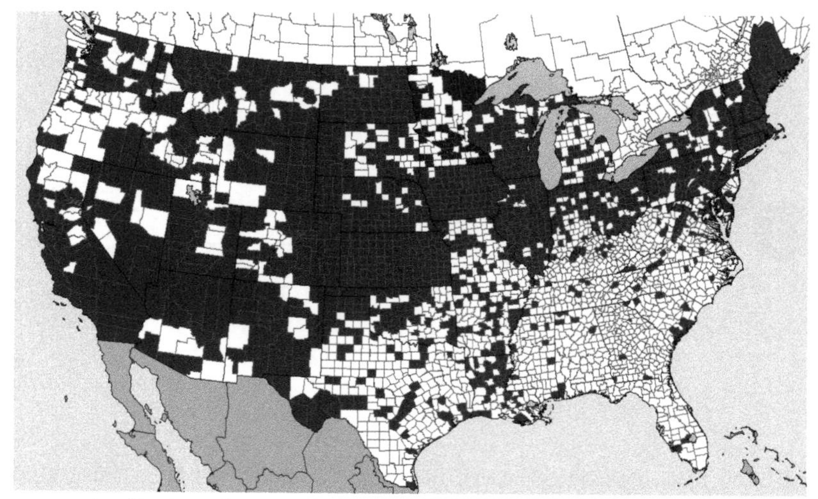

Distribution of **Amaranthus retroflexus**, REDROOT PIGWEED

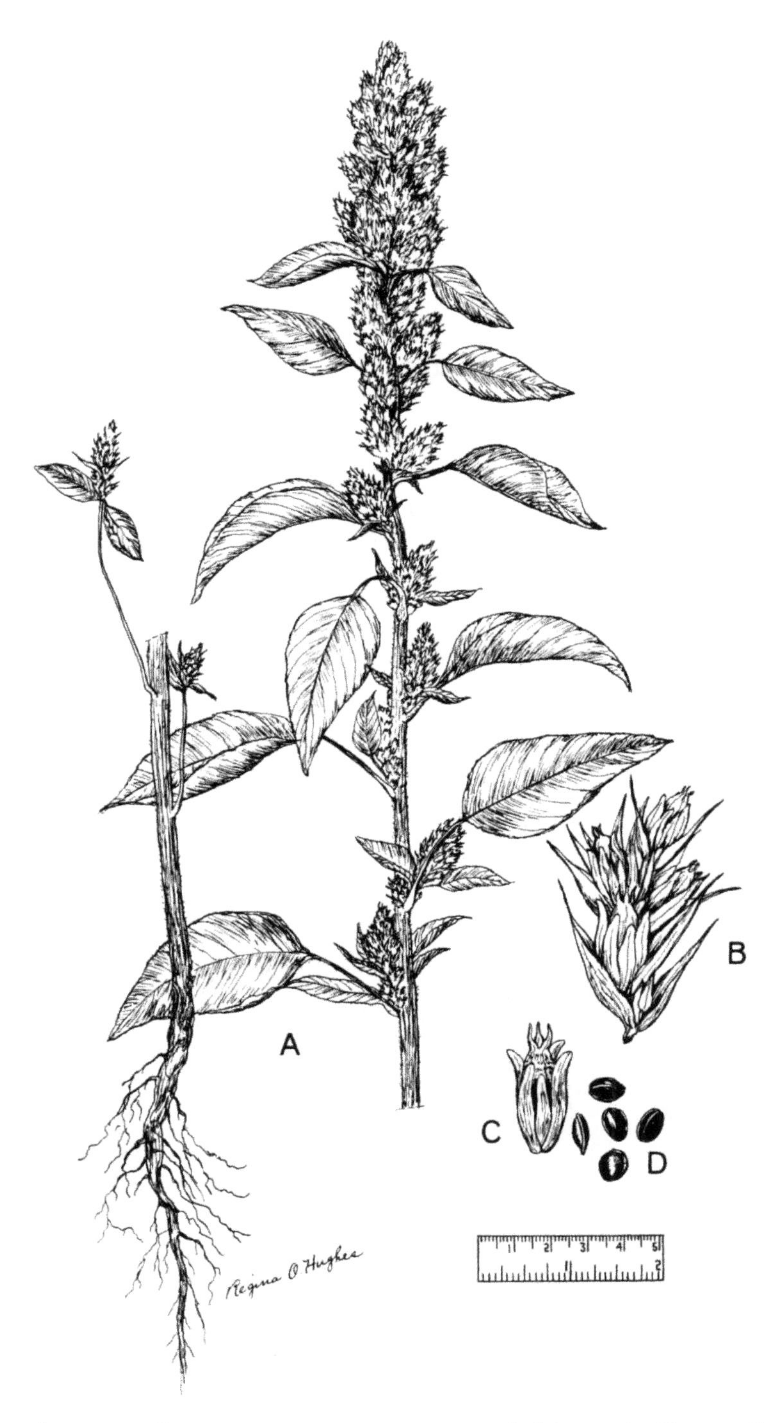

Amaranthus retroflexus, REDROOT PIGWEED. A, habit. B, pistillate spikelet. C, utricle. D, seeds.

Bassia scoparia (L.) A. J. Scott — KOCHIA, MEXICAN FIREWEED

Annual, with taproot, reproducing from seeds.

DESCRIPTION Stems erect, much-branched, 3–10 dm tall, smooth but usually hairy above. **Leaves** alternate, simple, pubescent to nearly glabrous, 2.5–5 cm long, lanceolate to linear, with hairy margins, without petioles. **Spikes** 5–100 mm long, hairy, left bracts 3–10 mm long. **Flowers** perfect, solitary or paired, small, greenish, without petals, in the axils of the upper leaves and in terminal panicles. **Calyx** 5–10-lobed, each lobe developing into a winglike appendage. **Seed** about 1.8 mm long, ovate, flattened with a groove on each side from the narrow end, finely granular, surface dull, brown with yellow markings, a fragile shell-like hull (calyx) may enclose seed.

FLOWERING July–September.

SYNONYMS *Kochia scoparia* (L.) Roth

WHERE FOUND Waste places, ballast grounds, dry pastures, rangeland, and cropland; a drought-resistant weed serious in the Plains States; introduced as an ornamental for its bright-red autumnal color (burning bush, summer cypress); often escaped from cultivation.

ORIGIN Naturalized from Eurasia.

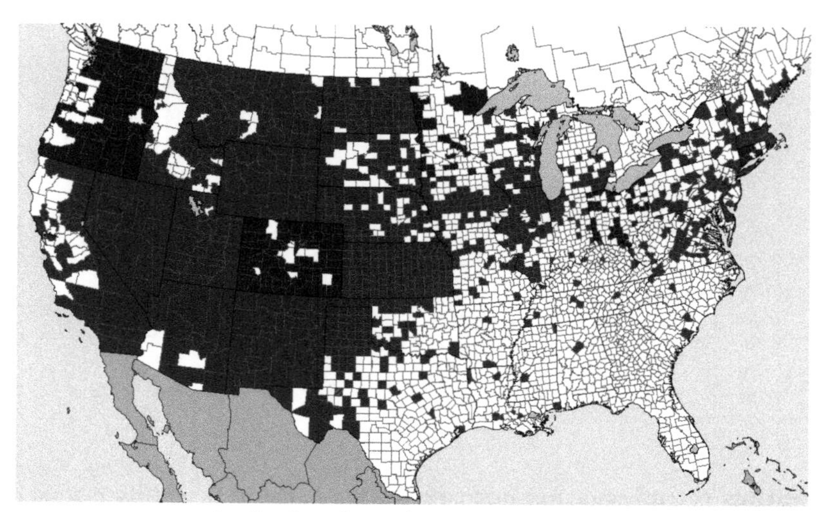

Distribution of **Bassia scoparia**, KOCHIA

Bassia scoparia, KOCHIA. A, branch of plant. B, magnified branchlet, showing flowers at different stages. C, rootstock. D, flower. E, fruits. F, seeds.

Chenopodium album L. COMMON LAMB'S-QUARTERS

Annual herb, reproducing by seeds. Taproot short and much-branched.

DESCRIPTION Stems erect, much-branched above, 0.2-2 m tall, glabrous, grooved, often with red or light-green streaks, branching varying from slight to much. **Leaves** alternate, simple, ovate to lanceolate, without stipules, the upper leaves sometimes linear and sessile, glabrous, usually white mealy-coated, especially on the underside and in the early stages, the edges with a few low broad teeth. **Inflorescence** in irregular spikes clustered in panicles at the ends of the branches and in the keeled and nearly covering the mature fruit; axils of leaves. **Flowers** perfect, small, sessile, green. Calyx of 5 sepals that are more or less keeled and nearly covering the mature fruit. Petals none. Stamens 6. Pistil 1, with 2 or 3 styles, ovary 1-celled, attached at right angles to the flower axis. **Fruit** a utricle (a seed covered by the thin papery pericarp). **Seed** shining, black, lens- to disk-shaped, 1.3-1.5 mm in diameter, with a marginal notch.

FLOWERING June-October.

WHERE FOUND Cultivated crops, gardens, grain fields, and waste ground.

ORIGIN Likely introduced and naturalized from Eurasia, but sometimes considered native or adventive in various regions of North America.

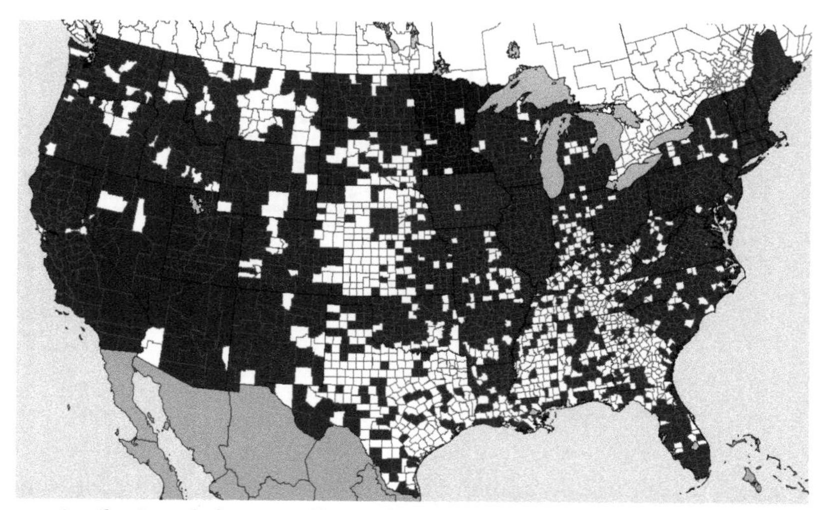

Distribution of **Chenopodium album**, COMMON LAMB'S-QUARTERS

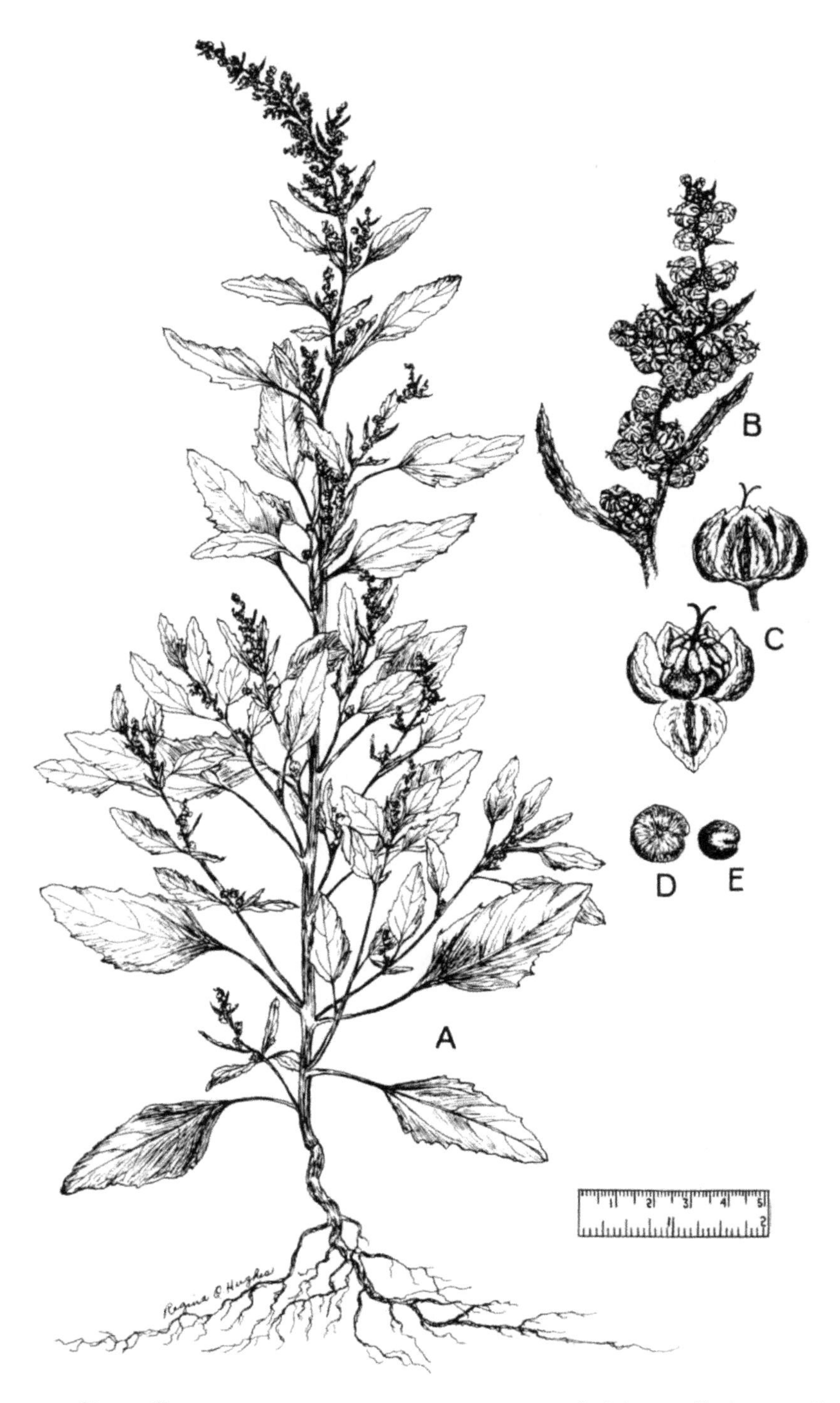

Chenopodium album, COMMON LAMB'S-QUARTERS. A, habit, small plant. B, floral spike. C, flowers. D, utricle. E, seed.

Halogeton glomeratus (M. Bieb.) C.A. Mey. HALOGETON, BARILLA

Annual herb, branching from the base.

DESCRIPTION Stems at first spreading, then usually becoming erect, 0.5–5 dm tall, often with numerous short lateral branches. Plant glabrous or pubescent, fleshy, bluish-green in early spring becoming yellow or red in late summer. **Leaves** alternate but in bunches along the stems, fleshy, round, abruptly ending in a slender needle-like hair, sessile, 6–20 mm long, with tufts of cottony hairs in the angles between the leaves and the stems. **Flowers** in compact clusters in the leaf axils, 2-bracted, greenish-yellow, inconspicuous, numerous, of 2 kinds: the larger flowers with wing-tipped sepals surrounding the seed cases, the smaller flowers with toothlike sepals at the apex. **Seed cases** 5; wide-winged bracts (sepals) on black seed produced during short days of late summer densely crowded on the stems and short branches at maturity giving an appearance of being in flower in autumn; the brown seeds produced during the long days of early summer encased in short adherent thickened bases of bracts (sepals). **Seed** numerous and of two types: black seed (very dark chocolate-brown) and brown seed (light-tan), more or less flattened, about 1 mm long, the spiral form of the embryo clearly evident. Black seed all germinating during first growing season after production. Brown seed dormant but remaining viable in soil for many years. Winged bracts on black seed facilitating spread by wind. Plants also become tumbleweeds, distributing the seeds.

FRUITING July–October.

WHERE FOUND Dry deserts and dry lakebeds. Barren eroded burned-over areas, overgrazed ranges, roadsides, alkaline and disturbed soils, and abandoned farmlands; and along railroad beds and sheep trails.

ORIGIN Introduced to the United States from Siberia in about 1930.

NOTE Poisonous to livestock, especially sheep and cattle, because of high oxalate content.

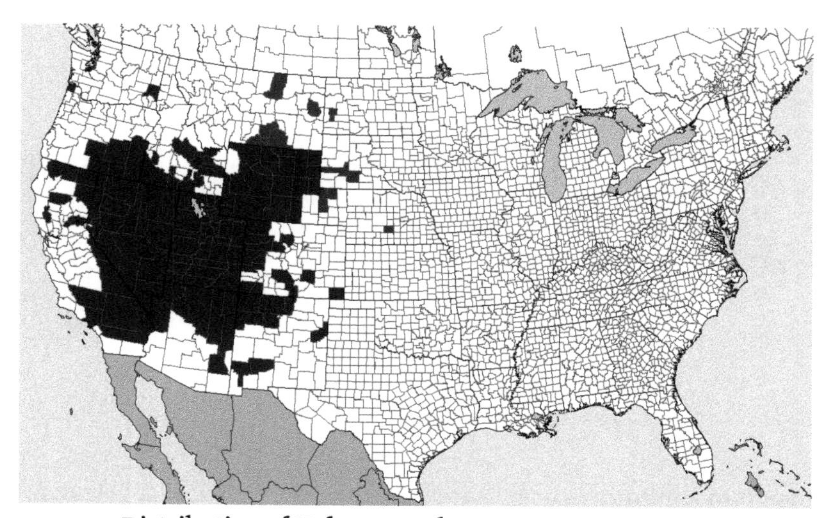

Distribution of **Halogeton glomeratus**, HALOGETON

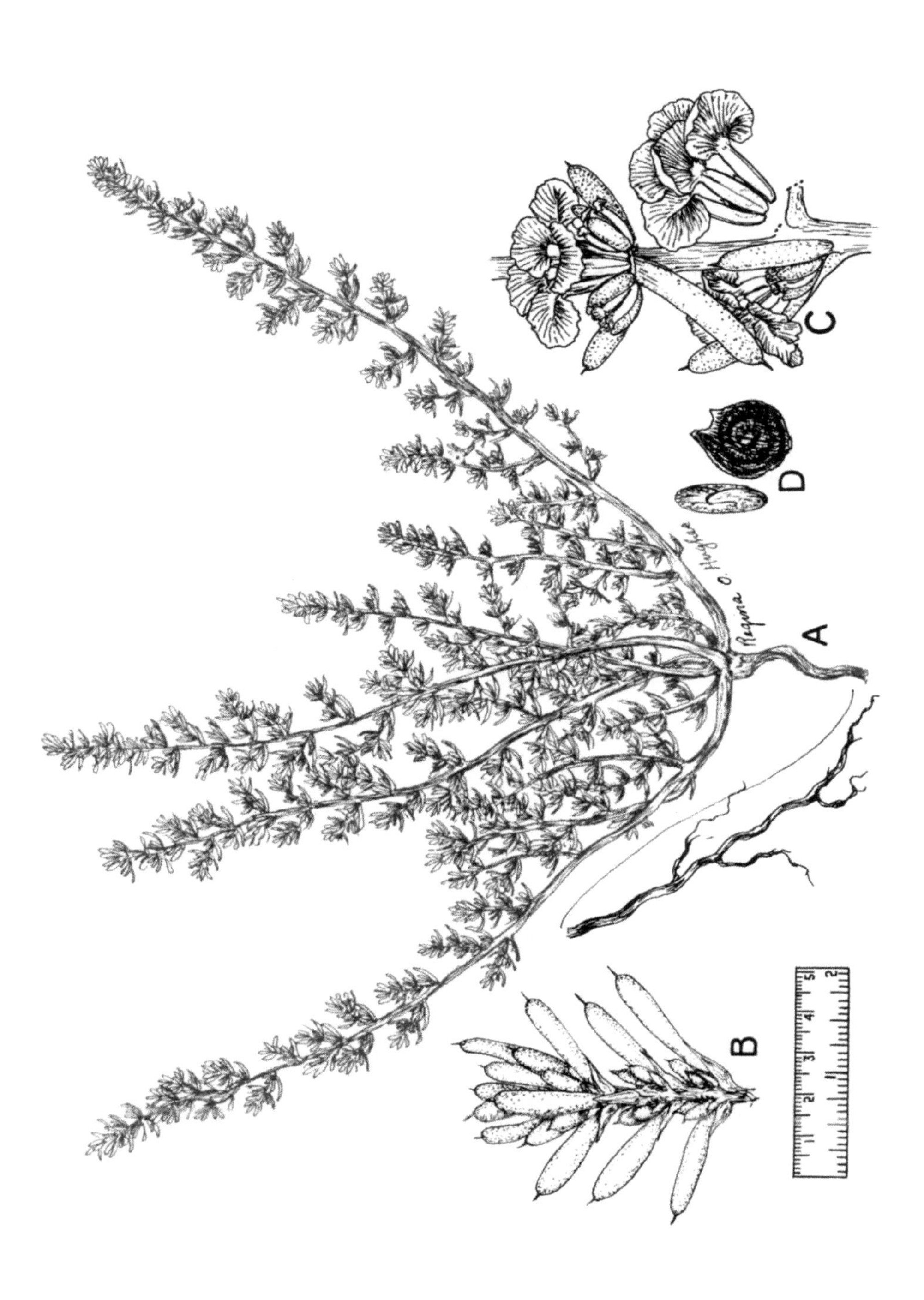

Halogeton glomeratus, HALOGETON. A, habit. B, enlarged leaves. C, flower. D, seed, 2 views.

Kali tragus (L.) Scop. RUSSIAN THISTLE

Annual herb, reproducing by seeds. Taproot spreading.

DESCRIPTION Stems bushy, much-branched, 1.5–12 dm tall, 3–15 dm in diameter, rigid, spiny, spherical, often reddish in age, young stems and leaves green and succulent. **Leaves** alternate, the first-formed leaves (in seedlings and young plants) fleshy, cylindrical or awl-shaped, 0.5 mm broad, 1.2–6.5 cm long, pointed at the tips, usually bearing a short shoot in the axil; the later-formed leaves shorter, stiff, dilated and thickened at the base, ending in a hard sharp spine. **Flowers** small, greenish, mostly solitary from the lowest to the uppermost leaf axils. Petals none. Sepals 5, papery and persistent. Stamens 5. Pistil 1, with a style. Bracts at the base of each flower 2, rigid, spine-tipped, resembling the leaf. **Fruit** surrounded by the 5 enlarged sepals, each developing a fan-shaped, strongly veined wing on its back, 3–9 mm broad, thus forming a beaked or conical cover over the 1-seeded fruit. **Seed** numerous (thousands per plant), top-shaped, about 2 mm broad, with a yellowish coiled embryo, visible through the thin gray wall of the fruit.

FLOWERING July until frost; fruiting from August into winter.

SYNONYMS *Salsola kali* L., *Salsola pestifer* A. Nels.

WHERE FOUND Disturbed areas, roadsides, ditchbanks, fallow abandoned grainfields, overgrazed ranges, and pastures. Old plants break loose to form "tumbleweeds."

ORIGIN Introduced from Eurasia; first reported in United States in South Dakota in the 1870s, in contaminated flaxseed from Russia.

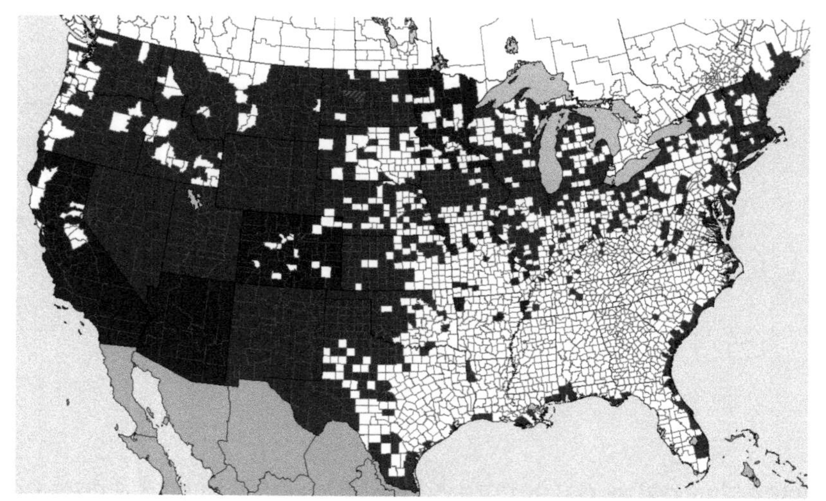

Distribution of **Kali tragus**, RUSSIAN THISTLE

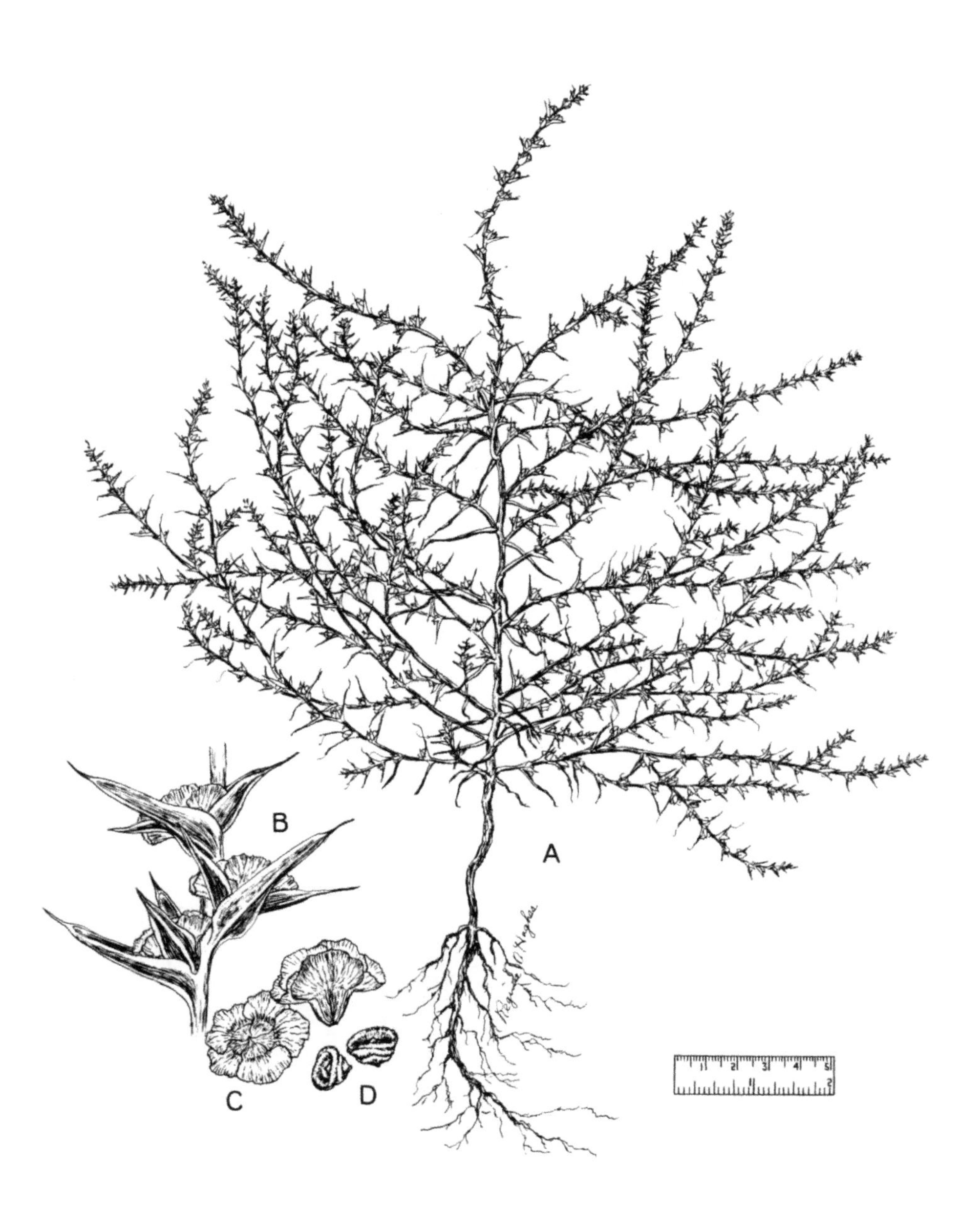

Kali tragus, RUSSIAN THISTLE. A, habit. B, flowering branch. C, fruiting calyces; D, seeds.

Rhus glabra L. SMOOTH SUMAC

Perennial sparsely branched shrub or small tree, to 6 m tall, reproducing by seeds and rootstocks.

DESCRIPTION Stems glabrous or merely soft pubescent, glaucous, often clumped forming thickets. **Leaves** compound, odd-pinnate, the leaflets 11–31, thin, much paler beneath, sharply serrate, lanceolate to narrowly oblong, 5–10 cm long, tapering to a point. **Flowers** in terminal dense panicles, open and up to 4.5 dm long, the perfect or pistillate flowers dense and smaller. Flowers greenish-white or yellow, petals 5. Calyx small, 5-parted. Stamens 5, inserted under the edge or between the lobes of a flattened disk in the bottom of the calyx. **Drupes** covered with minute, bright-red, acrid, appressed hairs (about 0.2 mm long), indehiscent.

FLOWERING June–July.

WHERE FOUND Dry soils, road embankments, thickets, dry wastes, upland soils, old fields, and margins of woods.

ORIGIN Native.

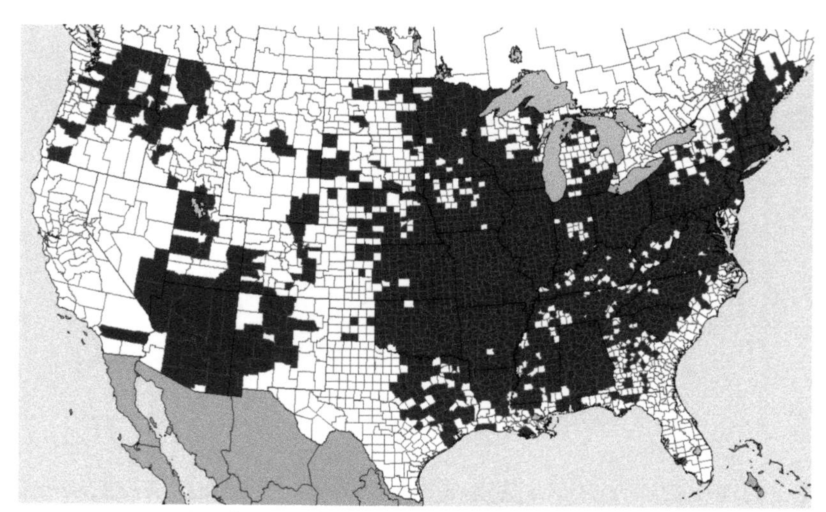

Distribution of **Rhus glabra,** SMOOTH SUMAC

Rhus glabra, SMOOTH SUMAC. A, habit. B, flowers; b, diagram of flower to show disk with stamens. C, fruit. D, seeds.

Toxicodendron diversilobum (Torr. & Gray) Greene

PACIFIC POISON OAK

Perennial woody shrub or vine.

DESCRIPTION Stems forming an upright shrub (common form), with many small woody stems rising from the ground, or attached to upright objects for support, becoming a vine, up to 8-10 m high; **Leaves** 3-parted, very irregular as to lobing, especially the two lateral ones, the margins either even or lobed, the surface of the leaves usually glossy and uneven, thus appearing thick and leathery. **Flowers** in clusters on slender stems diverging from the axis of the leaf, greenish-white, about 2-3 mm in diameter. **Drupes** spherical or somewhat flattened, greenish or creamy-white, with a smooth glossy surface, striped into segments, the fruits remaining on the plants fall and winter.

FRUITING mid-October.

SYNONYMS *Rhus diversiloba* Torr. & Gray

WHERE FOUND Roadsides, cultivated fields, and abandoned land. A skin irritant.

ORIGIN Native.

Toxicodendron pubescens P. Mill.

POISON OAK

Perennial shrub, reproducing from seeds and spreading by subterranean creeping rootstocks.

DESCRIPTION Stems slender, erect, woody for 0.5-6 dm, simple or with a few erect branches, not climbing, nor with aerial roots. **Leaves** 3-parted, on long erect velvety petioles, mostly near the top of the stem and often appearing falsely whorled. **Leaflets** elliptic, rhombic or obovate, hairy above, velvety beneath or eventually glabrous above, obtuse or rounded above, with 3-7 deep teeth or variously lobed, suggesting oak leaves, or unlobed, wedge-shaped to rounded at the base. **Fruit** (drupe) greenish to buff, about 5 mm in diameter, pubescent or glabrous.

FLOWERING May-June; fruiting August-November, sometimes all winter.

SYNONYMS *Rhus quercifolia* (Michx.) Steutl., *Rhus toxicodendron* L., *Toxicodendron quercifolium* Greene.

WHERE FOUND Dry barrens, sandy wastes, pinewoods, sandy woods. A skin irritant.

ORIGIN Native.

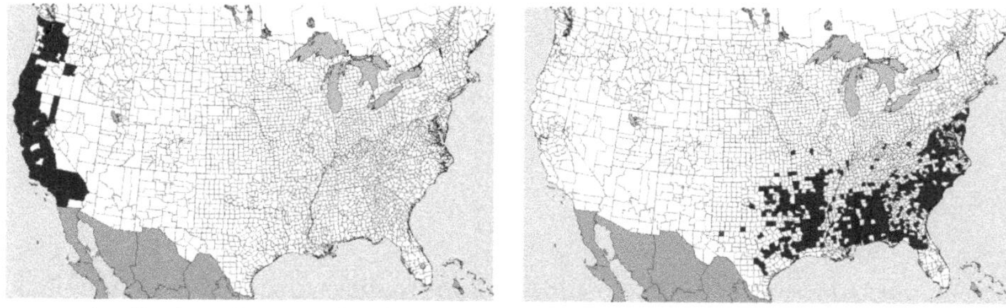

Distribution of **Toxicodendron diversilobum** (left), **T. pubescens** (right)

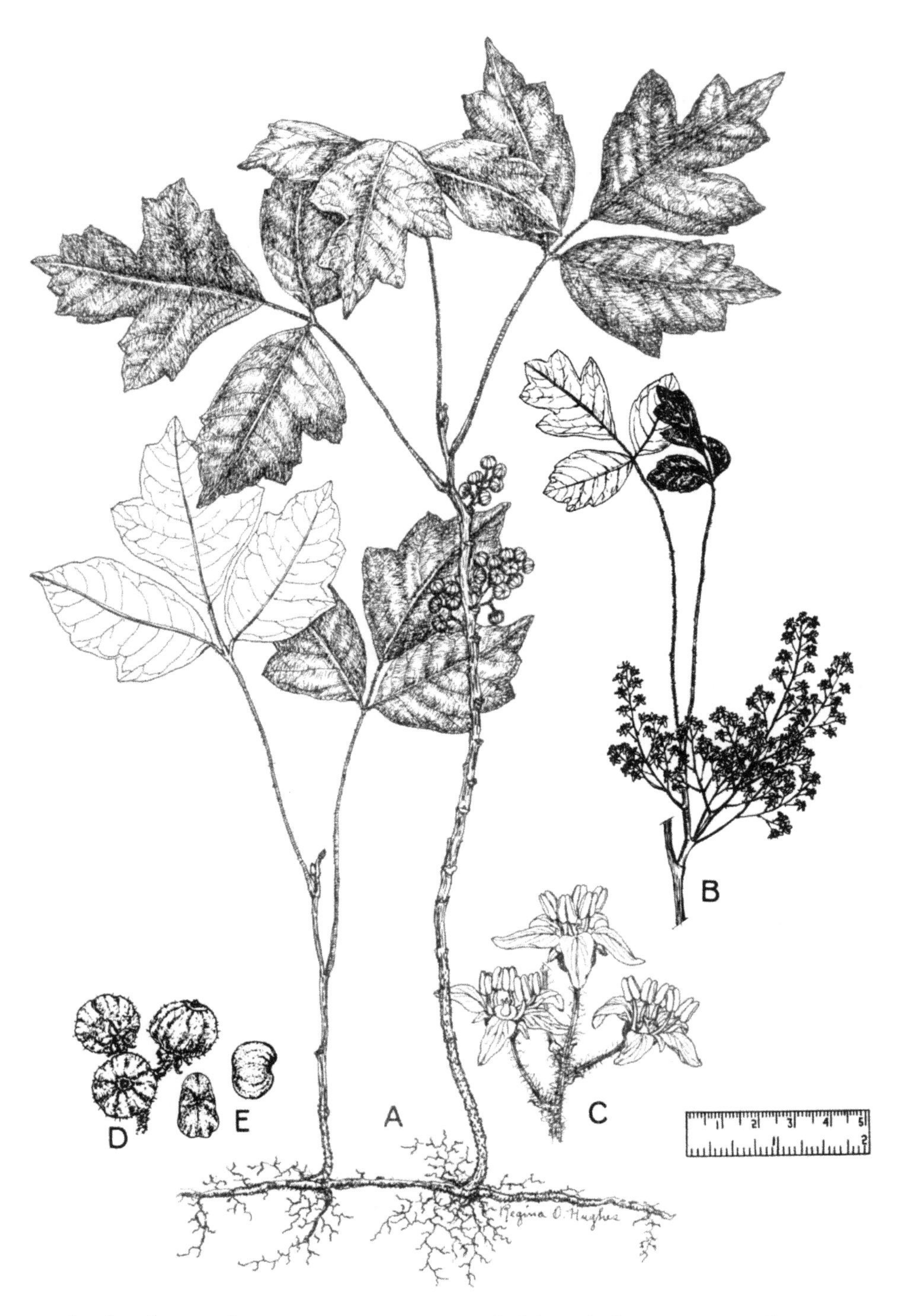

Toxicodendron pubescens, POISON OAK. A, habit. B, inflorescence. C, flowers. D, drupes. E, stones.

Toxicodendron radicans (L.) Kuntze POISON IVY

Perennial woody shrub or vine, reproducing by seeds and by creeping rootstocks from the basal stem nodes, sometimes running horizontally underground for several meters, sending up leafy shoots from their nodes.

DESCRIPTION Stems erect and shrubby or a vine climbing high into trees, on fences, the stem supported by aerial roots along the stem. **Leaves** alternate, quite variable in outline and in marginal cutting, ovate or elliptic, acute or tapering to a point, rounded to wedge-shaped at the base, entire to irregularly serrate or wavy, glabrous or thinly pubescent, compound with 3 large shiny leaflets, each 5-10 cm long, pointed at the tip, the terminal leaflet longer petioled than the lateral leaflets. **Panicles** up to 1 dm long, axillary or from axils of past years, ascending and divergent. Flowers small, yellowish-green, 5-petaled. **Drupes** small, 5-6 mm in diameter, grayish-white, nearly globose, hard, usually glabrous, with a grayish striped, 1-seeded stone about 3-4 mm in diameter.

FLOWERING June-July.

SYNONYMS *Rhus radicans* L.

WHERE FOUND Rocky fields, pastures, thickets, woods, and waste places, often climbing trees, fences, and dwellings; a ubiquitous weed.

ORIGIN Native.

NOTE All parts of the plant contain a poisonous principle that may cause blistering of the skin. A variable species as to habit in growth, leaflet shape, rooting habit, pubescence of the leaves, petioles, and fruit, leading to the naming of several named varieties and forms.

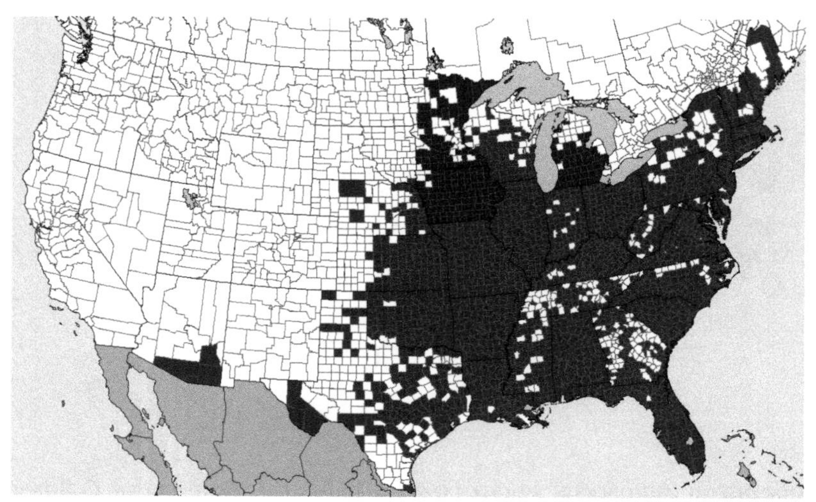

Distribution of **Rhus radicans**, POISON IVY

Rhus radicans, POISON IVY. A, habit. B, flower panicle. C, flowers. D, drupe. E, stones; F, aerial roots.

Cicuta maculata L. SPOTTED WATER-HEMLOCK

Perennial herb, reproducing by seed and by fleshy roots.

DESCRIPTION Roots elongated, 3-10 cm long, crowded or tufted. **Stems** erect, branching, stout, hollow, jointed, streaked with purple, ridged, glabrous, 1-2.2 m tall. **Leaves** alternate or basal, petioled, the 3 major divisions nearly compound with leaflets lanceolate to oblong-lanceolate, 3-12 cm long, pointed, deeply serrate, glabrous. **Flowers** in open spreading umbels having no involucre, but involucels of slender bractlets, pedicels of the smaller umbels very unequal. Calyx teeth prominent. Corolla white; stylopodium depressed. **Mericarps** broadly oval, about 2.5-3.5 mm long, flat on one side, rounded with 5 light ribs on the other side, oil tubes solitary between the ridges, brown with yellow ribs.

FLOWERING June-August.

WHERE FOUND Marshy ground, pastures, and wet meadows and along ditches and streams.

ORIGIN Native.

NOTE Roots poisonous if eaten.

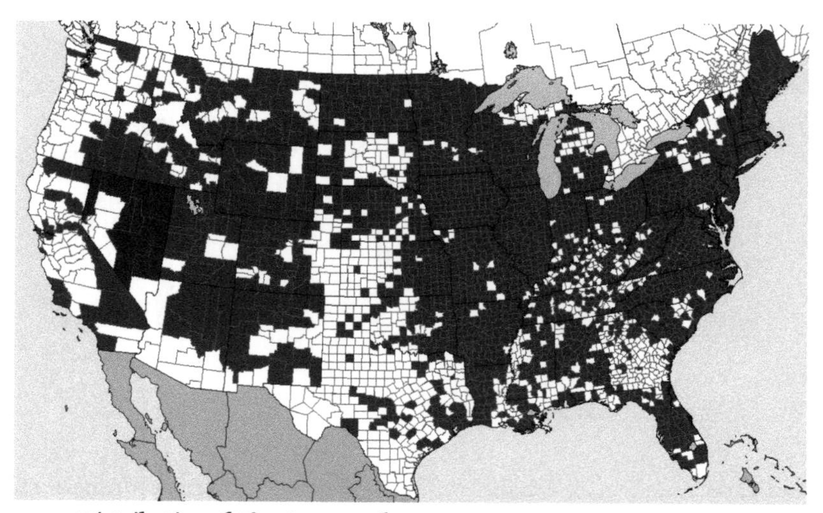

Distribution of **Cicuta maculata**, SPOTTED WATER-HEMLOCK

Cicuta maculata, SPOTTED WATER-HEMLOCK. A, habit. B, flower. C, schizocarp. D, mericarp.

Conium maculatum L. POISON HEMLOCK

Biennial herb, reproducing by seeds. Taproot long, white, often branched.

DESCRIPTION Stems erect, branching, stout, glabrous, purple-spotted, ridged, up to 3 m tall. **Leaves** alternate or basal, 2–4 dm long, petioled, broadly triangular-ovate in outline, 3–4 times pinnately compound, the leaflets lanceolate to ovate-oblong, dentate or finely cut, 4–10 mm long. **Inflorescence** in large open compound umbels, the umbels 4–6 cm wide, the terminal inflorescence blooming first but soon overtopped by the others. Involucral bracts entire. Corolla white. Schizocarp containing 2 mericarps. **Mericarps** broadly ovoid, about 2–3 mm long, granular, with conspicuous pale-brown wavy ribs, without oil tubes but with a layer of secreting cells next to the seed, grayish-brown; face of the seed deeply and narrowly concave.

FLOWERING June–September.

WHERE FOUND Borders of fields, pastures, meadows, roadsides, and waste places; on rich, gravelly, or loamy soils.

ORIGIN Introduced and naturalized from Eurasia; throughout most of the United States.

NOTE All parts of plant notoriously poisonous, fatal (given to Socrates in ancient Greece).

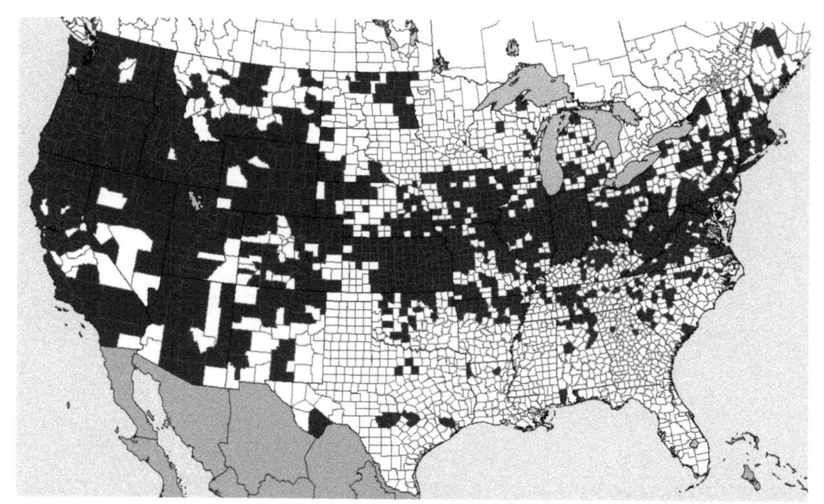

Distribution of **Conium maculatum**, POISON HEMLOCK

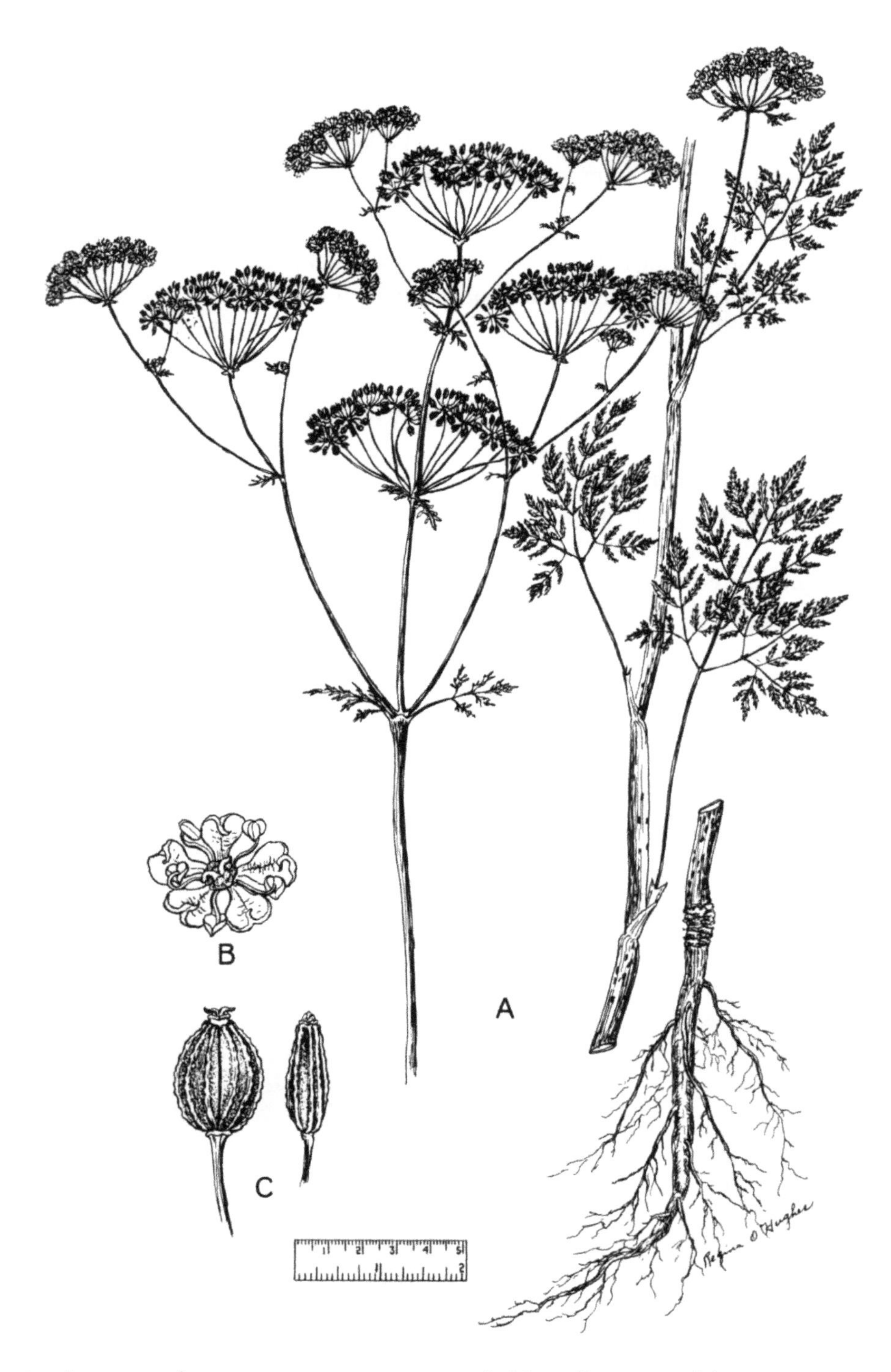

Conium maculatum, POISON HEMLOCK. A, habit. B, flower. C, schizocarps.

Daucus carota L. WILD CARROT, QUEEN ANNE'S-LACE

Biennial herb, reproducing by seeds. Taproot bearing a rosette of leaves the first season.

DESCRIPTION Stems erect, branching, slender, hollow, ridged, bristly-hairy, 3–16 dm high, bearing scattered stem leaves. **Leaves** alternate or basal, oblong in general outline, pinnately nearly compound, the ultimate segments linear, lanceolate or oblong, often lobed, somewhat hairy; stem leaves sessile, with a sheathing base. Basal leaves long-petioled. **Flowers** in flat-topped umbels, which become concave as the fruits mature, 6–15 cm broad. Involucral bracts cleft or pinnatifid into narrow segments. Corolla white to pinkish, usually with one purple or pink flower in the center. **Mericarps** oblong, 2–4 mm long, 1 side flattened, the other with 5 bristly primary ribs and 4 conspicuous secondary ribs that are winged and near a row of barbed prickles, oil tubes solitary under the secondary ribs and also 2 on the flat side, light grayish-brown.

FLOWERING May–October.

WHERE FOUND Dry fields, old meadows, pastures, and waste places; a pernicious weed.

ORIGIN Introduced from Eurasia.

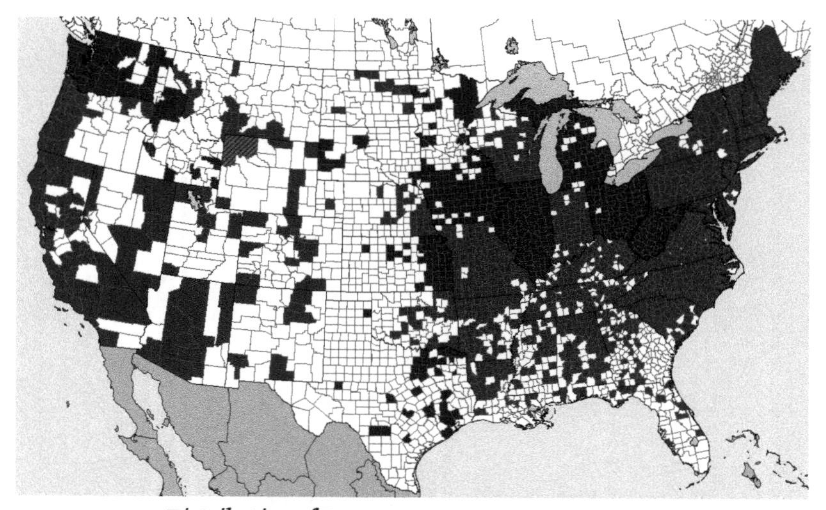

Distribution of **Daucus carota**, WILD CARROT

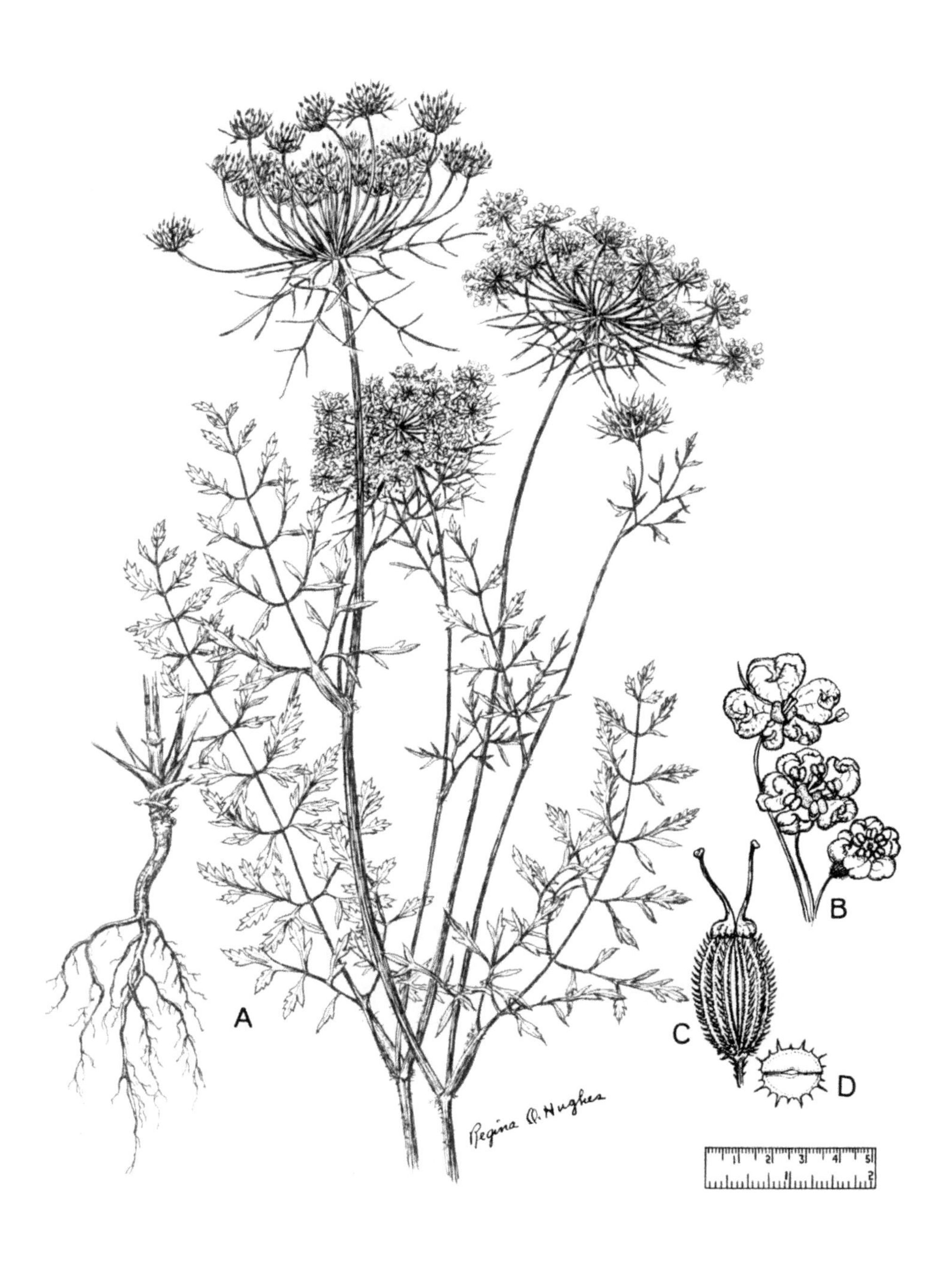

Daucus carota, WILD CARROT. A, habit. B, flowers. C, schizocarp. D, cross-section of fruit, showing 2 mericarps.

Apocynum cannabinum L. HEMP DOGBANE, INDIAN HEMP

Perennial herb, reproducing by seed and long, horizontal rootstocks.

DESCRIPTION Stems erect, 3-6 dm tall, from a woody base, exuding milky juice when broken, glabrous or nearly so. **Leaves** erect or ascending, blades 5-12 cm long, ovate to lanceolate, smooth-edged, glabrous to sparingly pubescent beneath, petiole 2-7 mm long to nearly sessile. **Cymes** terminal, of mostly ascending flowers with 5 greenish-white petals, 2-4 mm long, the lobes erect. Bracts dry and thin, linear-attenuate, not persisting. **Pod** (follicle) 1.2-2 dm long and slender, falcate. **Seed** 4-6 mm long, thin and flat, with a tuft of soft silky hairs at one end (coma 2.5-3 cm long).

FLOWERING June-August.

WHERE FOUND In old fields, wastelands, dumps, thickets, borders of woods, and open ground.

ORIGIN Native.

NOTE Poisonous to cattle.

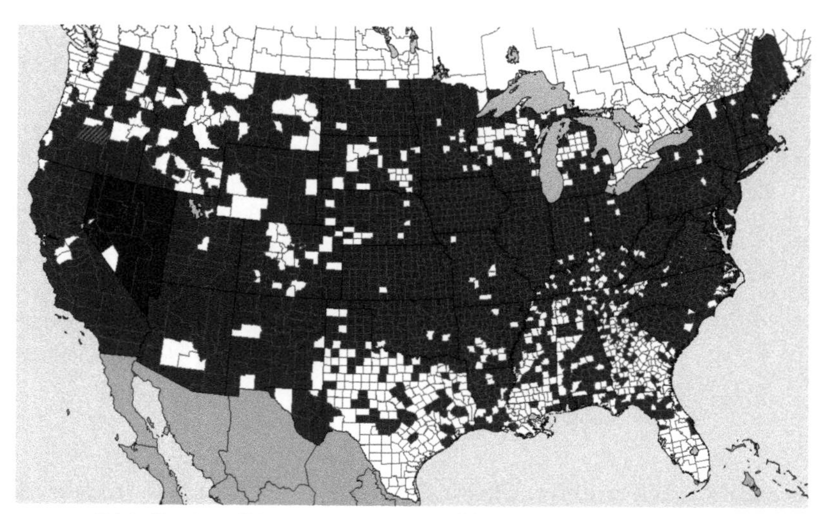

Distribution of **Apocynum cannabinum**, HEMP DOGBANE

Apocynum cannabinum, HEMP DOGBANE. A, habit. B, rootstock. C, flowers. D, follicles. E, seed with coma. F, seeds.

Asclepias syriaca L. COMMON MILKWEED

Perennial herb, reproducing by seeds and from long-spreading rhizomes.

DESCRIPTION Stems stout and erect, mostly simple, up to 2 m, covered with short, downy hairs, with milky juice. **Leaves** opposite, oblong, rounded, 1-2.6 dm long, 0.4-1.8 dm broad, with prominent veins, upper surface smooth, lower surfaces covered with short white hairs and with strong transverse nerves. **Flowers** sweet-smelling, pink to white, in large, many-flowered, bell-like clusters at the tips of the stems and in the axils of the upper leaves, corolla lobes 6-9 mm long, hoods 3-4 mm high. **Pod** (follicle) grayish, hairy and beset with soft spiny projections, 1-3 mm high, slenderly ovoid, 2.5-3.5 cm thick. **Seed** brown, flat, oval, 6 mm long, 5 mm wide, with a tuft of silky white hairs attached to the tip.

FLOWERING June-August.

WHERE FOUND Dry and cultivated fields, pastures, woods, roadsides, and thickets; also on prairies in the West.

ORIGIN Native.

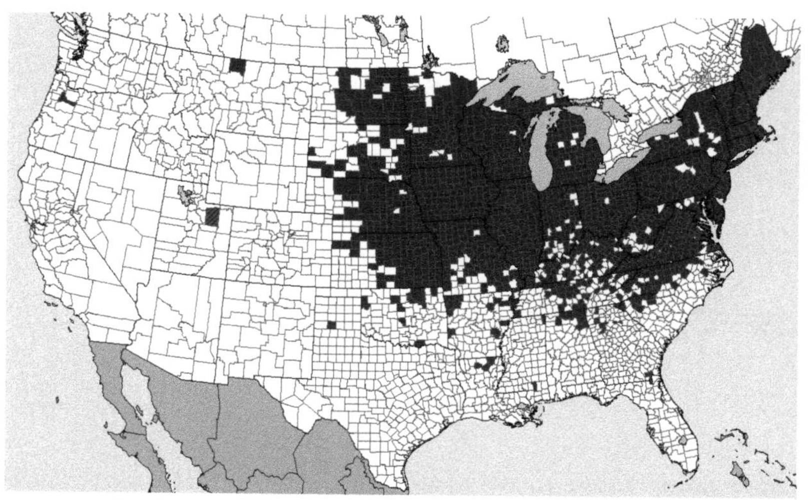

Distribution of **Asclepias syriaca**, COMMON MILKWEED

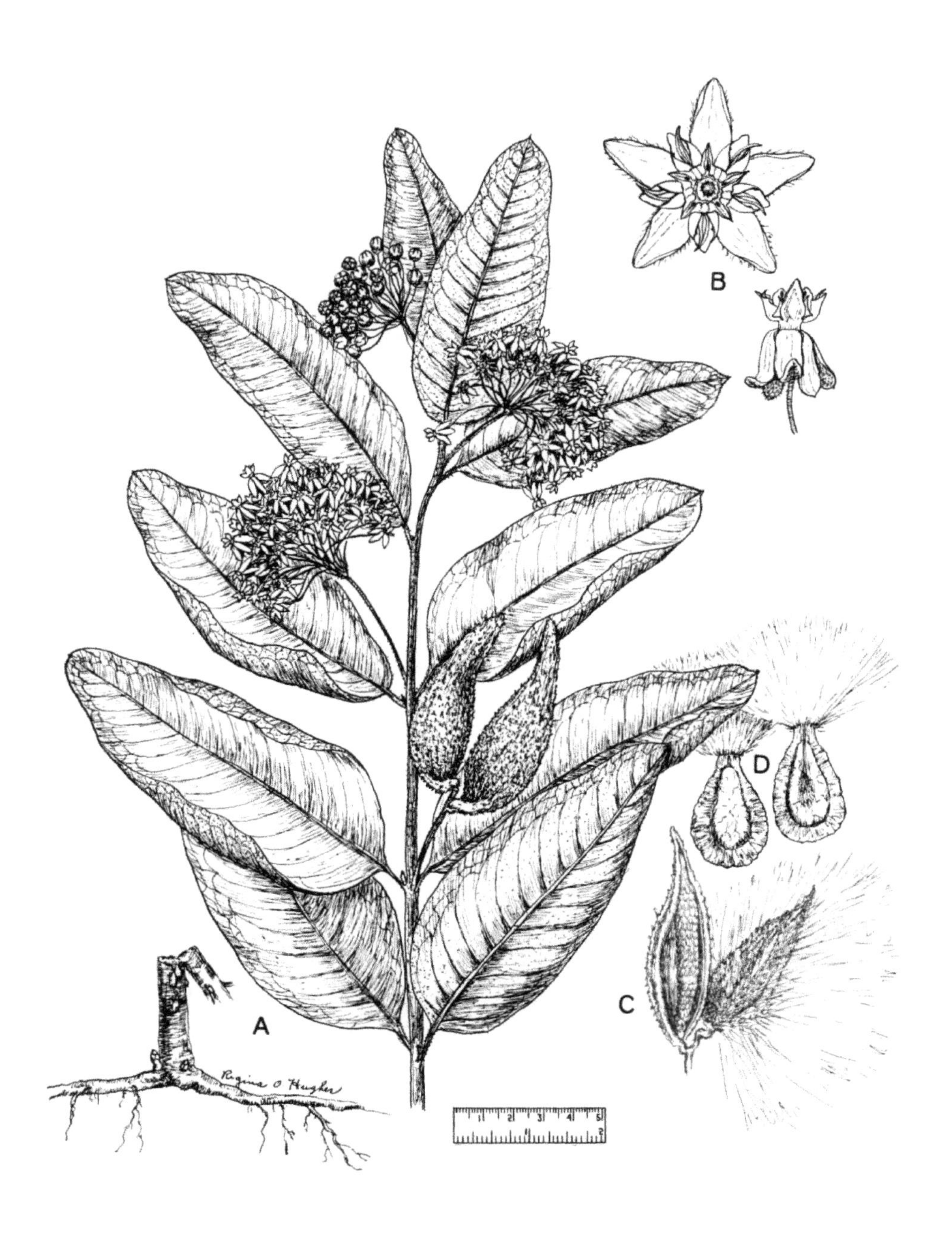

Asclepias syriaca, COMMON MILKWEED. A, habit. B, flower; upper view, side view. C, follicles. D, seeds with coma.

Asclepias verticillata L. WHORLED MILKWEED

Perennial herb, reproducing from seeds and creeping rootstocks.

DESCRIPTION Stems slender, erect, 3-9 dm high, smooth, branching at top of plant only, with milky juice. **Leaves** linear, light-green, arranged in groups of 3 or 7, in whorls around the stem, 2-5 cm long, 1-2 mm wide, rolled backwards and downward. **Flowers** in umbels, borne in clusters at the top of the stem or in the axils of the upper leaves, hoods roundish-oval, peduncles 1-3 cm long. Petals greenish-white, 5, ovate, 4-5 mm long. **Pod** (follicle) erect on erect stalk, 4-5 cm long, containing numerous seed. **Seed** flat, brown, with a tuft of fine hairs at the tip.

FLOWERING June-August.

WHERE FOUND Meadows, pastures, barrens, and waste and sterile places; seldom in cultivated fields.

ORIGIN Native.

NOTE Poisonous to livestock.

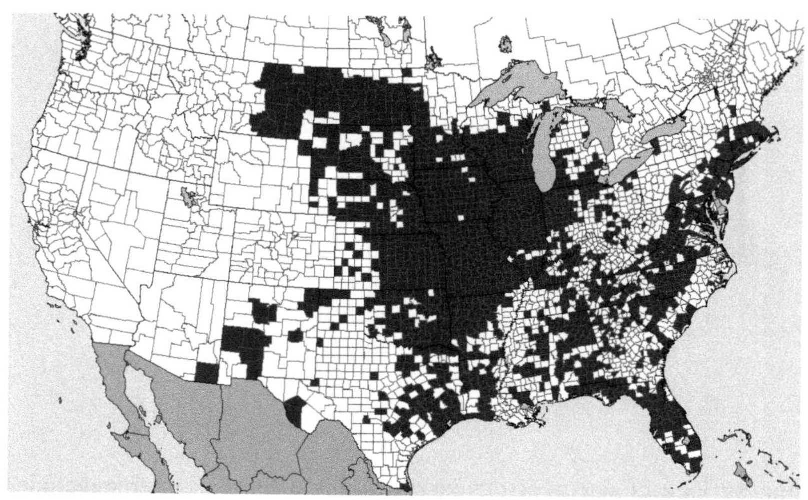

Distribution of **Asclepias verticillata**, WHORLED MILKWEED

Asclepias verticillata, WHORLED MILKWEED. A, habit. B, enlarged leaves, showing revolute margins. C, flower. D, seed; d, seeds with coma.

Achillea millefolium L. COMMON YARROW

Perennial herb, very variable, reproducing by seeds and underground rootstocks.

DESCRIPTION Stem simple or somewhat forked above, 3–10 dm high, web-hairy to smooth. **Stem leaves** 8–20 (or more), smooth to loosely pubescent, dissected into fine segments. **Corymbs** very compound, flattish-topped, 0.6–3 dm broad. Involucre slenderly cylindric, its scales pale, rarely dark-margined. Ligules usually whitish, passing to pink or deep rose-purple, short-oblong, 1.5–2.5 mm long. Receptacle greatly prolonged in fruit, the mature disk flowers becoming exserted. **Achene** tiny, flat, oblong, white or gray.

FLOWERING June–September.

WHERE FOUND Roadsides, meadows, and pastures; persists mainly on thin soil unfavorable for growth of more desirable plants; not common in cultivated fields.

ORIGIN Naturalized from Europe; throughout nearly all of the United States.

NOTE Plants have an offensive odor and a bitter taste.

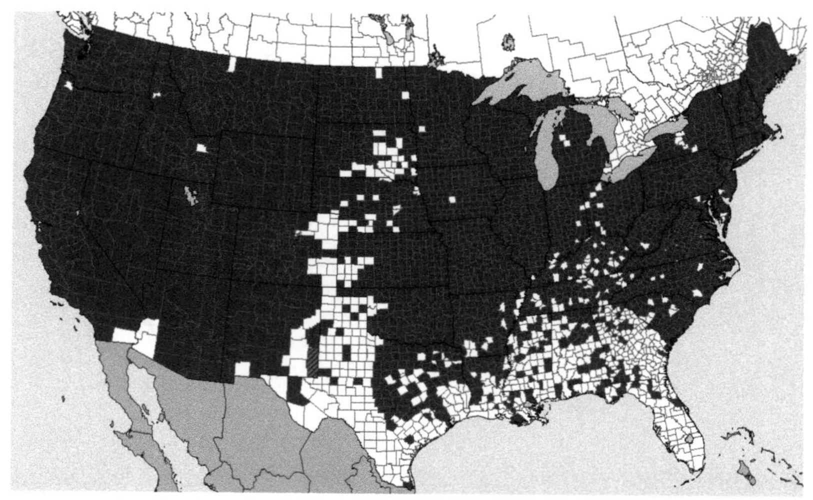

Distribution of **Achillea millefolium**, COMMON YARROW

Achillea millefolium, COMMON YARROW. A, habit. B, enlarged leaves and stem. C, flower head. D, female and male flowers. E, seeds.

Ageratina altissima (L.) King & H. E. Robins. WHITE SNAKEROOT

Perennial herb, reproducing by seeds and from the rhizome. Rhizome rough, knotty, fibrous-rooted.

DESCRIPTION **Stems** erect, firm, solitary or clustered, simple or branching above, 0.2–1.5 m tall, glabrous. **Leaves** opposite, simple, long-petioled, 3-nerved, glabrous, broadly ovate, the larger ones membranaceous flat blades, 5–18 cm long, 3–11 cm broad, coarsely and often sharply 9to 25-toothed on each margin, gradually tapering from above the base to a long tapering tip, often rounded or heart-shaped bases, the upper leaves gradually smaller. **Heads** in dense compound corymbs with 8–30 perfect disk flowers, in well-developed plants the corymbs open with loosely ascending branches from the upper axils, in smaller plants more compact and terminal. Involucral bracts (phyllaries) subequal, usually in 1 row, green, tapering to obtuse scarcely overlapping, 4–6 mm long, glabrous or short-hairy. **Flowers** 5–7 mm in diameter, with bright white petals 3–4 mm long. Receptacle flat and naked. **Achene** 2–2.5 mm long, linear, black or dark-brown, glabrous or nearly so.

FLOWERING Late July–October.

SYNONYMS *Eupatorium rugosum* Houtt., *Eupatorium urticaefolium* Reichard

WHERE FOUND Woodlands, thickets, clearings, damp and shady pastures, and fields; chiefly on gravelly and calcareous soils.

ORIGIN Native.

NOTE Poisonous to livestock; the poison transmissible to humans in milk.

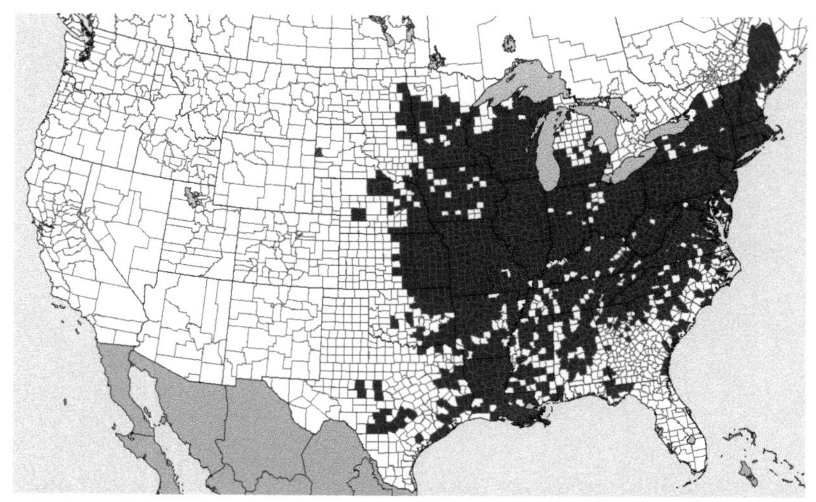

Distribution of **Ageratina altissima**, WHITE SNAKEROOT

Ageratina altissima, WHITE SNAKEROOT. A, habit. B, branch of corymb. C, achene.

Ambrosia artemisiifolia L. COMMON RAGWEED

Annual herb, shallow-rooted.

DESCRIPTION Stems glabrous or hairy, erect, simple or much branched, with one or both sexes, 0.2–2.5 m tall. **Leaves** nearly smooth, deeply cut into many lobes, being bipinnatifid or tripinnatifid with small segments, mostly alternate, some opposite below. **Flowers of two kinds**: the male flowers in small inverted racemes at the tips of branches, slender, becoming lax; the female flowers fewer, borne at the bases of leaves and in forks of the upper branches. Involucres saucer-shaped, scallop-margined, glabrous or hairy, indistinctly radiate-nerved. **Achene** about 4–5 mm long, enclosed in a woody hull, light-brown, top-shaped, pointed (the awl-shaped beak 1–2 mm long), bearing several longitudinal ridges ending in 4–7 short, spiny projections.

FLOWERING July–October.

WHERE FOUND Old pastures, wastelands, roadsides, vacant lots, stubble fields, cultivated lands, and sea-beaches. A variable and pernicious weed.

ORIGIN Native, but considered adventive in the western states.

NOTE Abundant pollen is a hazard to hay fever sufferers.

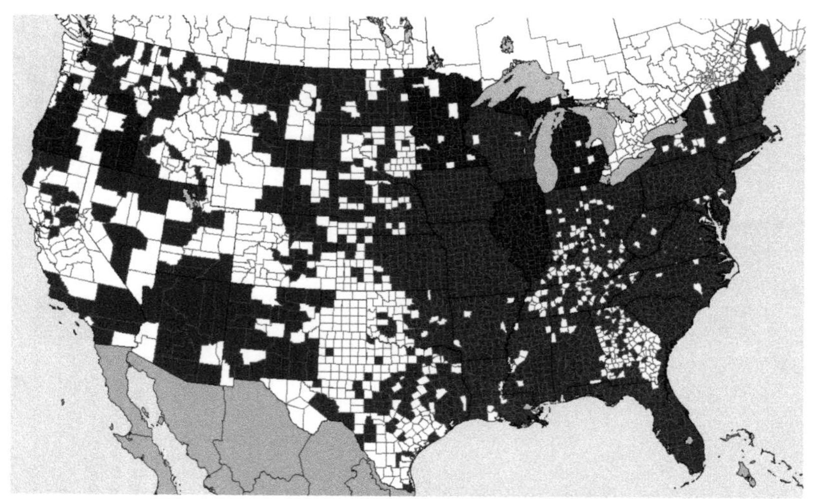

Distribution of **Ambrosia artemisiifolia**, COMMON RAGWEED

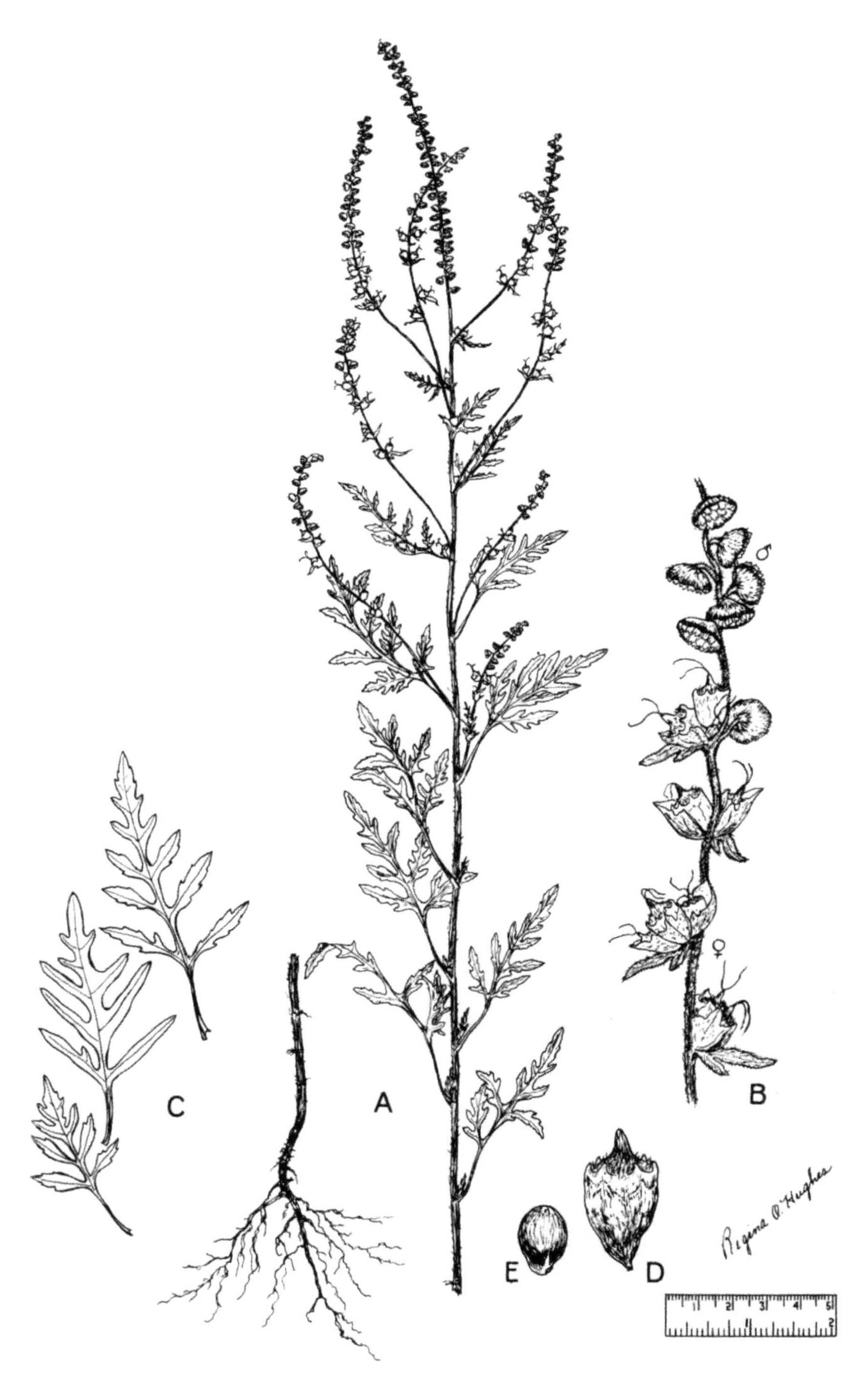

Ambrosia artemisiifolia, COMMON RAGWEED. A, habit. B, raceme with male heads (above) and female involucres (below). C, leaf variations. D, achene. E, seed.

Ambrosia psilostachya DC. WESTERN RAGWEED

Perennial or annual herb, reproducing by creeping roots and rootstocks and seeds.

DESCRIPTION Stem erect, 0.3–2.5 m tall, hairy, bushy, paniculately branched, often growing in dense patches. **Leaves** once-pinnatifid, with short petioles, alternate or opposite, ovate-lanceolate, with very deep lobes, sometimes compound, rough, nearly or quite sessile. **Flowers of two kinds:** the male flowers in clusters on ends of stems and branches, minutely roughened or pubescent; the female flowers few, without petals, in the axils of the upper leaves. **Achene** about 3 mm long, obovoid, in a woody hull having a pointed tip surrounded by 4 short blunt tubercles.

FLOWERING August–October.

WHERE FOUND Dry prairies, barrens, sands, openings, plains, and uncultivated places.

ORIGIN Native.

NOTE Abundant pollen is a hazard to hayfever sufferers.

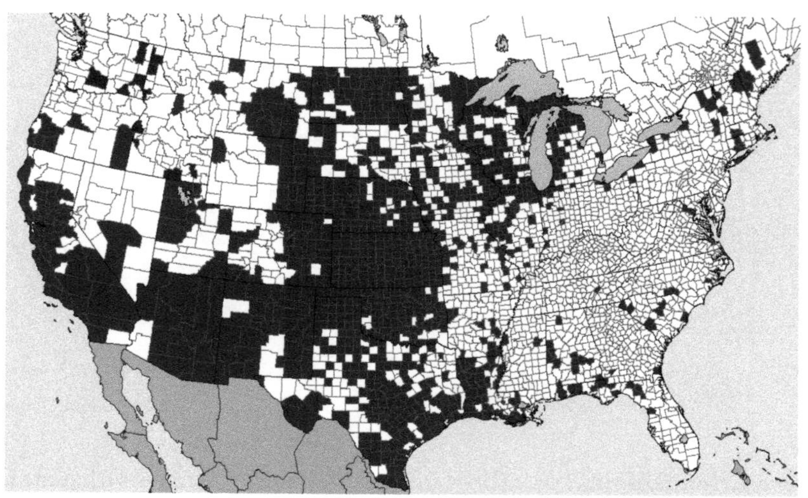

Distribution of **Ambrosia psilostachya**, WESTERN RAGWEED

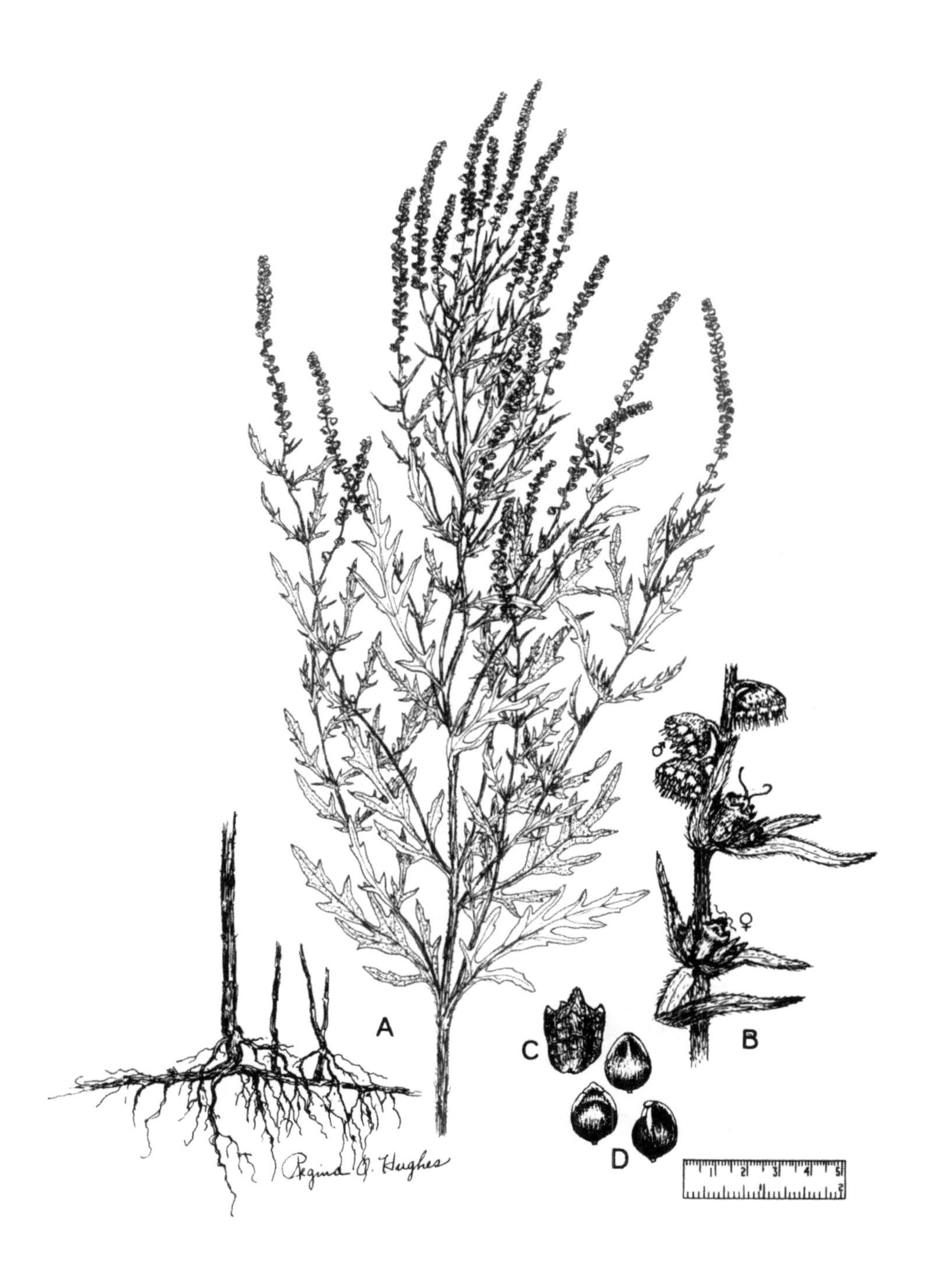

Ambrosia psilostachya, WESTERN RAGWEED. A, habit. B, raceme of male heads and female involucres. C, achene. D, seeds.

Ambrosia tomentosa Nutt. SKELETON-LEAF BURSAGE

Perennial herb, reproducing by seeds and by deep creeping rootstocks.

DESCRIPTION Stems erect or ascending, 2–6 dm tall, bushy, branching from the base, white-woolly to sparingly pubescent. **Leaves** alternate, mostly bipinnately lobed, white tomentose on the underside, green or grayish rough-haired to smooth above, 5–12 cm long, the lobes narrow, irregularly cut on the margin, petiole winged. **Flowers** small, male and female borne separately but on the same plant; the staminate heads with spathulate, 1-nerved chaff shorter than the corollas, in small drooping heads along the tips of the branches, solitary or in several terminal racemes; the pistillate flowers 1–2, in the axils of the upper leaves. **Burs** (mature involucres) 4–6 mm long, rough-haired, with 2 beaks, with 8–12 spines, about 1–1.5 mm long, conical. **Achene** 1–3 per bur, 4–6 mm long.

FLOWERING July–September.

SYNONYMS *Franseria discolor* Nutt.

WHERE FOUND Meadows, in moist cultivated fields, waste places, pastures, and irrigated fields, especially if poorly drained; also in dry regions and plains.

ORIGIN Native.

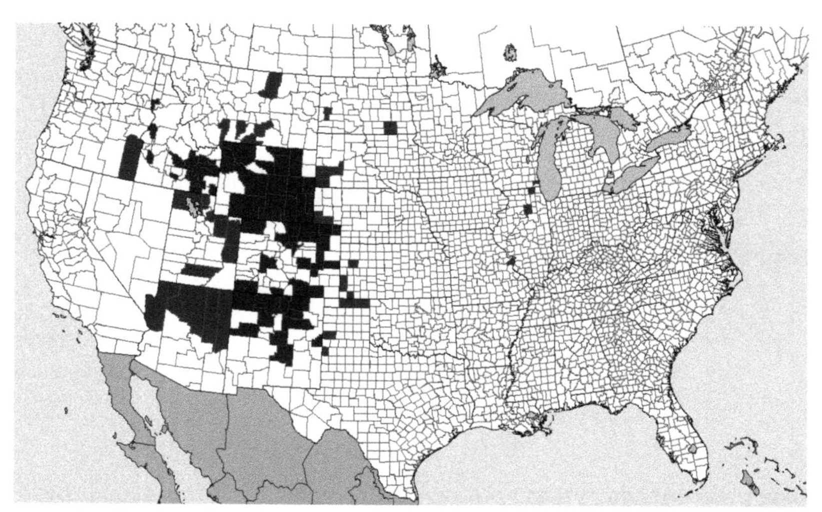

Distribution of **Ambrosia tomentosa**, SKELETON-LEAF BURSAGE

Ambrosia tomentosa, SKELETON-LEAF BURSAGE. A, habit. B, enlarged raceme, with staminate (a) and pistillate (b) flowers. C, fruit (burlike).

Ambrosia trifida L. GIANT RAGWEED

Annual herb, reproducing by seeds.

DESCRIPTION Stems coarse, rough-hairy, reaching a height of 4–6 m on fertile moist soils, 1–3 m on less fertile drier soils. **Leaves** all opposite, large, slightly hairy, entire or palmately cleft into 3 (occasionally 5) lobes, the lobes ovate-lanceolate, serrate, the petioles slightly margined. **Flowers of two kinds** and separated: the **male flowers** abundant in spikelike clusters (racemes) on the tips of branches and stems, the staminate involucres 3-ribbed on one side; the **female flowers** few, without petals, in the axils of the upper leaves, the female involucre (including the achene) 6–13 mm long, with a woody hull, the beak acute, conical, 2–4 mm long and 4–10 acute tubercules terminating the ribs.

FLOWERING Late June–September.

WHERE FOUND Alluvium, bottomlands, fertile moist soils, waste places, and thickets; most serious in corn, soybeans and other cultivated crops. Abundant pollen is a hazard to hay fever sufferers.

ORIGIN Adventive from Europe.

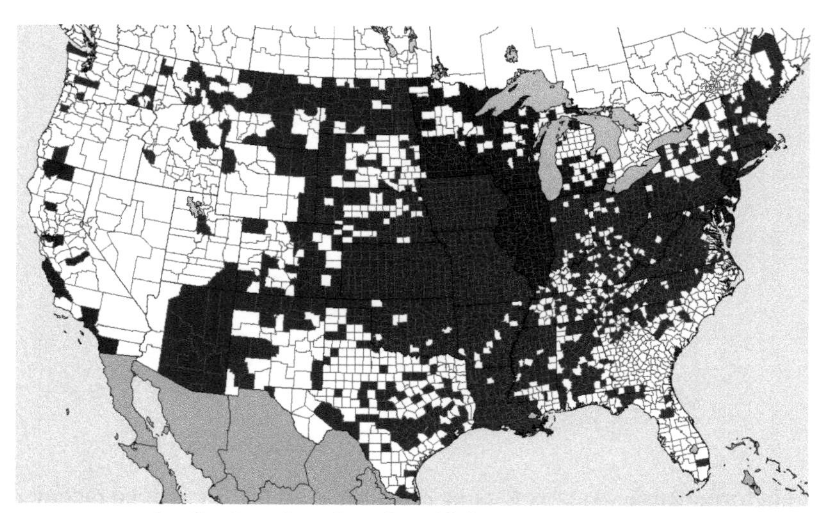

Distribution of **Ambrosia trifida**, GIANT RAGWEED

Ambrosia trifida, GIANT RAGWEED. A, habit, upper portion. B, portion of flowering raceme. C, achene.

Amphiachyris dracunculoides (DC.) Nutt. PRAIRIE BROOMWEED

Annual herb, reproducing by seeds.

DESCRIPTION Stems 3–10 dm tall, simple at the base, effusely branching above, paniculate. **Leaves** alternate, entire, blades of the upper leaves narrowly linear or thread-shaped, with a tendency to fall away early leaving the plant naked and wiry. **Flower heads** numerous, small with few yellow flowers. Involucres about 3 mm thick and 3 mm long. **Ray flowers** fertile, their pappus crown-like, of extremely minute, irregular, hyaline teeth in layers. **Disk flowers** abortive, their pappus a short papery cup, tipped with about 5 coarse, white or straw-colored fine bristles. **Achene** elongated, but never strongly flattened.

FLOWERING August–October.

SYNONYMS *Brachyris dracunculoides* DC., *Gutierrezia dracunculoides* (DC.) Blake, *Xanthocephalum dracunculoides* (DC.) Shinners)

WHERE FOUND Dry upland prairies, rocky open limestone barrens, roadsides, fallow fields, and along railroads.

ORIGIN Native.

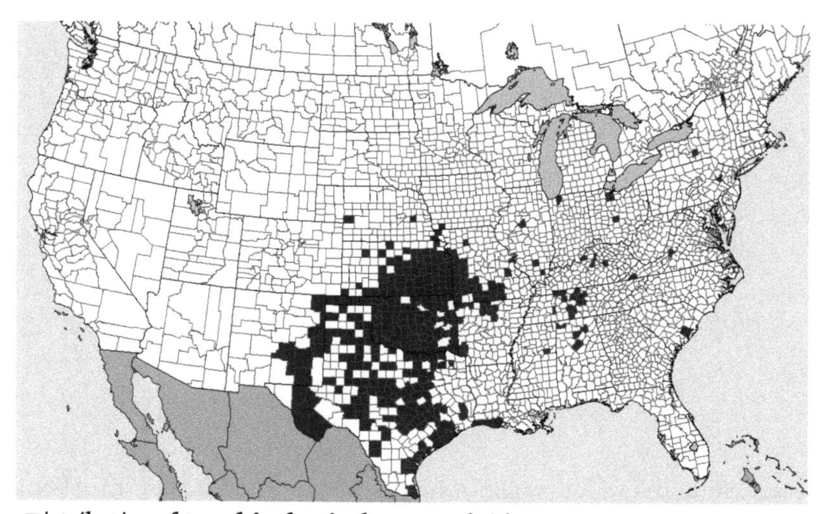

Distribution of **Amphiachyris dracunculoides,** PRAIRIE BROOMWEED

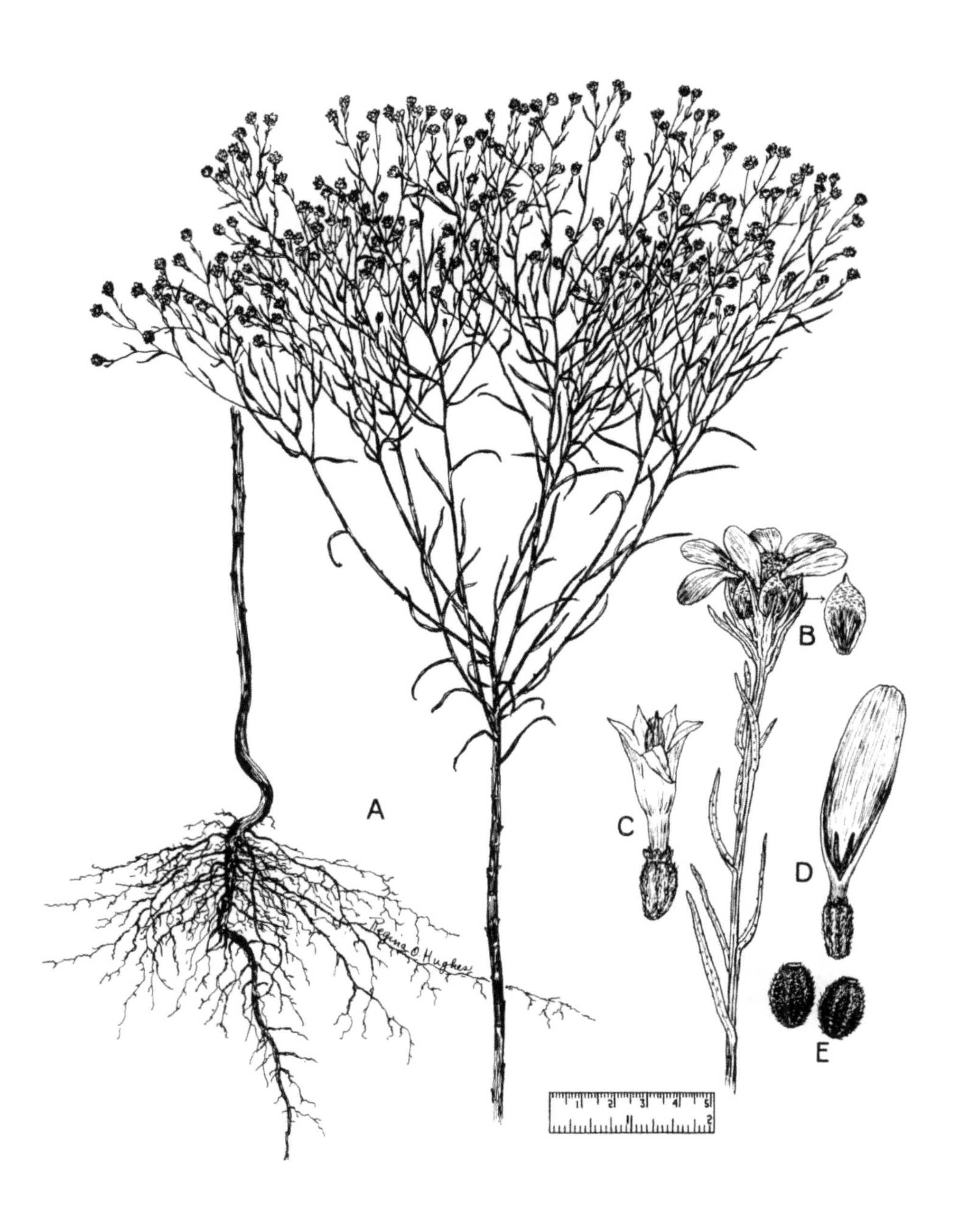

Amphiachyris dracunculoides, PRAIRIE BROOMWEED. A, habit. B, flower head. C, disk flower D, ray flower. E, achenes.

Anthemis cotula L. MAYWEED

Annual or winter-annual herb, with short thick taproot, reproducing only by seed.

DESCRIPTION Stems erect slender, much-branched, nearly smooth, 1-6 dm tall. **Leaves** about 2-6 cm long, finely divided, with narrow segments, with a strong disagreeable odor. **Flower heads** short-stalked, resembling those of the daisy, 1.3-2.5 cm wide, borne singly at the ends of branches, **ray flowers** white, 5-11 mm long, surrounding a mass of numerous small yellow **disk flowers**, the disk about 5-10 mm wide, becoming ovoid to short-cylindric at maturity. Involucre sparsely soft hairy. **Achene** brown, oblong, 10-ribbed, with glandular tubercles. **Pappus** none.

FLOWERING May-October.

WHERE FOUND In abandoned fields, waste places, and barnyards; a cosmopolitan weed.

ORIGIN Native of Europe; throughout the United States.

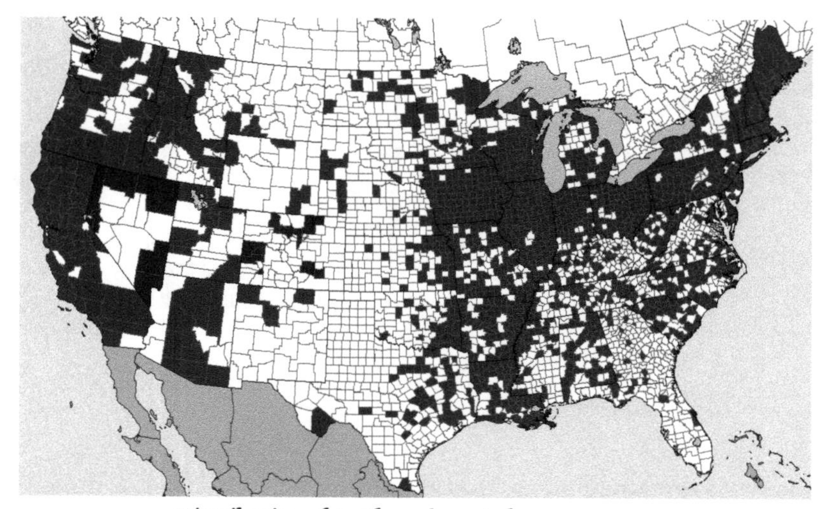

Distribution of **Anthemis cotula**, MAYWEED

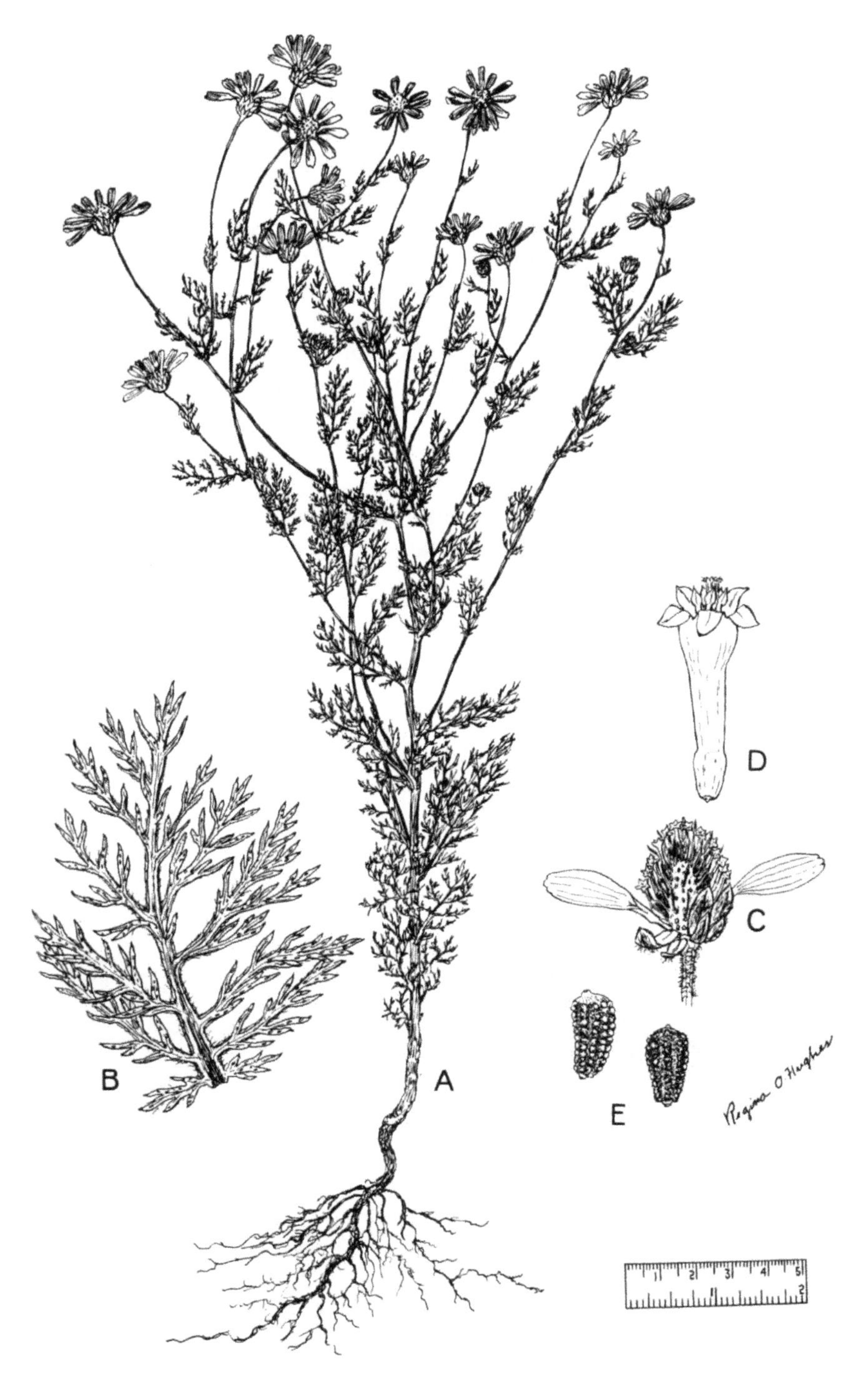

Anthemis cotula, MAYWEED. A, habit. B, enlarged leaf. C, flower head. D, disk flower. E, achenes.

Arctium minus (Hill) Bernh. COMMON BURDOCK

Biennial herb, reproducing by seed only. Taproot large, fleshy, living over one winter.

DESCRIPTION Stems a crown close to the soil surface the 1st year; the 2nd year much-branched, up to 1.5 m tall, hairy, somewhat grooved or angular. **Leaves** large, petiolate, the lower petioles hollow, the 1st year forming a dense rosette, the 2nd year distributed alternately on the stem with larger leaves toward the base, the blade narrowly to very broadly ovate, up to 5 dm long and 4 dm wide, thinly woolly, often eventually smooth beneath, nearly glabrous above. **Inflorescence** in raceme-like axillary clusters. **Flower heads** 1.5–3 cm in diameter, short-stalked or nearly sessile, glabrous or slightly glandular to sometimes webby-woolly, made up of numerous small red-violet **disk flowers**, surrounded by numerous hooked bracts that later form a bur about 1.4 cm in diameter. **Achene** rather rough, mottled dark-gray, about 0.7 cm long.

FLOWERING July–October.

WHERE FOUND Waste places where the soil is productive but undisturbed, neglected farmlands; not commonly found in cultivated areas.

ORIGIN Naturalized from Europe.

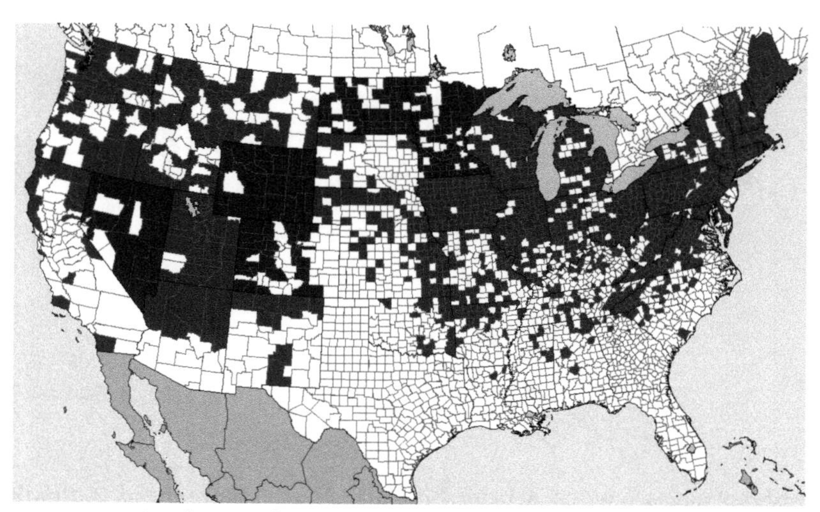

Distribution of **Arctium minus**, COMMON BURDOCK

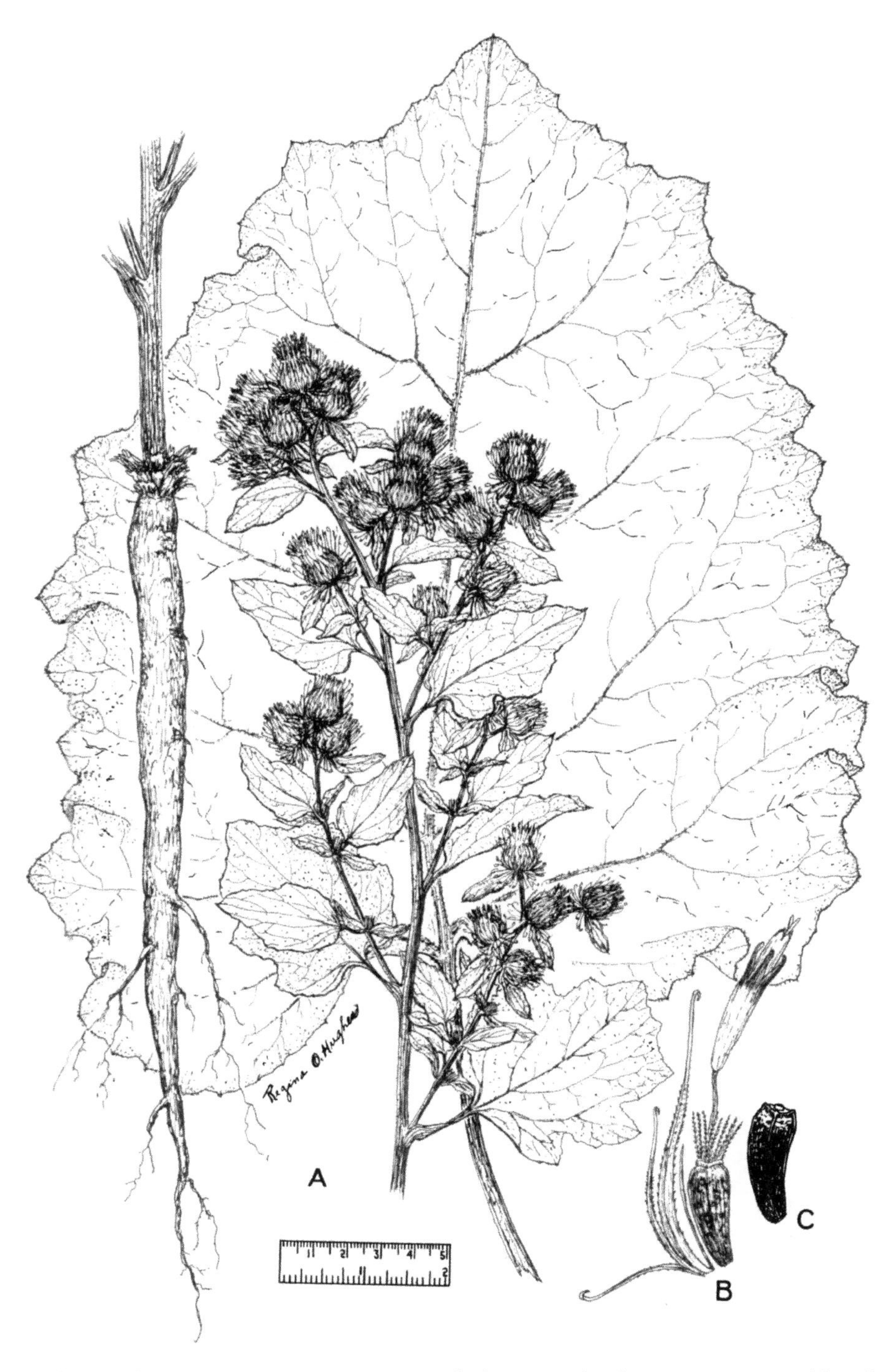

Arctium minus, COMMON BURDOCK. A, habit, root, leaf, upper raceme of heads. B, flower and phyllaries. C, achene.

Artemisia vulgaris L. MUGWORT

Perennial herb, with a short rhizome, aromatic.

DESCRIPTION Stems 0.5–1.5 m tall, simple or branched above, glabrous or nearly so below the inflorescence. **Leaves** green, glabrous above, densely white-woolly beneath, obovate or ovate, 5–10 cm long, 3–7 cm wide, the principal leaves cleft nearly to the midrib into ascending, acute, unequal segments that in turn are again toothed or more deeply cleft, frequently with 1 or more pairs of stipule-like lobes at the base. **Inflorescence** generally dense and leafy. Involucres mostly 3.4–4.5 mm high, more or less woolly. Receptacle without hairs. **Flowers** all fertile, the outer ones pistillate.

FLOWERING July–September.

WHERE FOUND Waste places, fields, and pastures; especially on limey soils.

ORIGIN Native of Europe.

NOTE Formerly used medicinally, but toxic properties that result from overdoses cause pain, spasms, and other disturbances.

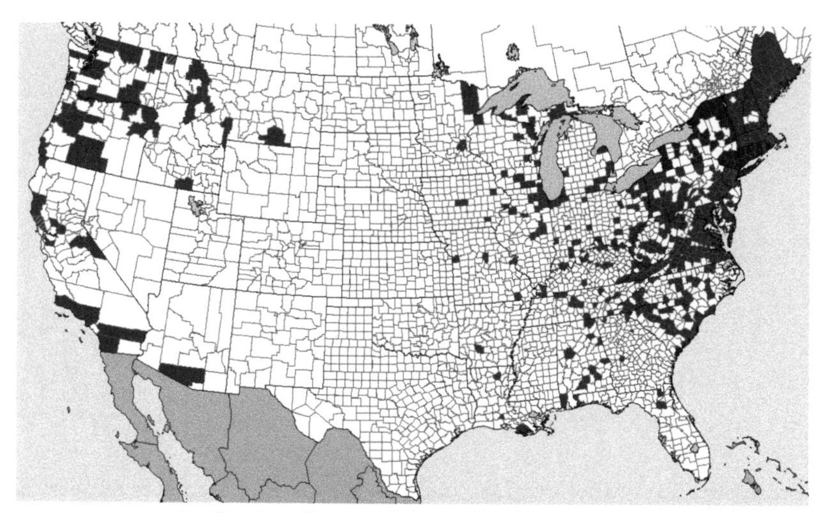

Distribution of **Artemisia vulgaris**, MUGWORT

Artemisia vulgaris, MUGWORT. A, habit. B, enlarged leaves. C, panicle. D, flower head. E, flowers. F, achenes.

Bidens bipinnata L. SPANISH-NEEDLES

Annual herb, reproducing by seeds. Taproot with numerous side branches.

DESCRIPTION Stems erect, square, glabrous or minutely hairy, 3–17 dm tall, branching in the upper portions. **Leaves** opposite on both stems and branches, 2–3 times pinnate, about 4–20 cm long, the ultimate segments tending to be rounded, the petioles about 2–5 cm long. **Heads** borne singly at the ends of long, slender, nearly leafless branches, narrow, discoid, the disk 4–6 mm wide at anthesis, the outer involucral bracts 7–10, linear, acute, not leafy, shorter than the inner; **ray flowers** pale-yellow and **disk flowers** yellow. **Achene** linear, black or dark-brown, attenuate above, often sparsely hairy, 10–18 mm long, the outer achenes often shorter. **Pappus** of 3–4 yellowish barbed awns.

FLOWERING August–October.

WHERE FOUND In cultivated fields, often in sandy soil, waste places, gardens, grain stubble fields, rocky woods, and moist or wet places.

ORIGIN Native or introduced northward; a semi-cosmopolitan weed.

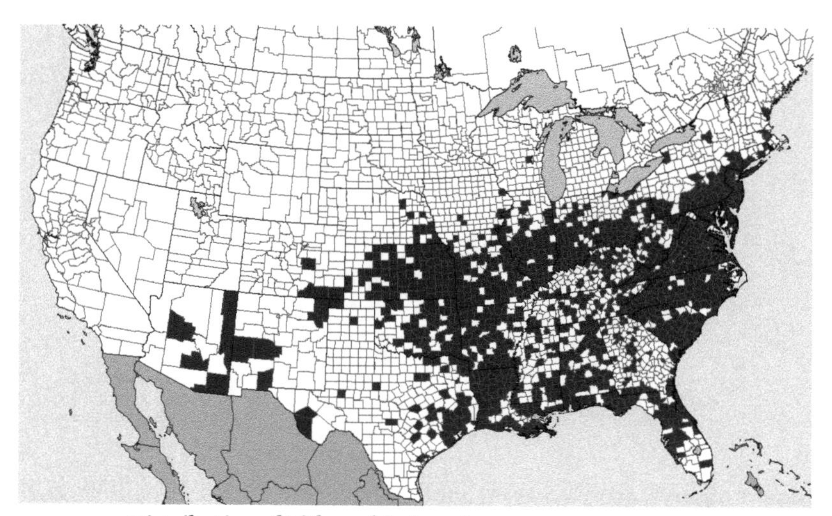

Distribution of **Bidens bipinnata**, SPANISH-NEEDLES

Bidens bipinnata, SPANISH-NEEDLES. A, habit. B, enlarged leaves. C, flower head. D, mature achenes. E, achene.

Bidens frondosa L. DEVIL'S-PITCHFORK

Annual herb, reproducing from seeds. Taproot shallow, much-branched.

DESCRIPTION Stems erect, glabrous or nearly so, 2–12 dm (occasionally up to 1 m) tall, somewhat 4-sided, branching near the top. **Leaves** opposite, 4-ranked, on petioles 1–6 cm long, pinnately compound, with 3–5 lanceolate, acuminate, serrate leaflets up to 10 cm long, 3 cm wide, sometimes sparsely short-hairy beneath, at least the terminal one slenderly stalked. **Heads** saucer-shaped to hemispheric or nearly globose, discoid or nearly so, the **ray flowers** orange-yellow, neutral, sometimes absent, the **disk corollas** 2.5–3 mm long, orange, equaling the disk, 5-toothed, the ligules golden-yellow, the disk up to 1 cm wide at flowering; the outer involucral bracts 5–10 (typically 8), green and more or less leafy, usually longer than the orange disk, hairy on the margins, at least toward the base. **Achene** flat, wedge-shaped, strongly 1-nerved on each face, dark-brown or blackish, nearly glabrous or appressed-hairy, mostly 5–10 mm long. **Pappus** of slightly divergent to erect barbed awns.

FLOWERING June–October.

WHERE FOUND In rich, moist soils, but occasionally in dry waste places; pastures, roadsides, gardens, and damp open habitats; in cultivated ground.

ORIGIN Native.

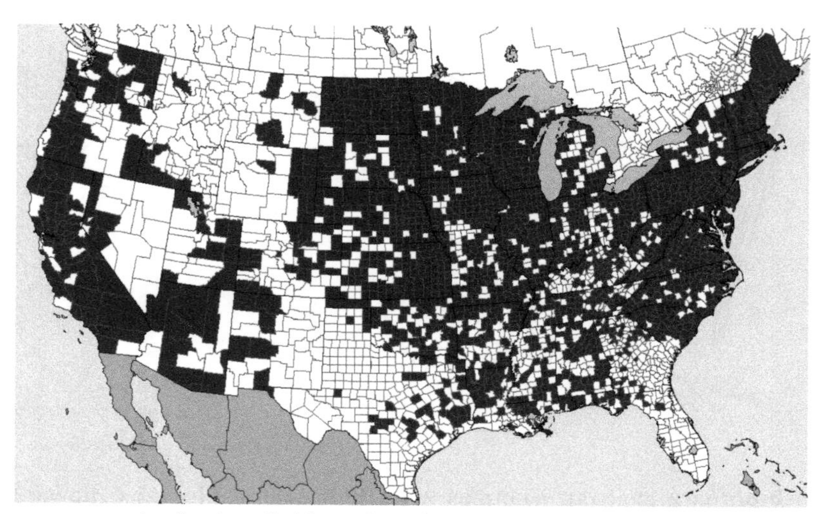

Distribution of **Bidens frondosa**, DEVIL'S-PITCHFORK

Bidens frondosa, DEVIL'S-PITCHFORK. A, habit. B, flower head. C, flower, female. D, achenes.

Centaurea diffusa Lam. DIFFUSE KNAPWEED

Biennial herb, pubescent becoming glabrate, rough, with an elongated taproot.

DESCRIPTION Stems erect, 5–8 dm tall, angled but not winged, branched near or above the base. **Leaves** alternate, the basal ones in a whorl, bipinnate to bipinnatifid, oblanceolate to oblong, up to 20 cm long and 5 cm wide, short-petioled, the ultimate segments narrowly oblong to elliptic, usually acute and wedge-shaped; the stem leaves sessile, the lower leaves bipinnate to bipinnatifid, the upper leaves much reduced and pinnately lobed, the uppermost leaves bract-like and entire or minutely lobed. **Flower heads** solitary, more or less clustered at the ends of the branches, 1.5 cm long. Involucre narrowly ovate or oblong, about 1 cm long, woolly becoming glabrate and granular, the phyllaries leathery, nerved, the outer and middle phyllaries broadly to narrowly ovate, pale yellowish-green with a light-brown margin, the upper part narrowed into a stiff spine; the inner phyllaries lanceolate, tipped by a papery or leathery fringed appendage, spiny or spineless. **Flowers** white, pink or lavender, the outermost flowers sterile, inconspicuous, with threadlike corolla lobes. **Achene** oblong, 2.5 mm long, dark-brown, marked with several conspicuous to faint, pale-brown or ivory lines. **Pappus** none or on the inner achenes as white chaffy scales less than 1 mm long.

FLOWERING June–September.

WHERE FOUND Waste grounds, fields, and roadsides; locally common.

ORIGIN Native in southeastern Europe and western Asia.

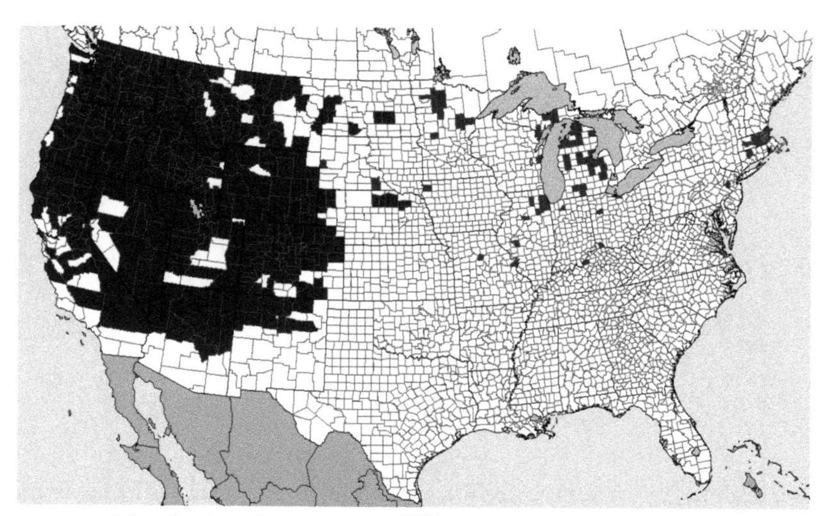

Distribution of **Centaurea diffusa**, DIFFUSE KNAPWEED

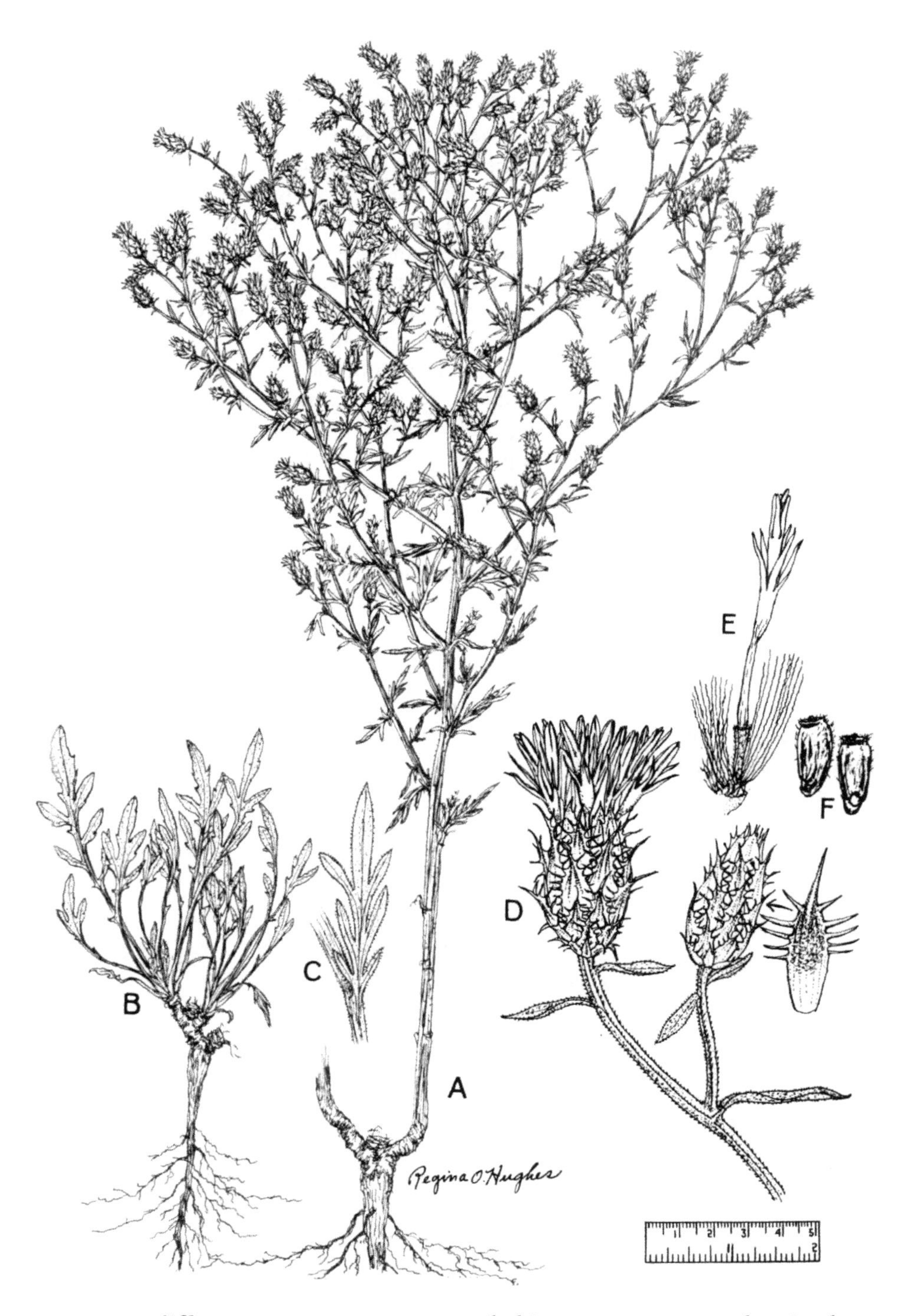

Centaurea diffusa, DIFFUSE KNAPWEED. A, habit. B, young rosette, showing long petioled basal leaves. C, cauline leaf, sessile. D, flower head. E, flower. F, achenes.

Centaurea repens L. RUSSIAN KNAPWEED

Perennial herb, reproducing by leafy shoots from the underground stems and by seed.

DESCRIPTION Stems corymbose-branched, bushy, 3–10 dm tall, from creeping horizontal and vertical underground stems (6–12 dm deep, the younger stems light-colored, the older stems dark-brown or black), the stem wingless, ridged, very leafy, covered with soft gray hairs when young, becoming velvety to nearly glabrous. **Leaves** alternate, simple, firm, variable in shape, the basal leaves deeply lobed or pinnatifid, 5–10 cm long, 1–2.5 cm broad, forming a quickly withering rosette; the lower stem leaves smaller, lobed or sharply toothed; the upper leaves entire, 1–3 cm long, linear to narrowly oblong, the tip sharp-pointed and the margins smooth or slightly toothed. **Flower heads** solitary, terminating leafy branchlets, small, cone-shaped, 0.8–1.3 cm in diameter just above the base, many-flowered. **Flowers** all tubular, lavender-blue to rose-pink or purple. Involucre slenderly ovoid, pale, about 1 cm high, phyllaries in many overlapping series, the outer phyllaries rounded-ovate with clear entire margins, the inner phyllaries oblong-acuminate and cut-margined, with very hairy, long, tail-like tips. **Achene** shortly obovoid, 2–3 mm long, grayish or ivory-colored, smooth with inconspicuous lines, the basal scar not oblique and not notched near the base. **Pappus** whitish, threadlike, in one series, deciduous (dropping off by achene maturity).

FLOWERING June–October.

SYNONYMS *Centaurea picris* Pall., *Rhaponticum repens* (L.) Hidalgo

WHERE FOUND Pastures, grainfields, cultivated fields, meadows, waste places, roadsides, and irrigation ditches; pest in corn, alfalfa, sugarbeets, and forage-seed crops. A noxious perennial weed, avoided by all livestock because of the bitter quinine-like taste.

ORIGIN Introduced and naturalized from the Caucasus in southern Russia and Asia.

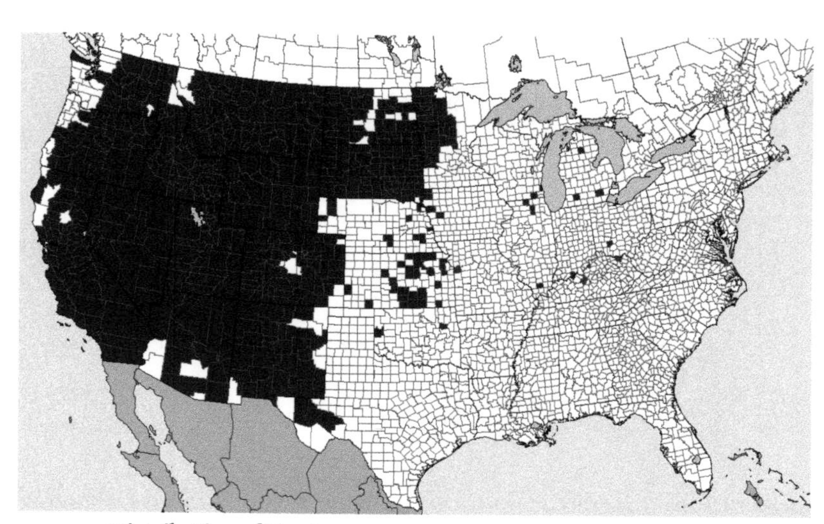

Distribution of **Centaurea repens**, RUSSIAN KNAPWEED

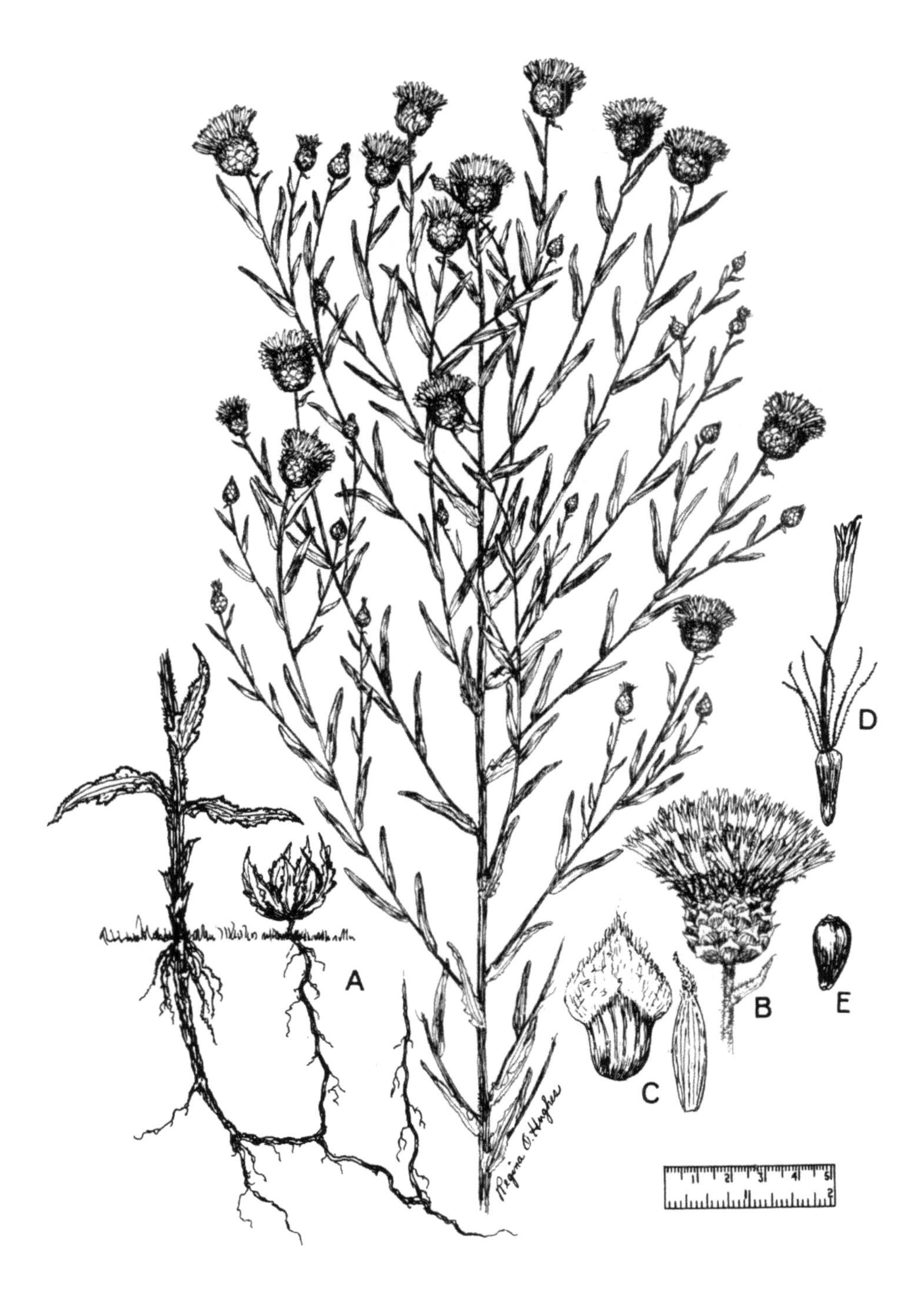

Centaurea repens, RUSSIAN KNAPWEED. A, habit. B, head. C, bracts, outer and inner. D, flower. E, achene.

Centaurea solstitialis L. YELLOW STAR-THISTLE

Annual herb, sometimes biennial, from a taproot, reproducing only by seed.

DESCRIPTION **Stems** rigid, branched from near the base, bushy, woolly, grayish, 1.5-7.5 dm tall, with narrow green wings. **Leaves** alternate, cottony-hairy; the basal leaves clustered on the ground, 5-15 cm long, usually petioled, deeply lobed with a much larger lobe at the tip; the stem leaves entire, stalked, narrow, sharp-pointed and lobeless, margins smooth, often wavy, the bases continuing down the stem as wings; the upper leaves greatly reduced. **Flower heads** terminal, urn-shaped with many inconspicuous bright-yellow, tubular flowers stiffly spreading above the narrow tip, the whole 1.8-2.5 cm high, without leafy bracts. Involucre globose, woolly. Involucral bracts numerous, closely overlapping, stiff and papery, the middle bracts ending in a stout, rigid, unbranched yellow spine, 1.2-2.5 cm long, with 1 or 2 pairs of very short spines at its base. Receptacle bristly, flat. **Achene** smooth, light-colored, often darker mottled, or dark-brown, 2-3 mm long, oblong, notched on one side just above the base. **Pappus** of many white, thin bristlelike scales of unequal length, 3-5 mm long, these lacking on the outermost seeds, persistent or deciduous.

FLOWERING May or June to frost.

WHERE FOUND In cultivated and fallow fields, pastures, rangelands, and waste places; mostly at low elevations, but sometimes up to 8,200 feet. A noxious weed in some areas; in fields of alfalfa and small cereals (wheat, oats, barley, and rice); a rangeland weed in Western States; an injurious pest because of the vicious spines.

ORIGIN Native in the Mediterranean region; also in western and central Asia.

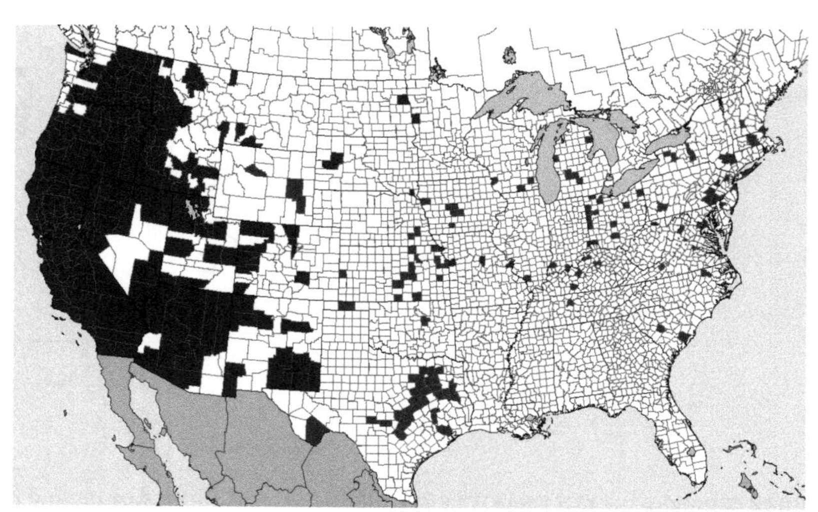

Distribution of **Centaurea solstitialis**, YELLOW STAR-THISTLE

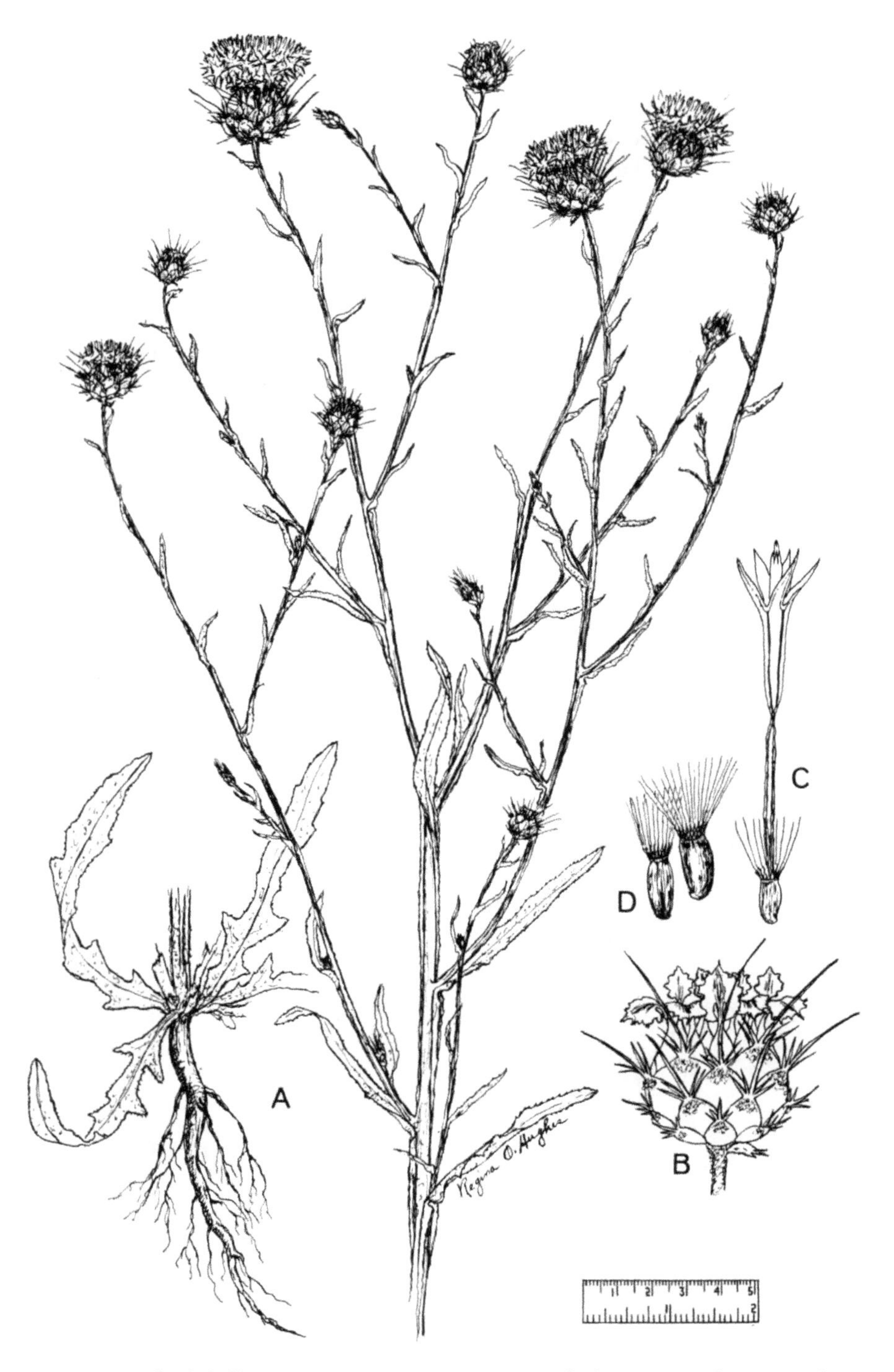

Centaurea solstitialis, YELLOW STAR-THISTLE. A, habit. B, involucre. C, flower. D, achenes.

Centaurea stoebe L. SPOTTED KNAPWEED

Biennial or short-lived perennial herb, reproducing by seeds.

DESCRIPTION **Stems** erect or ascending, with slender wiry branches, rough-pubescent, 3–10 dm high. **Leaves** alternate, pinnatifid, with narrow divisions, rough-pubescent, the upper leaves often linear. **Flower heads** terminal and axillary, numerous and clustered, 1.5–2.5 cm in diameter, many-flowered. Involucre pale, 1–1.4 cm high, its smooth strongly ribbed outer and median ovate phyllaries with firm points and 5–7 pairs of cilia, the dark tip 1–2 mm long, the innermost phyllaries entire or fringed. Receptacle bristly, flat. **Flowers** all tubular, whitish to pink or purple, perfect, the marginal flowers enlarged, falsely radiate, neutral. **Achene** brownish, about 2 mm long, notched on one side of the base. **Pappus** with a short tuft of bristles at tip end, 1–2 mm long, persistent.

FLOWERING August–September.

SYNONYMS *Centaurea maculosa* Lam.

WHERE FOUND Dry sterile, gravelly, or sandy pastures, old fields, and roadsides.

ORIGIN Introduced and naturalized from Europe.

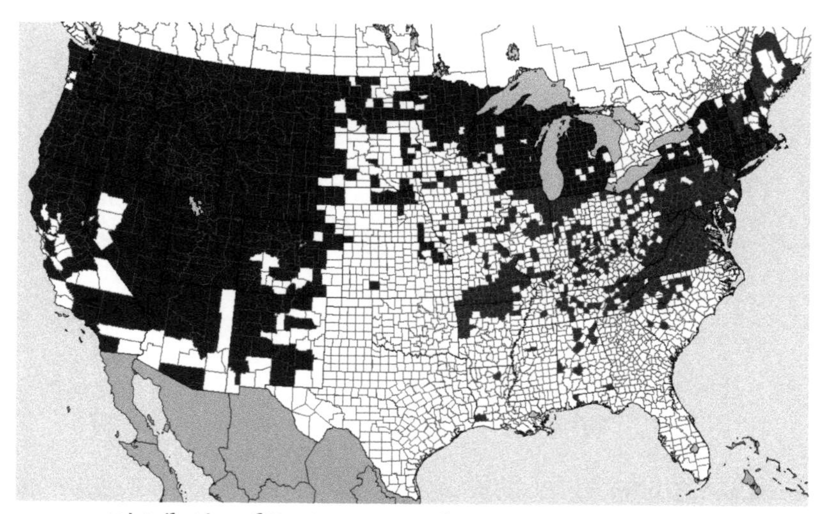

Distribution of **Centaurea stoebe**, SPOTTED KNAPWEED

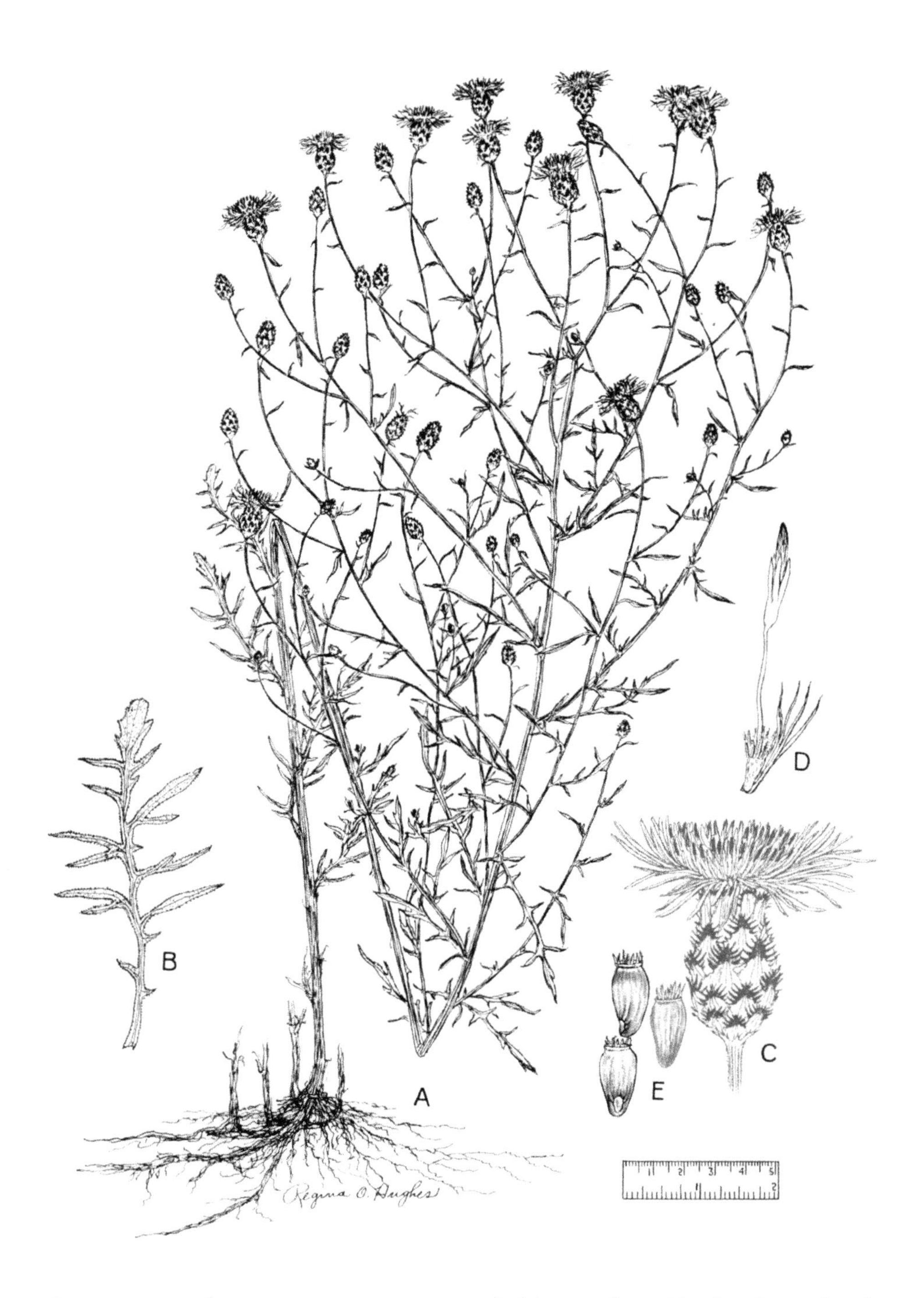

Centaurea stoebe, SPOTTED KNAPWEED. A, habit. B, enlarged leaf. C, flower head. D, disk flower. E, achenes.

Chrysothamnus viscidiflorus (Hook.) Nutt. DOUGLAS RABBITBRUSH

Perennial shrub, rounded, many upright branches, white-barked, up to 1 m tall, with brittle twigs, glabrous or minutely velvety with rather stiff spreading hairs.

DESCRIPTION Leaves linear to linear-lanceolate, flat or twisted, erect, spreading or reflexed, 1-6 cm long, 0.5-10 mm wide, 1to 5-nerved, glabrous or puberulent, sticky, sometimes with punctate glands ventrally. **Flower heads** about 5-flowered, in terminal broad cymes. Involucre 5-8 mm high, the phyllaries (about 15) linear-oblong to lanceolate, obtuse to acute, not keeled, strongly graduate but in obscure vertical ranks, papery. **Achene** densely hairy. **Pappus** brownish-white. A very polymorphic species, with several subspecies or varieties.

FLOWERING July-September.

WHERE FOUND Dry open places in valleys, plains, and foothills.

ORIGIN Native.

Distribution of **Chrysothamnus viscidiflorus**, DOUGLAS RABBITBRUSH

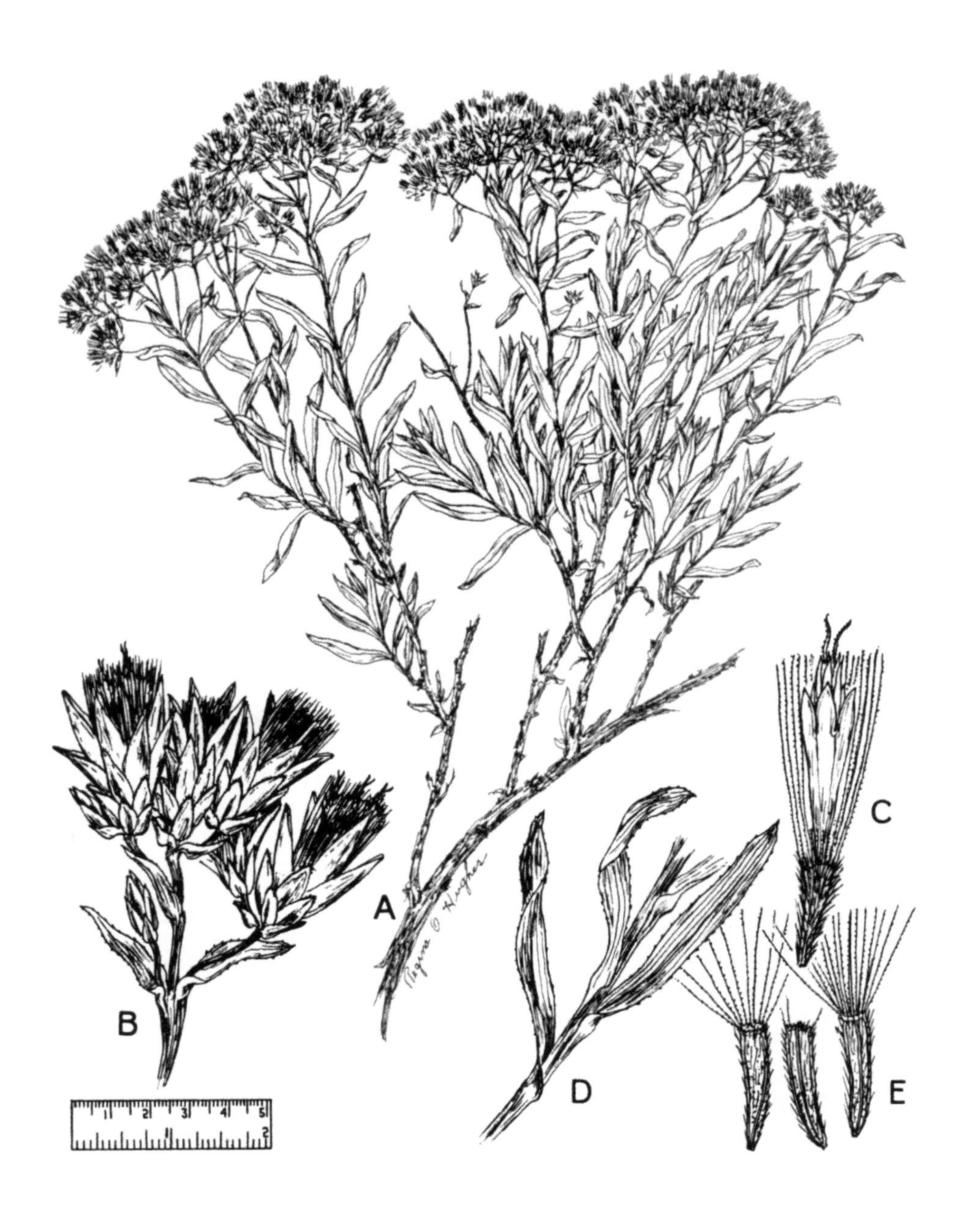

Chrysothamnus viscidiflorus, DOUGLAS RABBITBRUSH. A, habit (a branch) B, inflorescence. C, flower. D, enlarged leaves. E, achenes.

Cichorium intybus L. CHICORY

Perennial herb, reproducing by seeds and from roots below the crown. Taproot long, deep, branched, with a milky sap.

DESCRIPTION Stems hollow, 3–24 dm tall, often rough-hairy, becoming woody and reddish, the branches rigid, stiffly spreading. **Leaves** alternate, mainly clustered near the base, or forming a rosette at the ground, long-petioled, 1–2 dm long, up to 12 cm broad, irregularly toothed to deeply lobed, glabrous to rough-hairy. Upper leaves entire and dentate, oblong-lanceolate, greatly reduced (3–7 cm long), sessile, clasping the stem, the base extended into a pair of ear-like projections. **Flower heads** numerous, 2.5–3 cm in diameter, axillary, 1–4 together in sessile clusters along the rigid, nearly glabrous branches or at the tip of short, stiff branches that often have gland-tipped hairs. **Flowers** perfect, all strap-shaped ray flowers, sky-blue, sometimes white or rarely pink. Bracts surrounding the flower heads in 2 rows, the outer 5 about half as long as the 8–10 inner, thickened and yellowish at the base, sometimes with gland-tipped hairs, the margins minutely spiny. **Achene** 2–3 mm long, obovate, light-brown and darker mottled, finely granular, obscurely 4- to 5-angled, the tip blunt, beakless. **Pappus** a minute fringed crown of tiny bristle-like scales.

FLOWERING June–October, as early as March in the South and in the Pacific Northwest.

WHERE FOUND Along roadsides, grasslands, fence rows, and waste ground, preferring neutral or limestone (where most troublesome) soils; in lawns, fields of small grain and pastures; does not survive in cultivated land.

ORIGIN Native of the Mediterranean region, cultivated in Europe.

NOTE Roots used for a coffee substitute; roots eaten boiled; leaves cooked for greens or in salads; relished by all livestock but gives bitter taste to milk and butter.

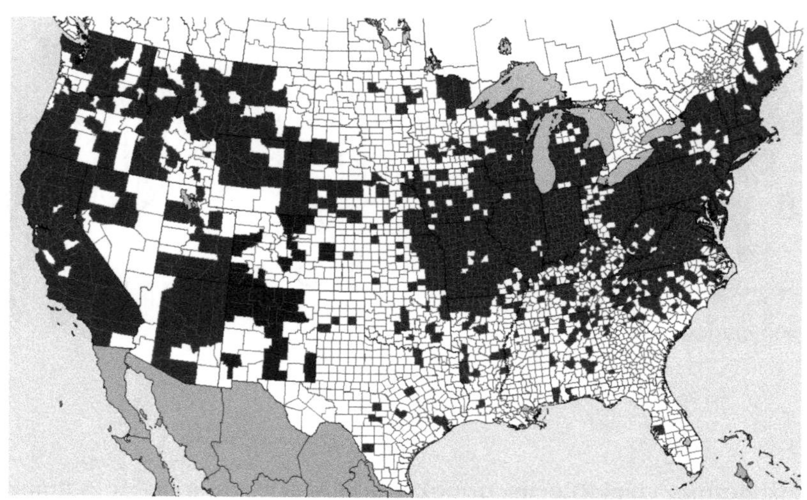

Distribution of **Cichorium intybus**, CHICORY

Cichorium intybus, CHICORY. A, habit. B, terminal portion of inflorescence. C, involucre. D, flower. E, achenes.

Cirsium arvense (L.) Scop. CANADA THISTLE

Perennial herb, reproducing by seeds and horizontal roots.

DESCRIPTION Roots extending several feet into the soil or extensively creeping horizontally. **Stems** erect, 4-12 dm tall, grooved, branching only at the top, nearly glabrous or slightly hairy when young, increasingly hairy with maturity. **Leaves** alternate, oblong or lanceolate, usually with crinkled edges and spiny margins, somewhat lobed, hairy beneath or often glabrous or nearly so at maturity, the upper leaves sessile and only slightly decurrent. **Flower heads** dioecious (male and female flowers usually in separate heads and borne on different plants), numerous, compact in corymbose clusters, terminal and axillary, 2-2.5 cm in diameter, with lavender, rose-purple, or white disk flowers only. Involucre 1-2 cm high, the bracts numerous, overlapping, spineless. Receptacle bristly, chaffy. **Achene** light to dark-brown, smooth oblong, 2.5-3.5 mm long, flattened, curved or straight, apex blunt with a tubercle in the center. **Pappus** tannish-brown, plumose, about 2 mm long, deciduous.

FLOWERING July-October.

WHERE FOUND In all crops, pastures, meadows, and waste places; in rich, heavy soils; a noxious weed.

ORIGIN Introduced and naturalized from Eurasia; not native of Canada.

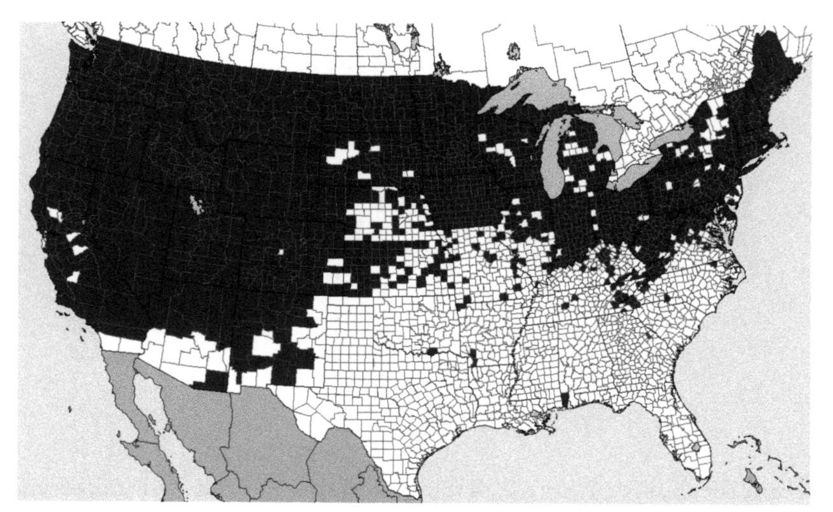

Distribution of **Cirsium arvense**, CANADA THISTLE

Cirsium arvense, CANADA THISTLE. A, habit. B, head. C, flower. D, achenes.

Cirsium vulgare (Savi) Tenore — BULL THISTLE

Biennial herb, reproducing by seeds, plants forming 1st year oblanceolate to elliptic coarsely toothed leaves in a rosette, from a large fleshy taproot.

DESCRIPTION Stems erect in 2nd year, 1-2 m tall, stout, often branched, more or less hairy, spiny-winged by the decurrent leaf bases. **Leaves** alternate lanceolate, pinnatifid, spiny with long needle-pointed tips, rough-spiny above, thinly white-woolly to green and merely hairy beneath. **Flower heads** compact, 2.5-5 cm in diameter, composed of many deep-purple or rose disk flowers. Involucre 2.5-4 cm high, its numerous bracts spine-tipped, without any well-developed sticky dorsal ridges. Receptacle flat, bristly. **Achene** straw-colored, striped with brown or black, ridged around one end, tipped with down, less than 4 mm long. **Pappus** plumose, about 2 mm long, deciduous.

FLOWERING June-October.

SYNONYMS *Cirsium lanceolatum* Hill

WHERE FOUND Fields, pastures, first-year meadows, and wastelands. An aggressive and bad weed, but will not survive in cultivated fields.

ORIGIN Introduced and naturalized from Eurasia.

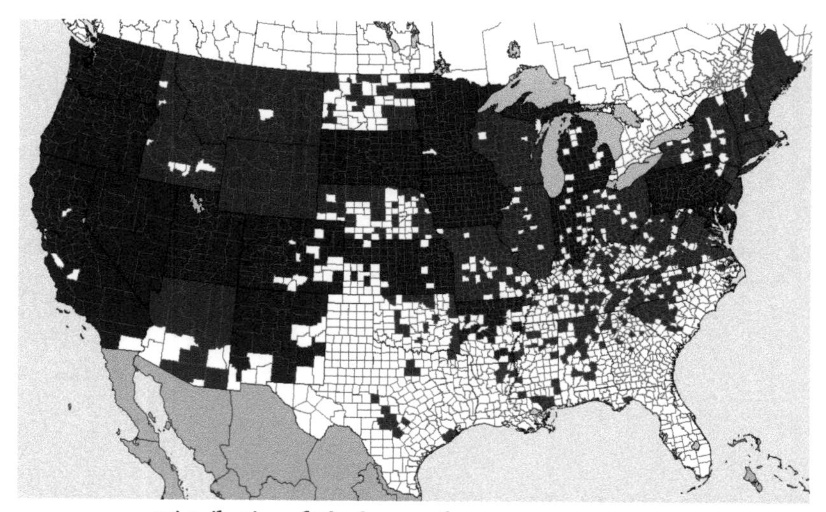

Distribution of **Cirsium vulgare**, BULL THISTLE

Cirsium vulgare, BULL THISTLE. A , habit. B, flower. C, immature fruit. D, achenes.

Cyclachaena xanthiifolia (Nutt.) Fresen. MARSH-ELDER

Annual herb, with a taproot, reproducing by seeds.

DESCRIPTION Stems 4–24 dm tall, coarse branching, grayish-green, glabrous below, becoming viscid-villous in the inflorescences. **Leaves** broad, mostly opposite, light grayish-green, ovate, covered with short hairs, 5–20 cm long and 2.5–15 cm wide, coarsely and often doubly serrate. **Inflorescence** large, in panicle shape. Flower heads small, numerous, nearly sessile, drooping in panicles at the top of the stem and in the axils of the upper leaves. **Flowers** either male or female, but borne in the same head, without petals, greenish-yellow. Involucre viscid-hairy or nearly glabrous, about 1.5–3 mm high, the 5 outer bracts larger than the 5 inner ones. **Achene** gray to black, triangular, somewhat flattened with a ridged surface, about 3 mm long, abundant.

FLOWERING August–October.

SYNONYMS *Iva xanthifolia* Nutt.

WHERE FOUND Bottomlands, along roadsides, ditches, pastures, and farmyards; infrequent in cultivated fields.

ORIGIN Native in an area extending from central Washington east to eastern Minnesota and south as far as New Mexico and central Texas and north into Manitoba; adventive in the northeast from New Jersey north through Maine and into Quebec and New Brunswick.

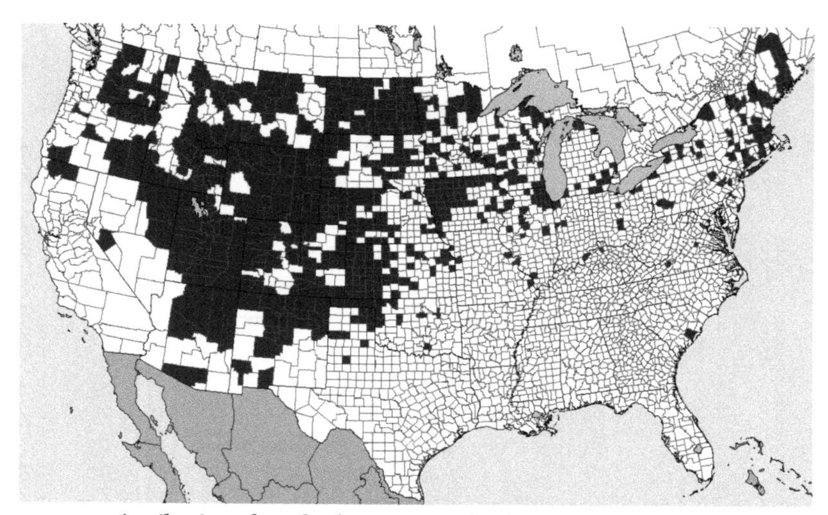

Distribution of **Cyclachaena xanthiifolia**, MARSH-ELDER

Cyclachaena xanthiifolia, MARSH-ELDER. A, habit. B, inflorescence, showing staminate and pistillate flowers in same head. C, achenes.

Ericameria nauseosa (Pallas ex Pursh) Nesom & Baird

RUBBER RABBITBRUSH

Perennial shrub, 3–20 dm tall, usually with several fibrous-barked main stems from the base, these much-branched.

DESCRIPTION Twigs ill-scented, erect, usually densely leafy, clothed with a persistent felt-like gray, white, or greenish wool. **Leaves** variable, linear-filiform to narrowly linear-oblanceolate, 2–7 cm long, 0.5–4 mm wide, 1- to 3-nerved, woolly to nearly glabrous, not much twisted. **Flower heads** in terminal rounded cymose clusters. Involucre 6–13 mm high, the phyllaries (20–25) usually 3- or 4-seriate, strongly graduate, mostly lanceolate or linear-lanceolate, not green-tipped, usually with resinous-thickened midrib. **Flowers** usually 5, yellow. Corolla 7–12 mm long. **Pappus** copious, dull-white.

FLOWERING August–October.

SYNONYMS *Chrysothamnus nauseosus* (Pall.) Britt.

WHERE FOUND In dry, open places, in valleys, plains, and foothills; also in mountains.

ORIGIN Native.

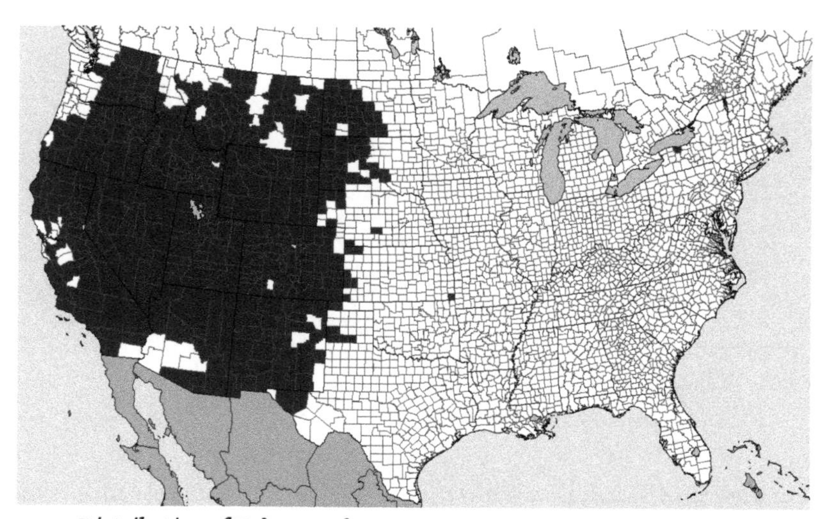

Distribution of **Ericameria nauseosa,** RUBBER RABBITBRUSH

Ericameria nauseosa, RUBBER RABBITBRUSH. A, habit. B, inflorescence. C, flower. D, enlarged leaves. E, achene.

Erigeron annuus (L.) Pers. ANNUAL FLEABANE

Annual or biennial herb, reproducing by seeds.

DESCRIPTION Stems erect, branched above, 3-15 dm tall, glabrous or with scattered spreading stiff hairs. **Leaves** alternate, simple, coarsely and sharply dentate, membranaceous, nearly glabrous, the lower leaves ovate, tapering into margined petioles, the upper leaves lanceolate, acute and entire at both ends, mostly sessile, the uppermost leaves often entire, linear. **Flower heads** in a corymbose cluster, numerous, 1.5-2 cm in diameter, many-flowered, on naked peduncles. Involucral bracts in 1- to 2-series, narrow, equal, slightly hairy. **Ray flowers** about 50-75, white or tinged purple or lavender, much longer than the disk flowers, 1 mm wide, pistillate. **Disk flowers** numerous, yellow, perfect. **Achene** about 1 mm long, obovate, flattened, with minute appressed hairs, straw-colored. **Pappus** double, the outer a crown of short scales, the inner of deciduous bristly hairs, usually absent in the ray flowers.

FLOWERING May-November.

WHERE FOUND Fields, meadows, pastures, and waste places; a cosmopolitan weed.

ORIGIN Native.

NOTE Used in herbal medicine as an astringent, tonic, and diuretic.

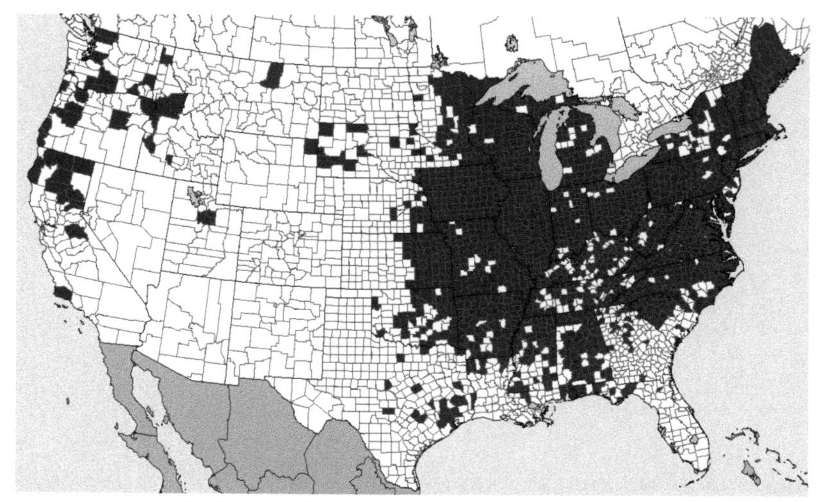

Distribution of **Erigeron annuus**, ANNUAL FLEABANE

Erigeron annuus, ANNUAL FLEABANE. A, habit. B, flower head. C, disk flower. D, ray flower. E, achenes.

Erigeron canadensis L. HORSEWEED

Annual herb, reproducing by seeds.

DESCRIPTION Stems erect, stout, unbranched at base, up to 1.5–2 m tall, with bristly hairs. **Leaves** alternate, simple, numerous, without petioles, dark-green, with scattered coarse white bristles; the lowest oblanceolate or spatulate, toothed or entire, tapering to petioles, 3–13 mm broad; the abundant stem leaves narrowly oblanceolate to linear, mostly entire and sessile. **Flower heads** numerous, racemose or cymose on the branches, forming elongate panicles. Involucres slenderly saucer-shaped, 2.5–5 mm long, the linear-tapering phyllaries with pale tips. **Ray flowers** greenish-white to lavender, inconspicuous, over 100 per head. **Disk flowers** numerous, perfect, yellow. Receptacle flat, 1.2–2.5 mm broad when bare. **Seed** about 1 mm long with numerous slender white bristles at one end.

FLOWERING July–November.

SYNONYMS *Conyza canadensis* (L.) Cronq.

WHERE FOUND In pastures, roadsides, wastelands, cultivated fields, and gardens; mostly on rather dry soils. A semi-cosmopolitan weed.

ORIGIN Introduced; a common weed.

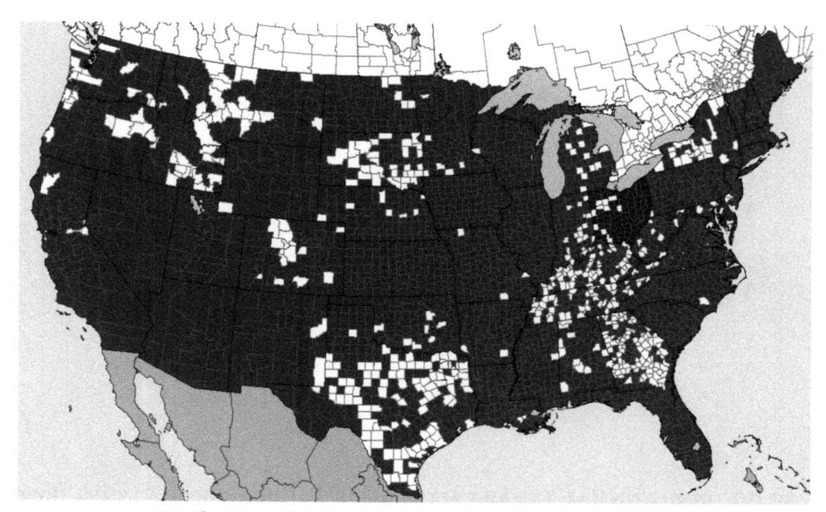

Distribution of **Erigeron canadensis**, HORSEWEED

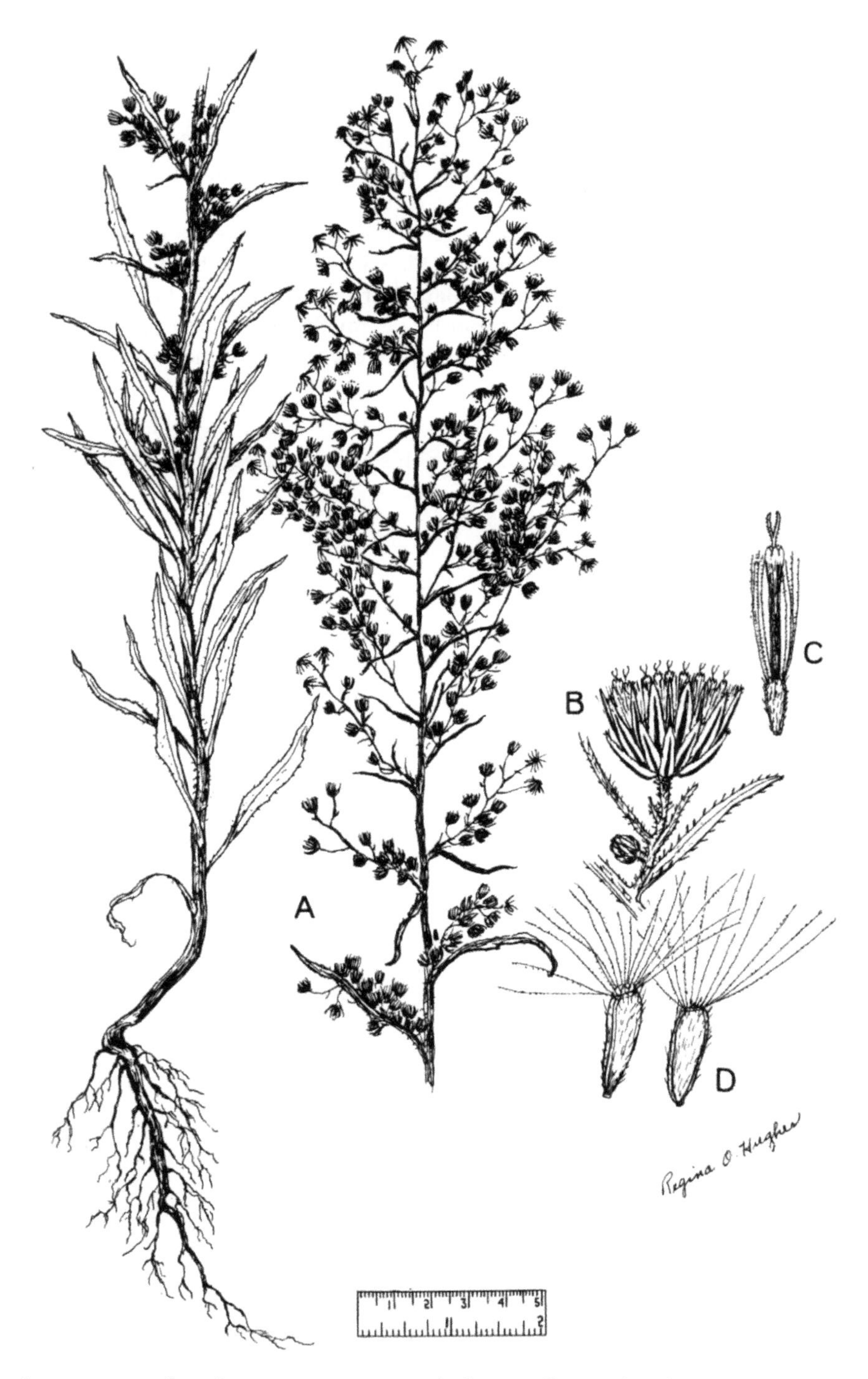

Erigeron canadensis, HORSEWEED. A, habit. B, flower head. C, disk flower. D, achenes.

Eupatorium capillifolium (Lam.) Small DOG-FENNEL

Annual herb.

DESCRIPTION Stems slender, several from a stout woody caudex, 5–30 dm tall, downy or smooth below, much-branched. **Leaves** once or twice pinnately divided or dissected into fine, linear divisions, those of the inflorescence mostly simple, glandular-punctate, glabrous, the lowermost leaves opposite, the others alternate, mostly 2–10 cm long, often with axillary clusters. **Flower heads** green to bronze, very numerous in an elongate, much-branched panicle, the lower or outer heads blooming first. Involucre 2–3.5 mm high, the inner bracts much longer than the outer ones, usually with a sharp tip or abruptly pointed. **Flowers** 3–6 in each head, corolla greenish-white, 1.5–3 mm long. **Achene** glabrous.

FLOWERING September–October.

WHERE FOUND Borders of woods and old fields and pastures; in wet or dry soil in open places; aggressive and weedy.

ORIGIN Native.

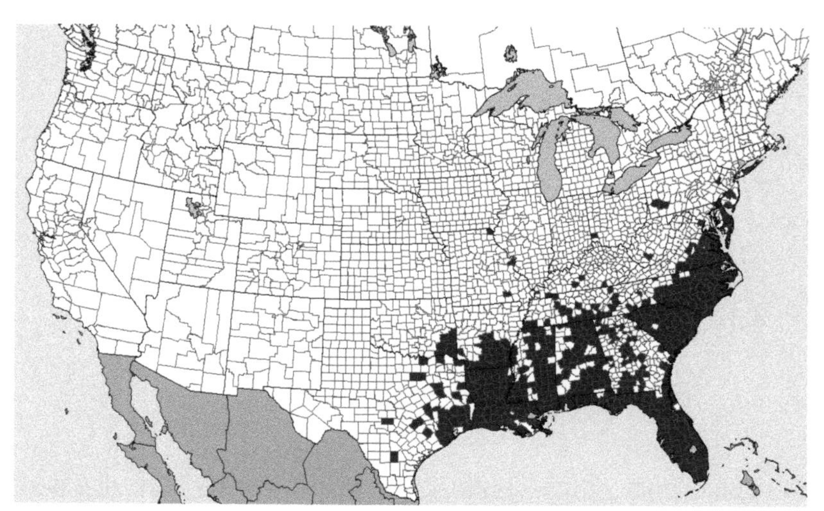

Distribution of **Eupatorium capillifolium**, DOG-FENNEL

Eupatorium capillifolium, DOG-FENNEL. A, habit. B, flower panicle. C, enlarged leaves. D, disk flower. E, achenes.

Galinsoga parviflora Cav. SMALL-FLOWER GALINSOGA

Annual herb, reproducing by seeds.

DESCRIPTION **Stems** erect or spreading, much branched, slender, 3-7 dm tall, glabrous or sparsely pubescent. **Leaves** opposite, ovate to lance-ovate, pointed at the tip, thin, 2-7 cm long, 1-4 cm wide, serrulate or crenulate, glabrous or sparsely appressed-hairy. **Flower heads** small, numerous, scattered at the ends of the branches, in leafy cymes. **Ray flowers** very small, white, 4-5 in number, surrounding the small yellow **disk flowers.** Pappus of the disk flowers without awns, equaling or longer than the corolla. **Achene** about 1.5 mm long, wedge-shaped, 4-sided, dark-brown to black, with a fringe of tiny scales at one end, or glabrous.

FLOWERING June-November.

WHERE FOUND Weedy gardens, dooryards, lowland fields, and waste places, especially in damp areas with rich soil. A cosmopolitan weed.

ORIGIN Naturalized from Mexico and South America.

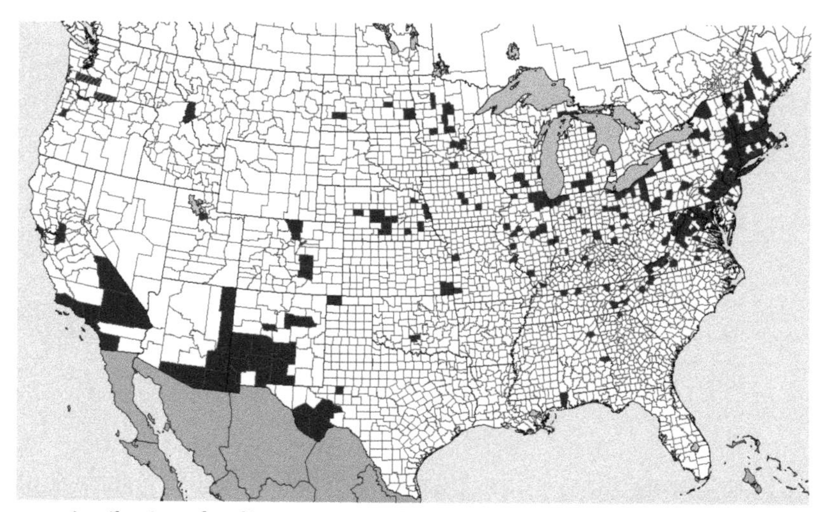

Distribution of **Galinsoga parviflora**, SMALL-FLOWER GALINSOGA

Galinsoga parviflora, SMALL-FLOWER GALINSOGA. A, habit. B, enlarged flowering branch. C, flower head. D, ray flower E, disk flower. F, achene with pappus.

Helenium amarum (Raf.) Rock — BITTER SNEEZEWEED

Annual herb, with short branching taproot, reproducing by seeds.

DESCRIPTION Stems smooth, erect, 1–8 dm tall, branching in the upper portion, very leafy. **Leaves** alternate, simple, numerous, crowded along the main stem and branches, sessile, linear-filiform, 1.5–8 cm long, up to 2 mm wide, glabrous and densely glandular-punctate. **Flower heads** 1.5–2 cm in diameter. **Ray flowers** yellow, pistillate and fertile, with 3-toothed tips, surrounding a dome-shaped mass of yellow perfect **disk flowers.** Pappus ovate, long-awned, about as long as the achene. **Achene** reddish-brown, 1–1.5 mm long, hairy along the edges, wedge-shaped, with bristle-tipped scales at the top.

FLOWERING August–October.

SYNONYMS *Helenium tenuifolium* Nutt.

WHERE FOUND Wastelands, old feed lots, pastures, idle lands, roadsides, and yards; does not persist under cultivation.

ORIGIN Native to Southeastern United States, considered adventive elsewhere.

NOTE Causes bitter, unmarketable milk produced by cows grazing heavily infested pastures, especially in the South.

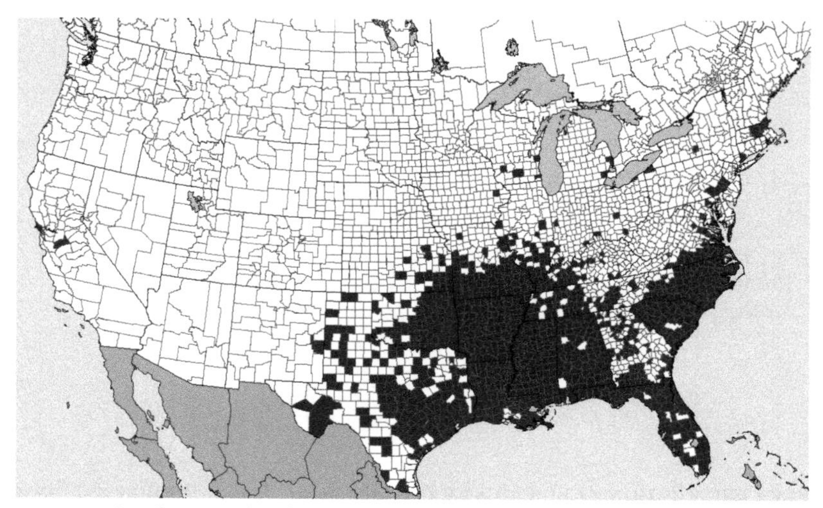

Distribution of **Helenium amarum,** BITTER SNEEZEWEED

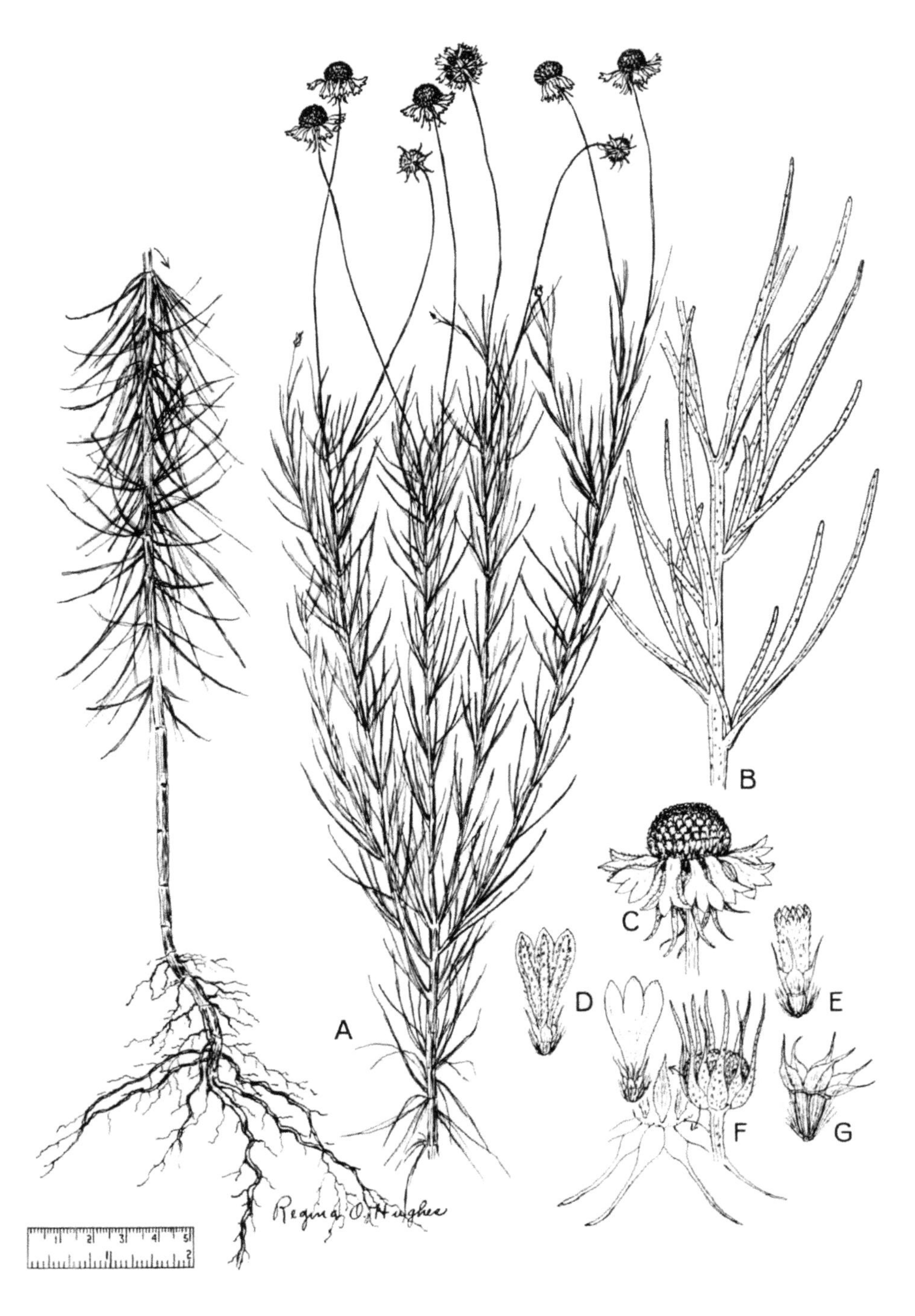

Helenium amarum, BITTER SNEEZEWEED. A, habit. B, leaf detail. C, flower head. D, ray flower, back and face. E, disk flower. F, involucre. G, achene.

Helianthus annuus L. COMMON SUNFLOWER

Annual herb, reproducing only by seeds; root system fibrous.

DESCRIPTION Stems erect, usually branching above, stout, coarse, rough and pubescent, 0.5–3 m or more high. **Leaves** mostly alternate, simple, petioled, with 3 main veins, the blade 0.2–3 dm or more wide, ovate or the lower ones cordate, rough on both surfaces, the margins serrate. **Heads** solitary and terminal, or also axillary, 6–12.5 cm in diameter. Involucral bracts overlapping, oblong-lanceolate to oval or broadly ovate, long-ciliate, with pubescent to glabrescent backs and tapering tips. Receptacle nearly flat, chaffy, 4 cm or more in diameter. **Ray flowers** mostly neutral, bright-yellow. **Disk flowers** numerous, perfect, with tubular brownish corolla. **Achene** about 9 mm long by 4–8 mm broad, glabrous except for the sparsely hairy summit, ovate to wedge-shaped, slightly 4-angled and flattened, white, gray, or dark-brown with light stripes or gray-mottled. **Pappus** of 2 thin scales, chaffy, deciduous.

FLOWERING July–November.

WHERE FOUND Plains, in cultivated fields, waste places, grainfields, pastures, fence rows, roadsides, and ballast lots.

ORIGIN Native; throughout the United States.

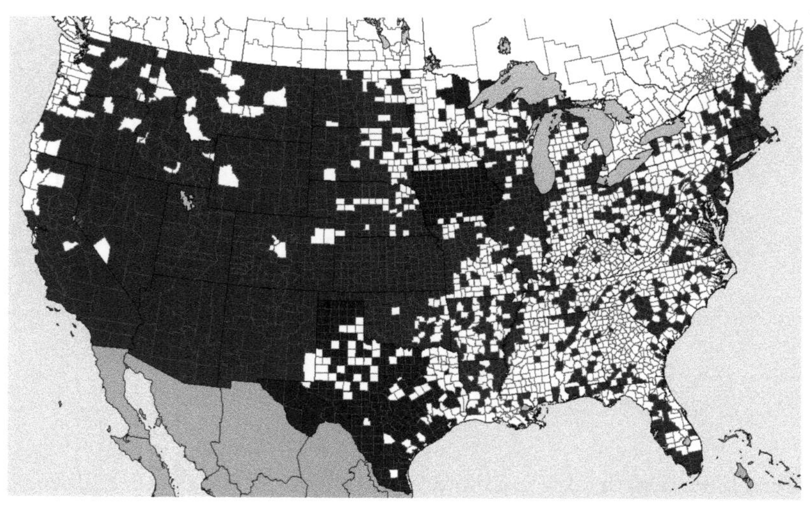

Distribution of **Helianthus annuus**, SUNFLOWER

Helianthus annuus, SUNFLOWER. A, habit. B, ray flower. C, disk flower. D, achenes, 2 views.

Hymenoxys odorata DC. BITTER RUBBERWEED

Annual herb, reproducing by seeds.

DESCRIPTION Stems low, spreading, many-branched, varying from 3-60 cm tall, the foliage with a bitter taste and an aromatic odor when crushed. **Leaves** somewhat fleshy, sparingly hairy to almost glabrous, divided into 3-13 very narrow segments about 1.6 mm broad, the surface covered by minute depressed glandular pits; a basal rosette of leaves 2.5-10 cm long formed first, soon withering, the stem leaves alternate, numerous, 1.8-7 cm long. **Flower heads** yellow, solitary at the tips of the stiff leafless stalks, 2.5-15 cm long, small, 0.8-3.3 cm wide, including the 8-13 golden-yellow ray flowers and the many tubular disk flowers in the center (as many as 3,000 heads per plant). Involucral bracts surrounding each head in 2 rows, erect, greenish-yellow, somewhat glandular, the outer 8-10 bracts 0.3-0.5 cm long, united and thickened at the base, the tips pointed, the inner bracts not united, free at the base, slightly longer than outer bracts and pressed inward in fruit. **Achene** numerous, 50-75 per head, narrowly cone-shaped, 1.6-2.2 mm long, indistinctly 4-angled and covered with silky grayish hairs. **Pappus** of 5 (rarely 6) dry, colorless or tawny scales, sharp-pointed or tipped with a short bristle, the scales about the same length as the achenes.

FLOWERING March-June, beginning in January in warmer areas.

WHERE FOUND Plains and drainage areas, barren disturbed soil around watering areas, and overgrazed rangelands and pastures; abundant where water collects along roadsides, lake beds, and bottomlands near rivers.

ORIGIN Native.

NOTE Heavy infestations may cause death to sheep and occasionally to cattle.

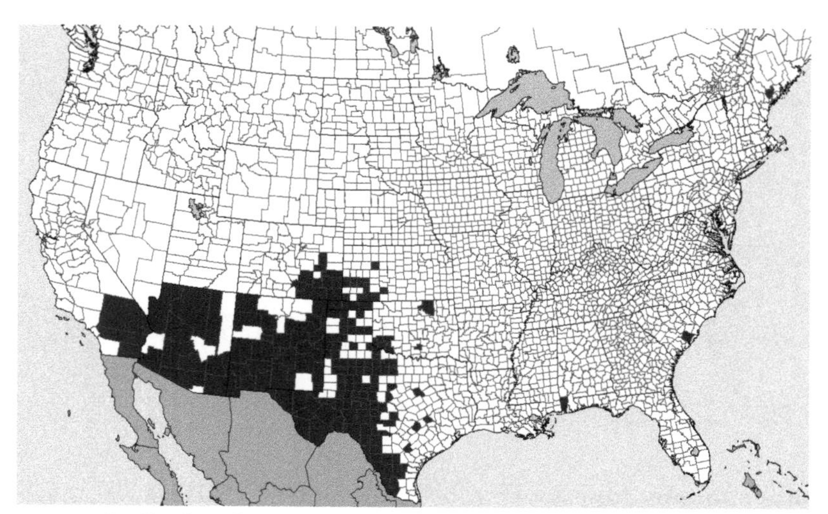

Distribution of **Hymenoxys odorata**, BITTER RUBBERWEED

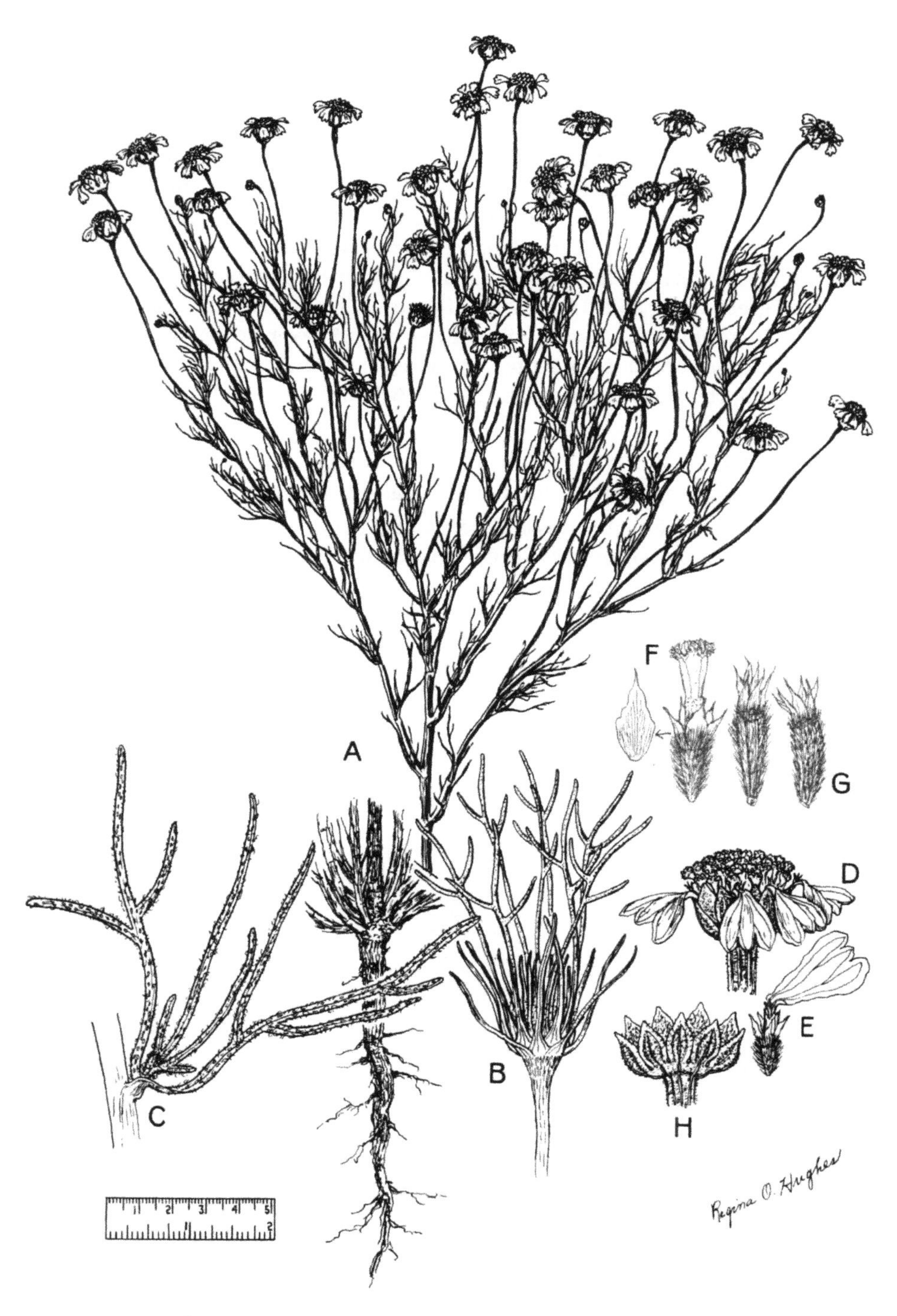

Hymenoxys odorata, BITTER RUBBERWEED. A, habit. B, young plant. C, enlarged leaf detail. D, flower head. E, ray flower; F, disk flower. G, achenes. H, involucre.

Isocoma tenuisecta Greene JIMMYWEED

Perennial, woody half-shrub, to 1 m wide and 1 m tall. Taproot up to 5 m deep, with a few shorter lateral roots.

DESCRIPTION Stems densely branched at the base, 5-7 dm tall (rarely taller), gray-green to brighter green above, forming a nearly spherical bush. **Leaves** pinnatifid, glandular, resinous and sticky, 20-35 mm long, about 1 mm broad, oblong or spathulate-oblong with narrow (2-22 mm long by 1 mm wide) linear lobes extending from each side. **Flower heads** discoid, with 8-12 flowers; each involucral bract with a thickened green warty apical spot, this with minute resin glands, the apical portion of the involucre at fruiting time 3-4 mm in diameter. **Flowers** yellow, borne in small dense heads, the corolla lobes acute, not more than 1 mm long. Style appendages triangular, shorter than the stigmatic region. **Achene** with simple pappus of 30-60 bristles.

FLOWERING September-October.

SYNONYMS *Haplopappus tenuisectus* (Greene) Blake ex Benson

WHERE FOUND On alluvial plains of desert and in semi-desert grasslands; invades depleted rangelands. Poisonous to cattle and to humans through the milk of cows.

ORIGIN Native; Arizona and New Mexico; Mexico.

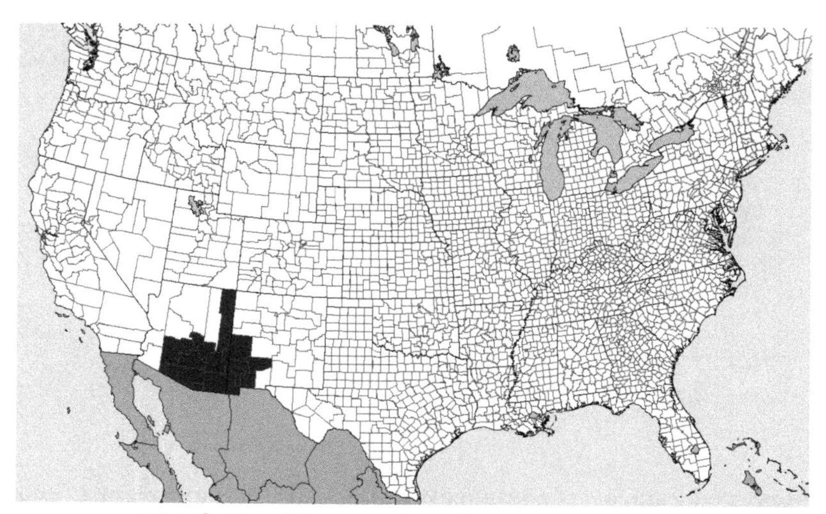

Distribution of **Isocoma tenuisecta**, JIMMYWEED

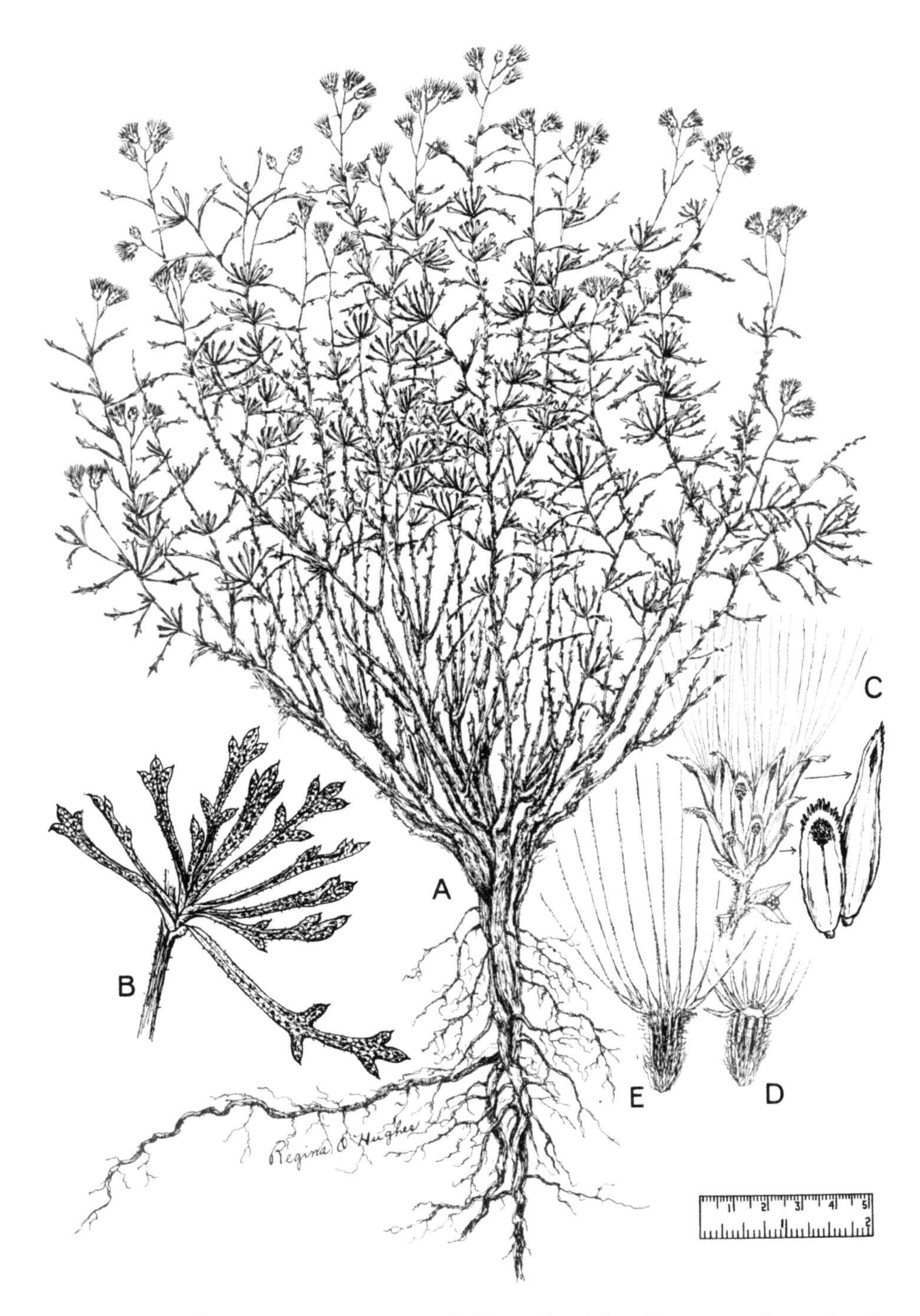

Isocoma tenuisecta, JIMMYWEED. A, habit. B, fascicle of leaves. C, flower head. D, disk flower. E, achene.

Jacobaea vulgaris Gaertn. — TANSY RAGWORT

Biennial herb, or a short-lived perennial. Taproot poorly to well-developed.

DESCRIPTION Stems solitary or several, erect, simple up to the inflorescence, 2–10 dm tall, pubescence floccose-tomentose, usually absent by flowering time. **Leaves** alternate, mostly 4–20 cm long, 2–6 cm wide, mostly 2 to 3 times pinnatifid, the lower leaves stalked and often deciduous, the upper leaves becoming sessile. **Flower heads** several or rather numerous, in a short broad inflorescence. Disk usually 7–10 mm wide. Involucre about 4 mm high, its bracts about 13, over 1 mm wide, generally dark-tipped. Rays commonly about 13, mostly 4–10 mm long. **Disk flowers** perfect and fertile. **Ray flowers** pistillate and fertile. **Achene** of the disk flowers minutely pubescent, achene of the ray flowers glabrous.

FLOWERING July–September.

SYNONYMS *Senecio jacobaea* L.

WHERE FOUND Pastures and disturbed situations.

ORIGIN Native of Europe.

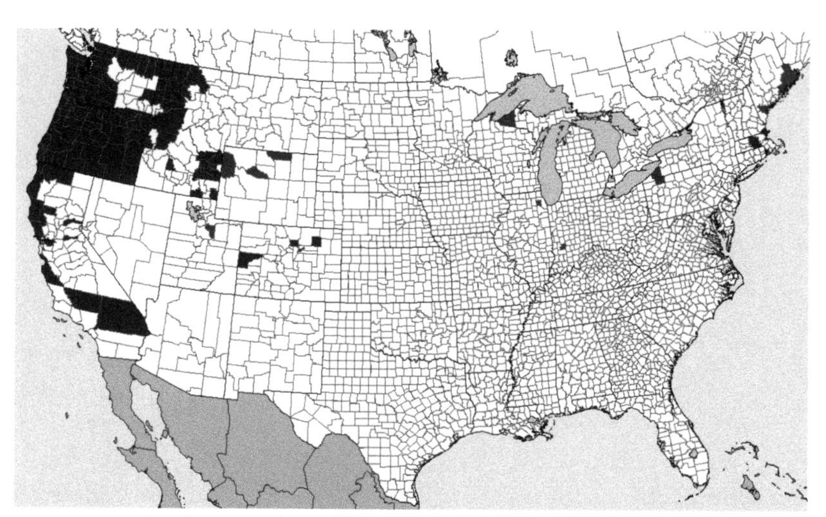

Distribution of **Jacobaea vulgaris**, TANSY RAGWORT

Jacobaea vulgaris, TANSY RAGWORT. A, habit. B, flower head. C, ray flower. D, disk flower. E, achenes, ray flower. F, achenes, disk flower.

Lactuca serriola L. — PRICKLY LETTUCE

Annual, winter annual, or biennial herb, with milky juice, from a large deep taproot; reproducing also by seeds.

DESCRIPTION Stems pale-green or straw-colored, 4.5–18 dm tall, the lower stem often prickly. **Leaves** alternate, bluish-green, the edges prickly toothed, stalkless, clasping the stem, the base extended into a pair of arrowhead-shaped projections; the lower leaves 5–30 cm long, 2.5–10 cm wide, the midrib usually with short spines on the underside; the principal leaves with 1 to several irregular lobes along each side, or the leaves unlobed and merely prickly toothed. **Flower heads** small, only 8–10 mm in diameter in flower, with very narrow buds, numerous in the large many-branched flowering part. **Flowers** all strap-shaped, petal-like ray flowers pale-yellow, often drying blue. Bracts surrounding the flower heads in 3 or 4 lengths, 9–16 mm high. **Achene** flattish, pale-brown to gray, 3–3.5 mm long, with 5–7 lengthwise ribs on each side, these and the 2 margined edges short-bristly near the summit, oblong, broader above, the tip sharp-pointed, from which arises a very slender, white, once-bent beak, slightly longer than the achene body, and bearing a parachute-like tuft of white hairs (pappus).

FLOWERING July–September.

WHERE FOUND Dry soils; along roadsides, railroads, and sidewalks; vacant lots, dumps, fence rows, run-down pastures; cultivated fields, orchards, and vineyards.

ORIGIN Native of Eurasia to the Himalayas; introduced from Europe.

NOTE In Wyoming, pulmonary emphysema develops in cattle feeding exclusively on this weed.

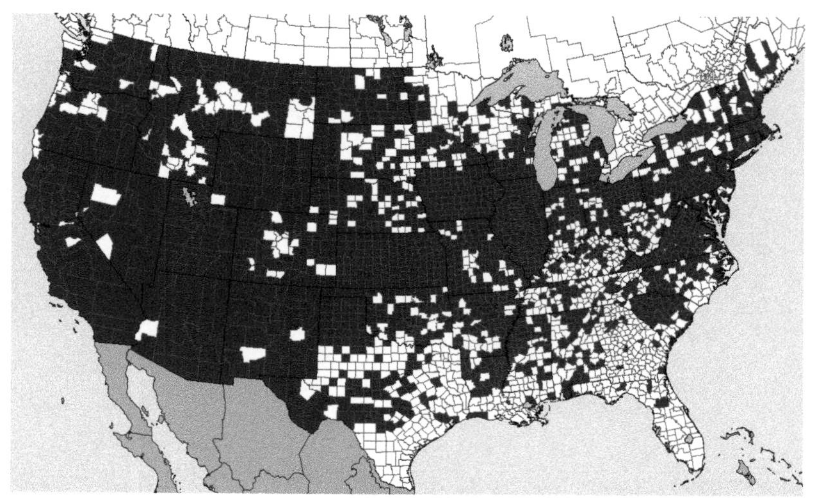

Distribution of **Lactuca serriola**, PRICKLY LETTUCE

Lactuca serriola, PRICKLY LETTUCE. A, habit; a, upper part of plant; b, lobed lower leaves; c, linear upper leaves; d, root. B, flower heads. C, achenes.

Lactuca tatarica (L.) C.A. Mey. BLUE LETTUCE

Perennial herb, reproducing by seeds and from a deep taproot or spreading horizontal roots that produce new shoots.

DESCRIPTION Stem 3-9 (rarely 12) dm tall, very leafy up to the flowering portion, the lower stem not spiny. **Leaves** alternate, thickish, all stalkless, or the lower leaves more or less narrowed into a winged stalk; the principal leaves 5-15 (occasionally to 20) cm long, slightly smaller upward, all unlobed or divided into 2 to many backward-pointing projections, these varying from mere notches to segments 2.5 cm or more long, the upper half of the leaf usually undivided; the upper leaves numerous, narrow lance-shaped, mostly unlobed. **Flowers heads** few, 2-3 cm in diameter in flower, their stalks with minute scaly bracts. **Flowers** all ray flowers, petal-like, blue or purple, drying whitish, much longer than the surrounding bracts. Bracts surrounding the flower heads 13-20 mm high, often purplish. **Achene** dark red-brown or sometimes slate-gray, 2-3 mm long, tapering above into a short firm beak, 0.5-3 mm long, which bears a tuft of white hairs (pappus) at the top, with 3-7 lengthwise ribs on each side.

FLOWERING Usually July-September; flowering June-October in the Pacific States.

SYNONYMS *Lactuca pulchella* (Pursh) DC.

WHERE FOUND Native of the grassland and sagebrush plains, alkali flats, wet meadows, and stream and river valleys; along roadsides, railroads, and ditches; irrigated pastures, cultivated lands, and grainfields.

ORIGIN Native, but considered adventive in eastern states.

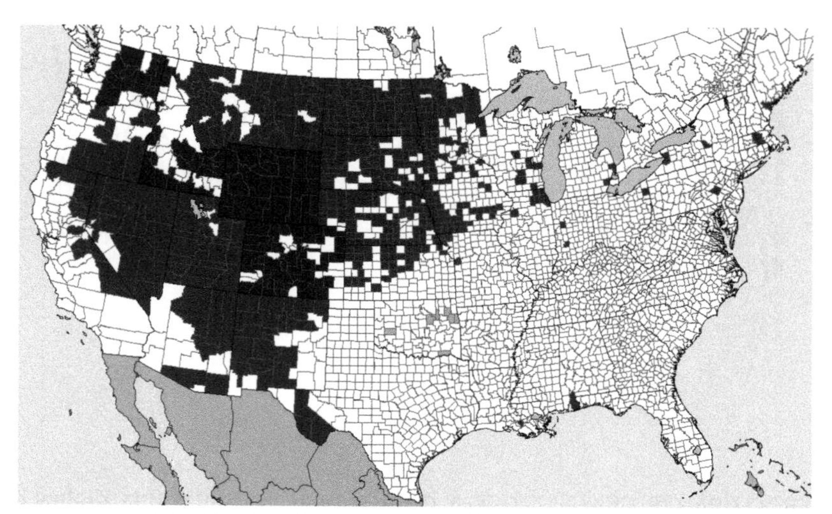

Distribution of **Lactuca tartarica**, BLUE LETTUCE

Lactuca tartarica, BLUE LETTUCE. A, habit. B, involucre and flowers. C, achene.

Leucanthemum vulgare Lam. FIELD OXEYE-DAISY

Perennial herb, reproducing by seed and by short to wide creeping rhizomes, often forming patches.

DESCRIPTION Stems erect, 0.3-1 m tall, simple or forked toward the top, glabrous. **Basal leaves** in rosettes, pinnatifid, subpinnatifid or coarsely and irregularly toothed; the middle and upper **stem leaves** alternate, simple, sessile, tooth-margined, narrowly oblong or oblanceolate, conspicuously nearly pinnatifid at the base, glabrous. **Heads** terminal, 2-6 cm in diameter, solitary at the tops of the stem and of long branches, many-flowered. Involucral bracts overlapping, numerous, narrow, brown-margined. **Ray flowers** 20-30, white, pistillate, 10-15 mm long. **Disk flowers** numerous, yellow, perfect. **Achene** narrowly obovate, 1-1.5 mm long, bearing a tubercle at the apex, black with 8-10 light-gray ribs. **Pappus** absent.

FLOWERING June-August.

SYNONYMS *Chrysanthemum leucanthemum* L.

WHERE FOUND Fields, wastes, meadows, and roadsides; a familiar weed.

ORIGIN Naturalized from Europe.

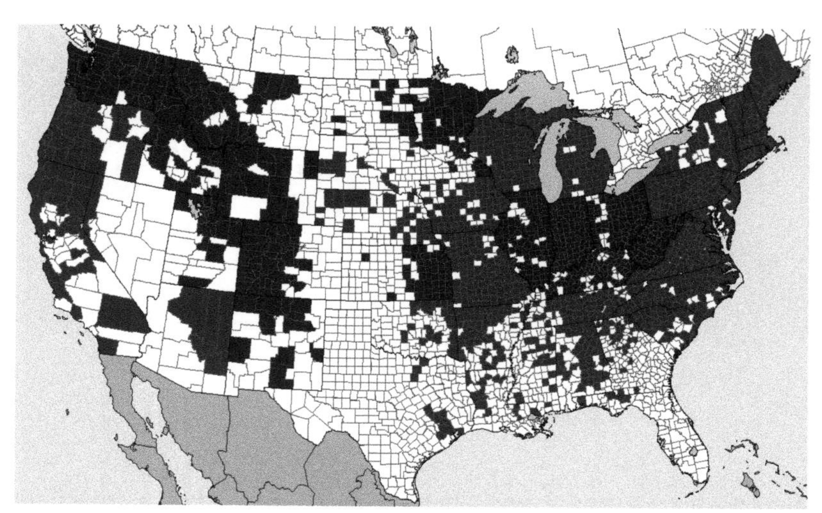

Distribution of **Leucanthemum vulgare**, FIELD OXEYE-DAISY

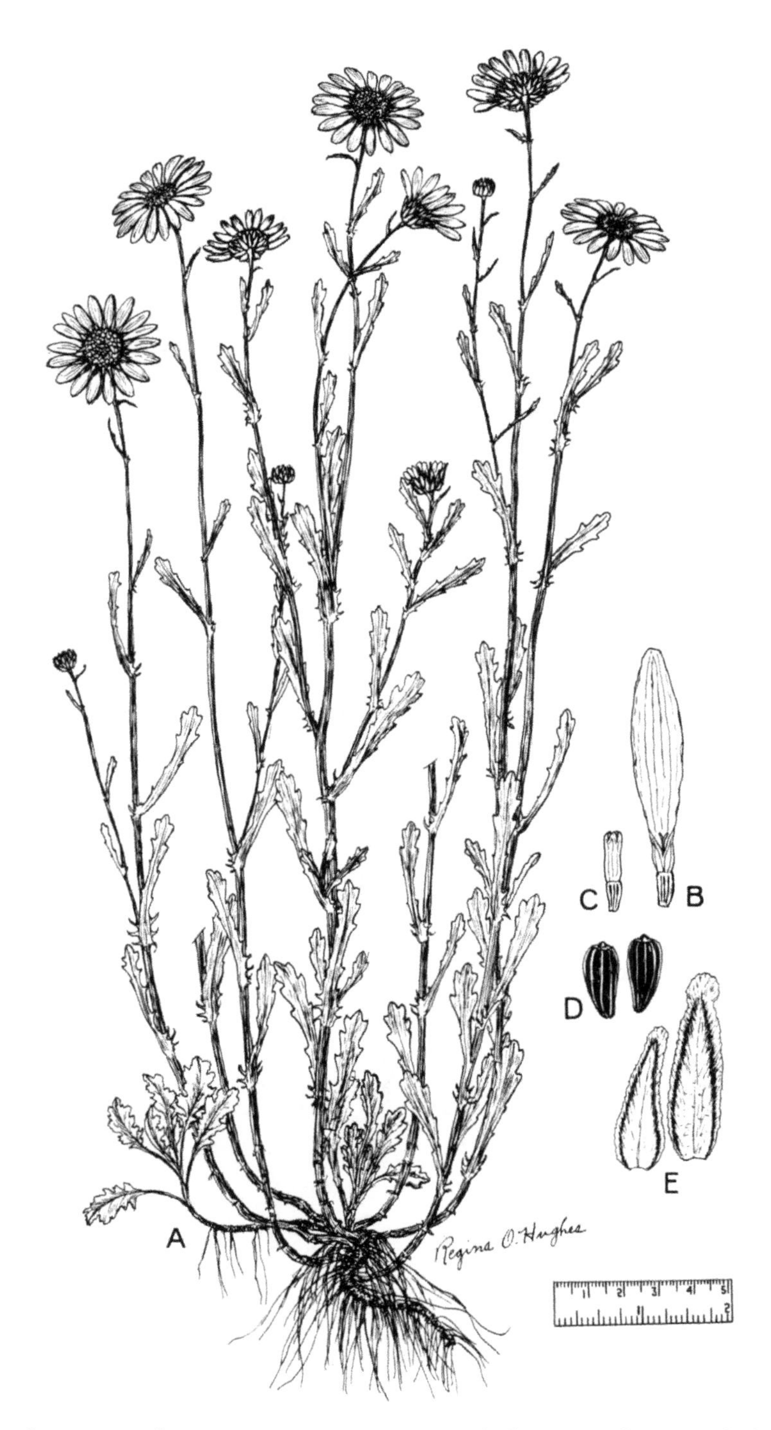

Leucanthemum vulgare, FIELD OXEYE-DAISY. A, habit. B, ray flower. C, disk flower. D, achenes. E, involucral bracts.

Solidago canadensis L. CANADA GOLDENROD

Perennial herb, reproducing from seed and from creeping rhizomes, without any well-developed caudex.

DESCRIPTION Stems clustered or solitary, about 3-15 dm tall, glabrous at base, more or less puberulent above the middle. **Leaves** alternate, thin, sharply serrate to nearly entire, glabrous or slightly roughened above, glabrous to puberulent on the midrib and main veins beneath; the basal leaves lacking, or similar to the lower stem leaves, reduced and soon deciduous, the other leaves numerous and crowded, being gradually reduced upwards, lance-linear to narrowly lance-elliptic, long acuminate, tapering to the sessile base, 3-nerved, the larger leaves 6-13 cm long, 5-18 mm wide. **Inflorescence** a terminal, broadly pyramid-like panicle, 5-40 cm high, with conspicuously recurved branches. Involucres about 2-3 mm high, their bracts (phyllaries) overlapping in 3 to 4 series, thin and slender, acute or acuminate, yellowish to yellowish-green, without well-defined green tips. **Ray flowers** mostly 10-17, sometimes only 7, minute, about 1-1.5 mm long. **Disk flowers** fewer, thin corollas, 2.4-2.8 mm long. **Achene** short-hairy.

FLOWERING July-October.

WHERE FOUND Open places, both moist and dry.

ORIGIN Native.

NOTE About 100 species of goldenrod are known, mainly from North America. In addition to *Solidago canadensis*, other common and sometimes weedy species are:

S. gigantea Ait.—Damp thickets.
S. graminifolia (L.) Salisb.—Damp or dry thickets and shores.
S. missouriensis Nutt.—Dry prairies, gravels, and rocky slopes.
S. nemoralis Ait.—Dry sterile open soils and thin woods.
S. occidentalis Nutt.—Moist ground.
S. rigida L.—Dry or gravelly open woods, thickets, and prairies.
S. rugosa Mill.—Damp open soil, thickets, and borders of woods and streams.

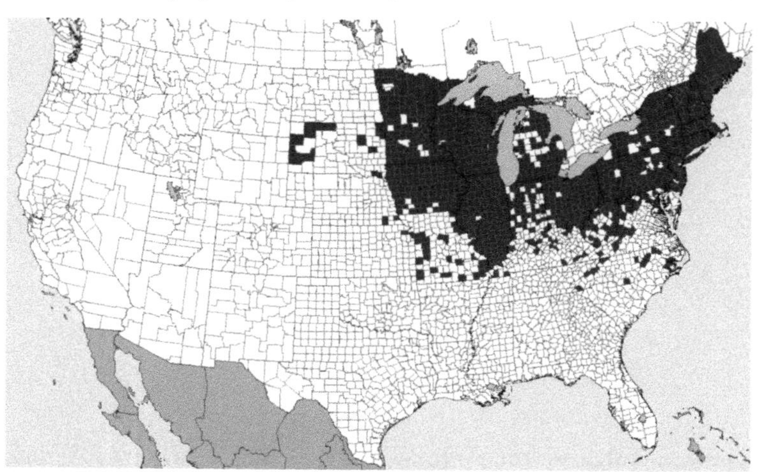

Distribution of **Solidago canadensis**, CANADA GOLDENROD

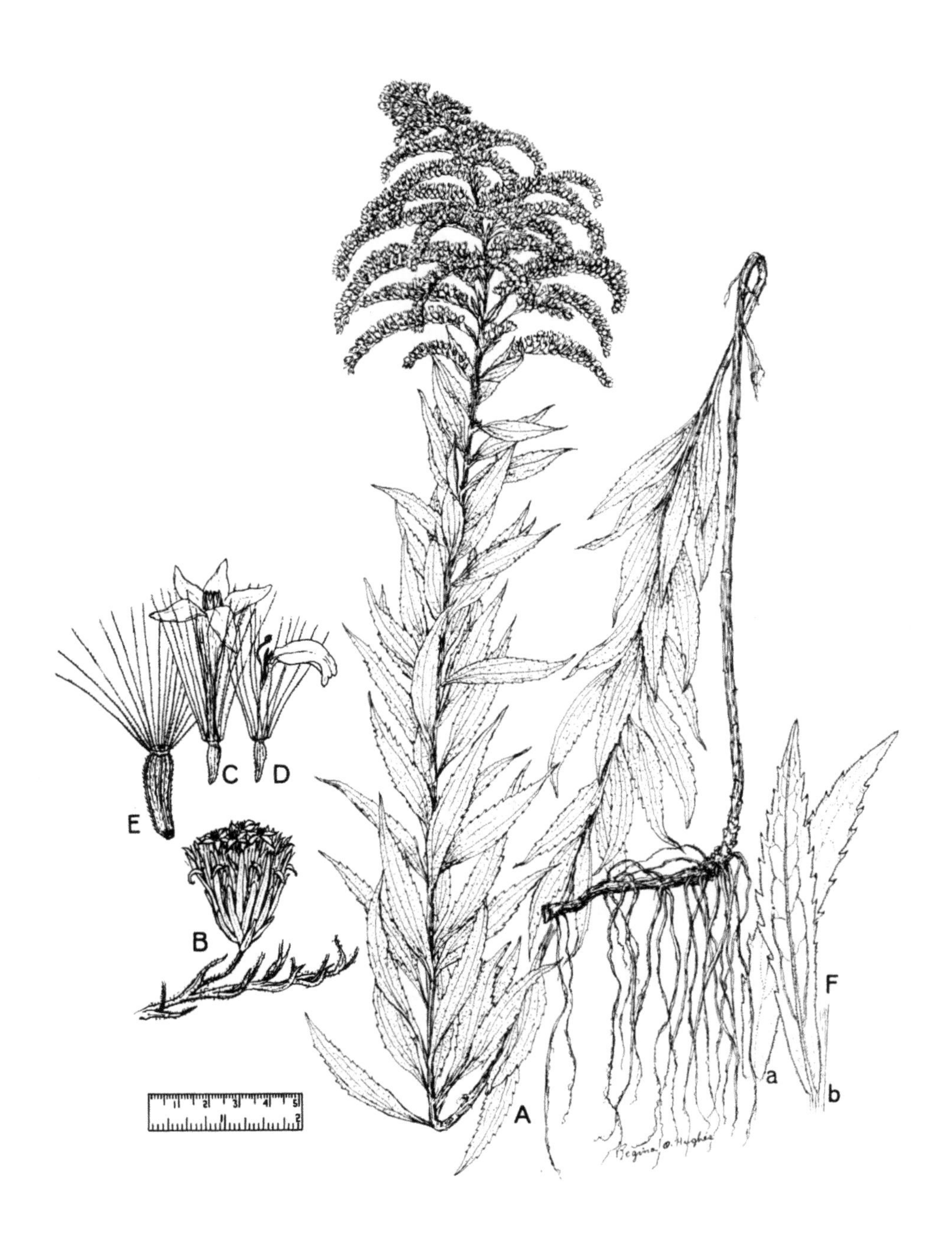

Solidago canadensis, CANADA GOLDENROD. A, habit. B, flower head. C, disk flower. D, ray flower. E, achene. F, leaves; a, upper surface; b, lower surface.

Sonchus arvensis L. PERENNIAL SOW-THISTLE

Perennial herb, reproducing by seeds, or from deep-seated (2–3 m deep) vertical roots and wide-spreading, horizontal, plant-producing roots from root buds to 60 cm deep, thus establishing large colonies.

DESCRIPTION Stems stout, large, hollow, containing milky juice, with conspicuous, stalked, gland-tipped hairs occurring on the upper stems, flowering stalks, particularly on the bracts surrounding the flower heads (glands lacking in ssp. *uliginosus* (Bieb.) Nyman, the common form in the northern United States). **Leaves** alternate, crowded on the lower half of the stem, the upper leaves scarce, greatly reduced and often unlobed; the principal leaves divided into 2–5 (occasionally 7) lobes along each side, usually with the tip lobe longer or broadly triangular, or with all the leaves mostly unlobed or merely toothed; the ear-like projections of the clasping leaf bases small and rounded at the tips. **Flower heads** large, 3–5 cm in diameter in flower, the flowers orange-yellow. Bracts surrounding the flower heads 14–25 mm high. **Achene** dark reddish-brown, 2–3.5 mm long at maturity, oblong, slightly narrowed at each end, 5–7 distinct, lengthwise ribs on each side, the ribs strongly cross-ridged, but not the furrows in between.

FLOWERING June–October, as early as April in warmer regions.

WHERE FOUND Cultivated lands, wastelands, oak woods, and grainfields.

ORIGIN Native of western Asia and Europe.

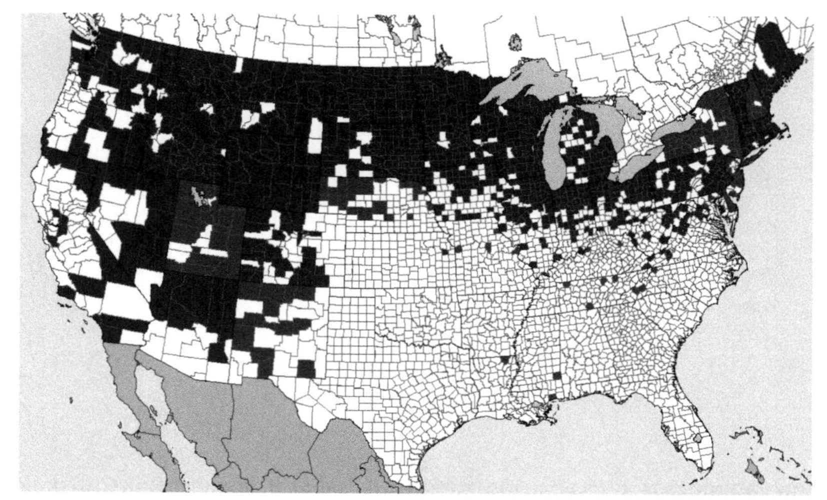

Distribution of **Sonchus arvensis,** PERENNIAL SOW-THISTLE

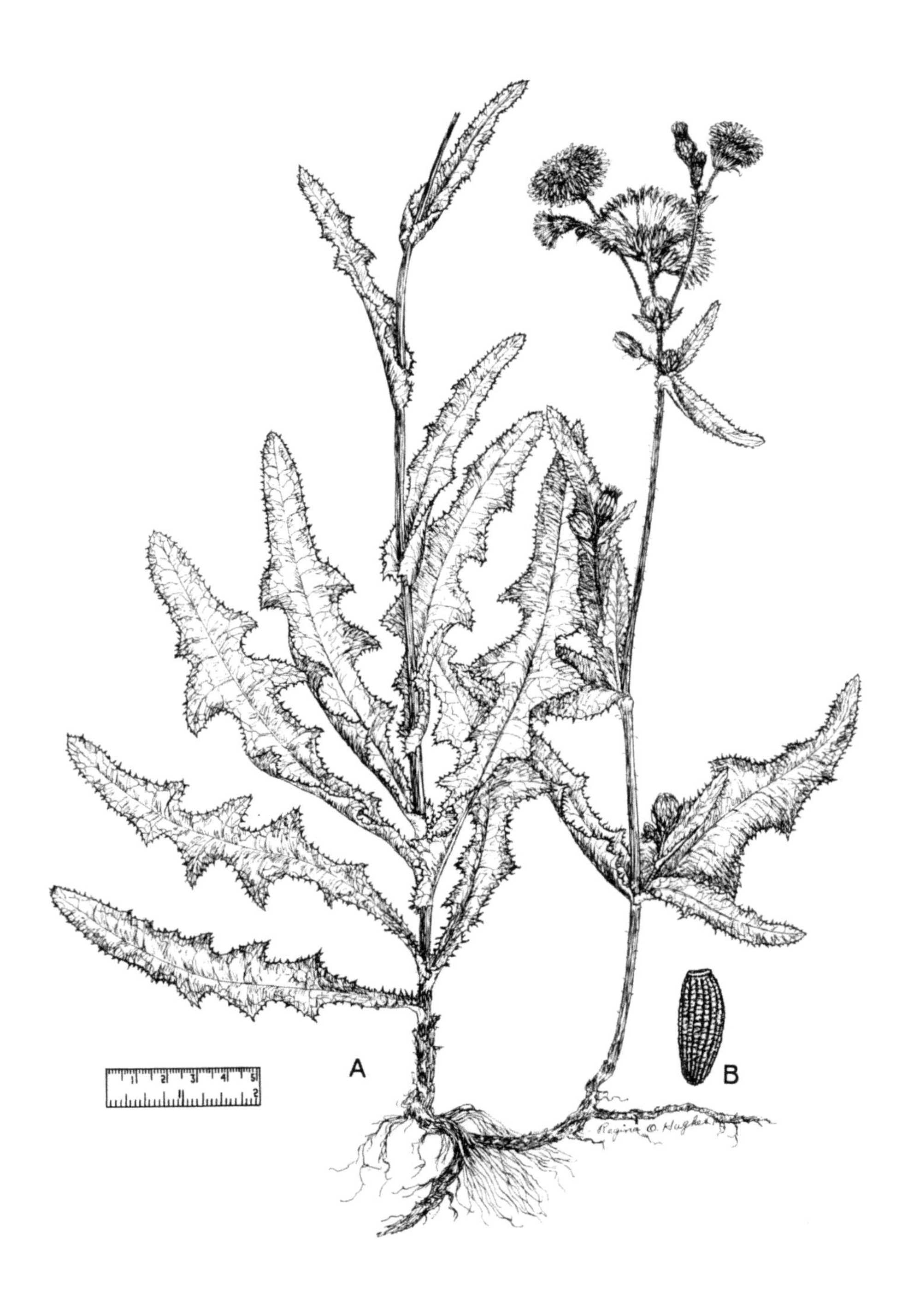

Sonchus arvensis, PERENNIAL SOW-THISTLE. A, habit. B, achene.

Sonchus asper (L.) Hill SPINY SOW-THISTLE

Annual herb, rarely biennial, reproducing by seeds only. Taproot short.

DESCRIPTION Stem erect, often reddish, very leafy up to the flowering part, gland-tipped hairs lacking or abundant on the upper stems and flower stalks. **Leaves** alternate, crowded along the stem, but progressively smaller and fewer lobes above; the main leaves many-lobed (5–11 lobes per side), usually with the tip lobe not longer or broadly triangular, to forms being scarcely lobed, margins with very long and stiff spines; the ear-like projections at the leaf bases rounded. **Flower heads** small, 1.2–2.5 cm in diameter in flower, the flowers pale-yellow. Bracts surrounding the flower heads only 9–16 mm high. **Achene** at maturity orange-brown, 2–3 mm long, thin and flat, broad through the center, the edges thinner than the body and wing-like, usually with 3 distinct central ribs on each side (rarely 4 or 5), not cross-ridged, but the edges sometimes minutely serrate or cross-marked.

FLOWERING Late June to frost; flowering year around in Florida, Texas, and California.

WHERE FOUND Winter vegetable crops, orchards, grainfields, alfalfa, and lawns.

ORIGIN Native of western Asia, northern Africa, and Europe. Throughout the United States.

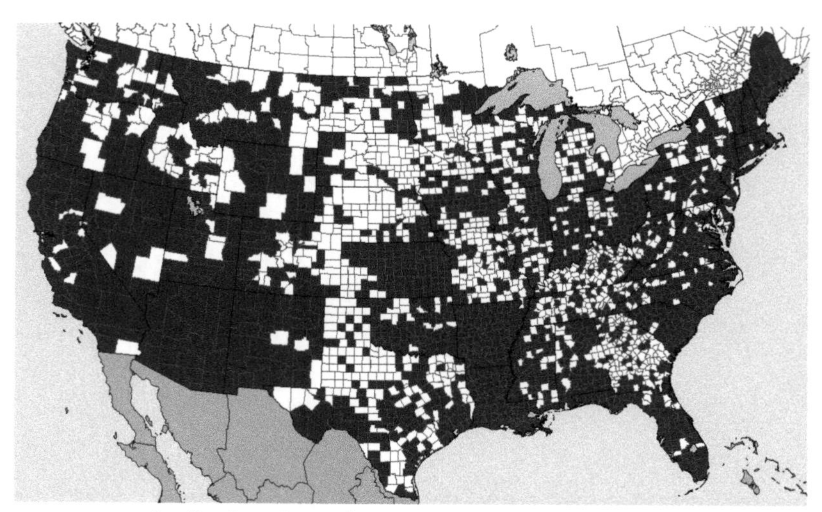

Distribution of **Sonchus asper**, SPINY SOWTHISTLE

Sonchus asper, SPINY SOWTHISTLE. A, habit. B, flower head. C, single flower. D, achenes.

Sonchus oleraceus L. ANNUAL SOW-THISTLE

Annual herb, rarely biennial, reproducing by seeds.

DESCRIPTION Stems from a taproot, erect, branched, stout, glabrous and glaucous, 3-20 dm tall, with a milky juice. **Leaves** alternate, crowded on the lower stem, fewer on the upper stem, often some of the principal leaves, as well as the basal leaves, stalked, the principal leaves deeply cut into 1-3 lobes along each side, with the tip lobe broadly triangular or rounded and cut nearly to the midvein; or the leaves sometimes with just 1 large lobe at the tip and a long winged stalk; or the leaves many-lobed with the tip lobe not enlarged; ear-like projections of clasping leaf bases not rounded, but sharp-pointed or tapering to a point; the margins with weak prickly teeth. **Flower heads** 2-3 cm broad, the **ligulate flowers** pale-yellow, the hairs (when present) on the bracts surrounding the flower heads, not gland-tipped. **Achene** 2.3-3 mm long, broadest toward the top, tapering to a narrow base, the 5-7 lengthwise ribs not prominent and distinct, often running together, the ribs not strongly cross-ridged.

FLOWERING July-September.

WHERE FOUND Cultivated fields, lawns, gardens, vineyards, orchards, roadsides, waste lots, railroad yards, beaches, grainfields, and ditchbanks.

ORIGIN Native of North Africa, western Asia, and Europe. Widely distributed in the United States, most common in the Pacific States.

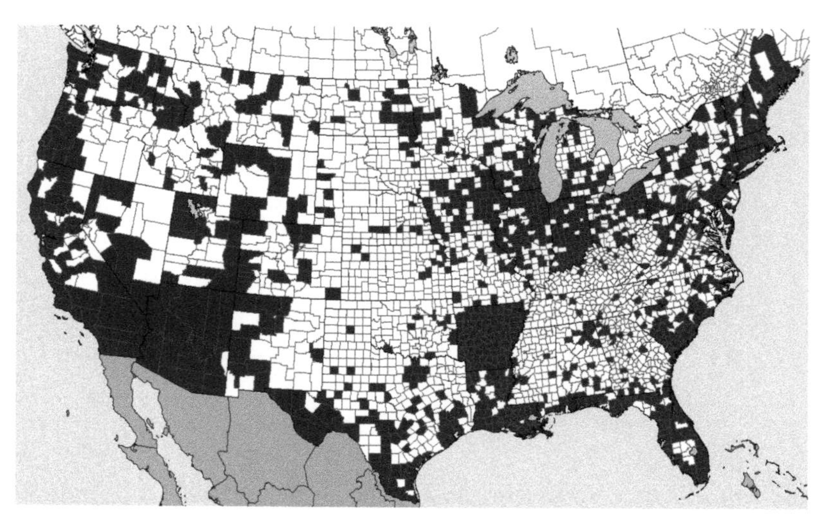

Distribution of **Sonchus oleraceus,** ANNUAL SOW-THISTLE

Sonchus oleraceus, ANNUAL SOW-THISTLE. A, habit. B, flower head. C, ray flower. D, achenes with pappus.

Taraxacum officinale G. H. Weber ex Wiggers — DANDELION

Perennial herb, from a thick taproot, often several feet deep, with many-branched crowns, with a milky juice, reproducing by seeds and by new shoots from the root crowns.

DESCRIPTION Stems very short and wholly underground, producing a rosette of leaves at the ground surface. **Leaves** 5–40 cm long, variable in shape from lobeless or entire to being divided into many shallow to deep-cut lobes with long soft points and intermediate small teeth, a larger lobe at the tip, or the edges merely toothed, narrowed at the base into a short hollow petiole, usually pubescent. **Flower heads** 2–5 cm in diameter in flower, solitary at the end of a naked hollow stalk 5–75 cm long. Receptacle flat or convex, naked. **Flowers** all strap-shaped ray flowers, golden-yellow, 5-notched at the tip, 100–300 per head. Bracts (phyllaries) green to brownish, surrounding the flower heads in 2 rows, the outer row hanging down and one-third to one-half as long as the inner, erect row. **Achene** yellowish to greenish-brown, 3–4 mm long, 5- to 8-ribbed on each side with minute curved spines on the rib margins of the upper half of the seed. Beak threadlike, 2–4 times longer than the body of the seed, topped by a tuft of whitish hairs (pappus), 3–4 mm long, persistent.

FLOWERING AND FRUITING from March to frost, or throughout the year in warmer areas.

WHERE FOUND Nearly a ubiquitous weed in waste areas, lawns, over-grazed pastures and meadows, open fields, and roadsides from sea level up to about 12,000 ft. elevation. Mainly a lawn pest, but also in hayfields and pastures; often an impurity in Kentucky bluegrass and forage grass seeds.

ORIGIN Introduced and naturalized from Eurasia; perhaps also native of northern North America. Throughout most of the United States.

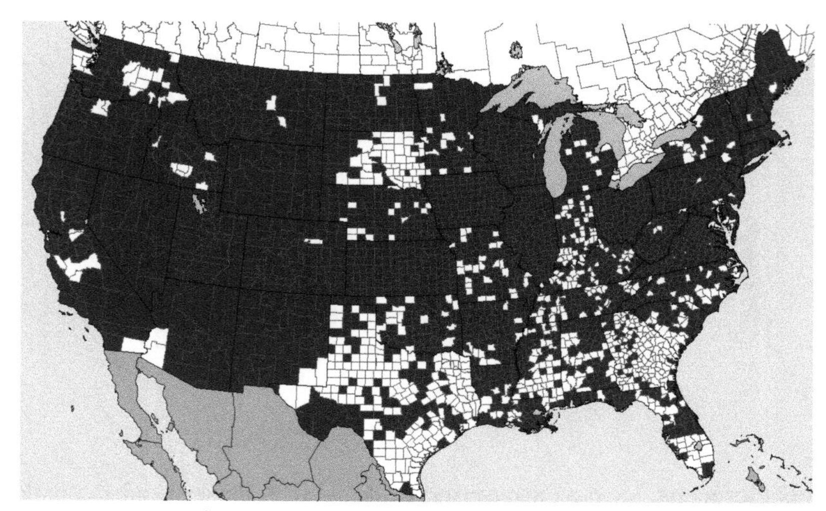

Distribution of **Taraxacum officinale**, DANDELION

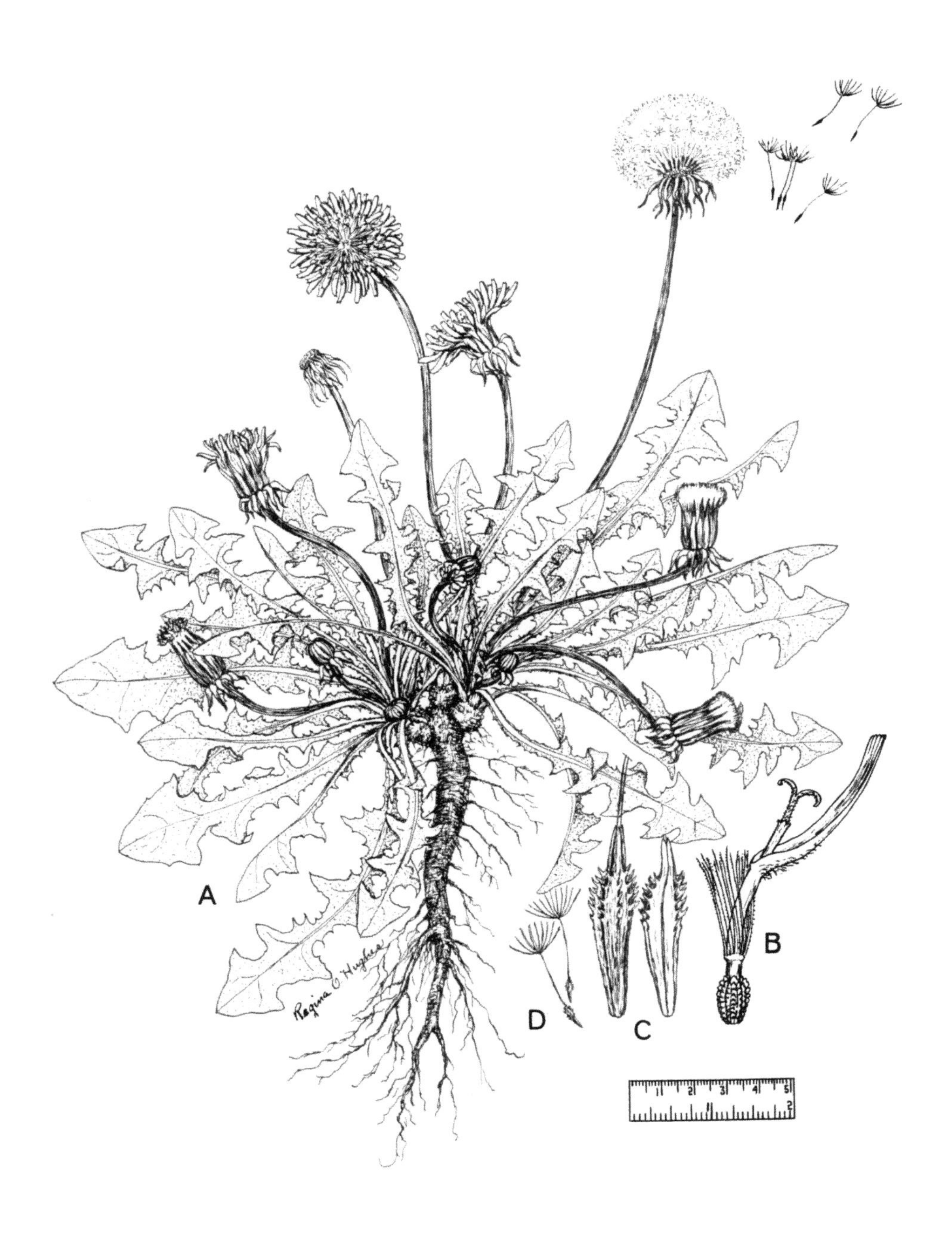

Taraxacum officinale, DANDELION. A, habit. B, flower. C, achenes. D, achenes with pappus.

Vernonia baldwinii Torr. WESTERN IRONWEED

Perennial herb.

DESCRIPTION Stems 6–20 dm tall, velvety. **Leaves** alternate, oblong or ovate-lanceolate, acute or acuminate, sharply cut on the margins, rough-bristly above, velvety beneath, 1–2 dm long. **Heads** 18–to 34-flowered. Involucre 6–8 mm broad, thick-cylindric to saucer-shaped, its tapering to awl-shaped phyllaries (bracts) spreading in all directions or recurving at the tip, pubescent within and their midribs prominent. **Inflorescence** corymb-like to panicle-like. Receptacle flat. **Achene** about 3 mm long.

FLOWERING July–September.

WHERE FOUND Prairies, pastures, open ground, and woods.

ORIGIN Native.

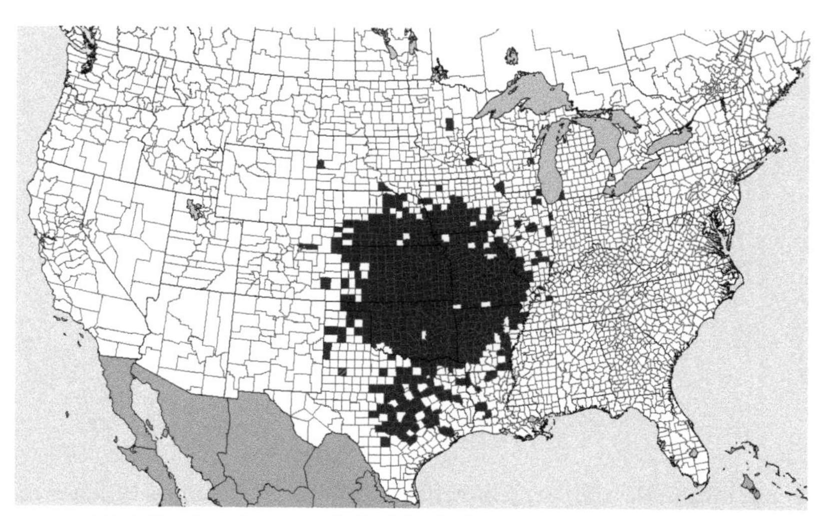

Distribution of **Vernonia baldwinii**, WESTERN IRONWEED

Vernonia baldwinii, WESTERN IRONWEED. A, habit. B, tomentose underside of leaf, enlarged. C, flower head. D, single flower. E, achenes.

Vernonia gigantea (Walt.) Trel. ex Branner & Colville TALL IRONWEED

Perennial herb, reproducing by seeds and rhizomes. Roots strong, fibrous.

DESCRIPTION Stem 1–2 m tall, widely branched in the upper region. **Leaves** alternate, loosely ascending to spreading, lance-oblong, or narrowly ovate, sharply toothed, mostly 3–8 cm broad, the lower surface minutely pubescent, rarely dotted, with cottony hairs along the midrib. **Flowers** in cymes, 1–5 dm broad, open, rarely dense. **Heads** saucer-shaped, 13-to 30-flowered. Involucres 4–6 mm high, their flattened ovate phyllaries obtuse, acute, or short-pointed. **Disk flowers** reddish-purple. **Achene** oblong, with rough bristly hairs along the ribs.

FLOWERING August–October.

SYNONYMS *Vernonia altissima* Nutt.

WHERE FOUND Damp rich soils, meadows, pastures, and wastelands.

ORIGIN Native.

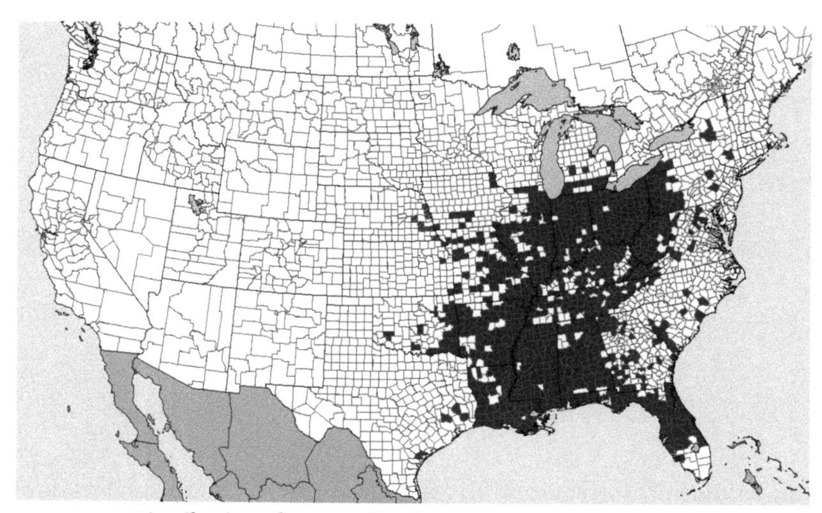

Distribution of **Vernonia gigantea**, TALL IRONWEED

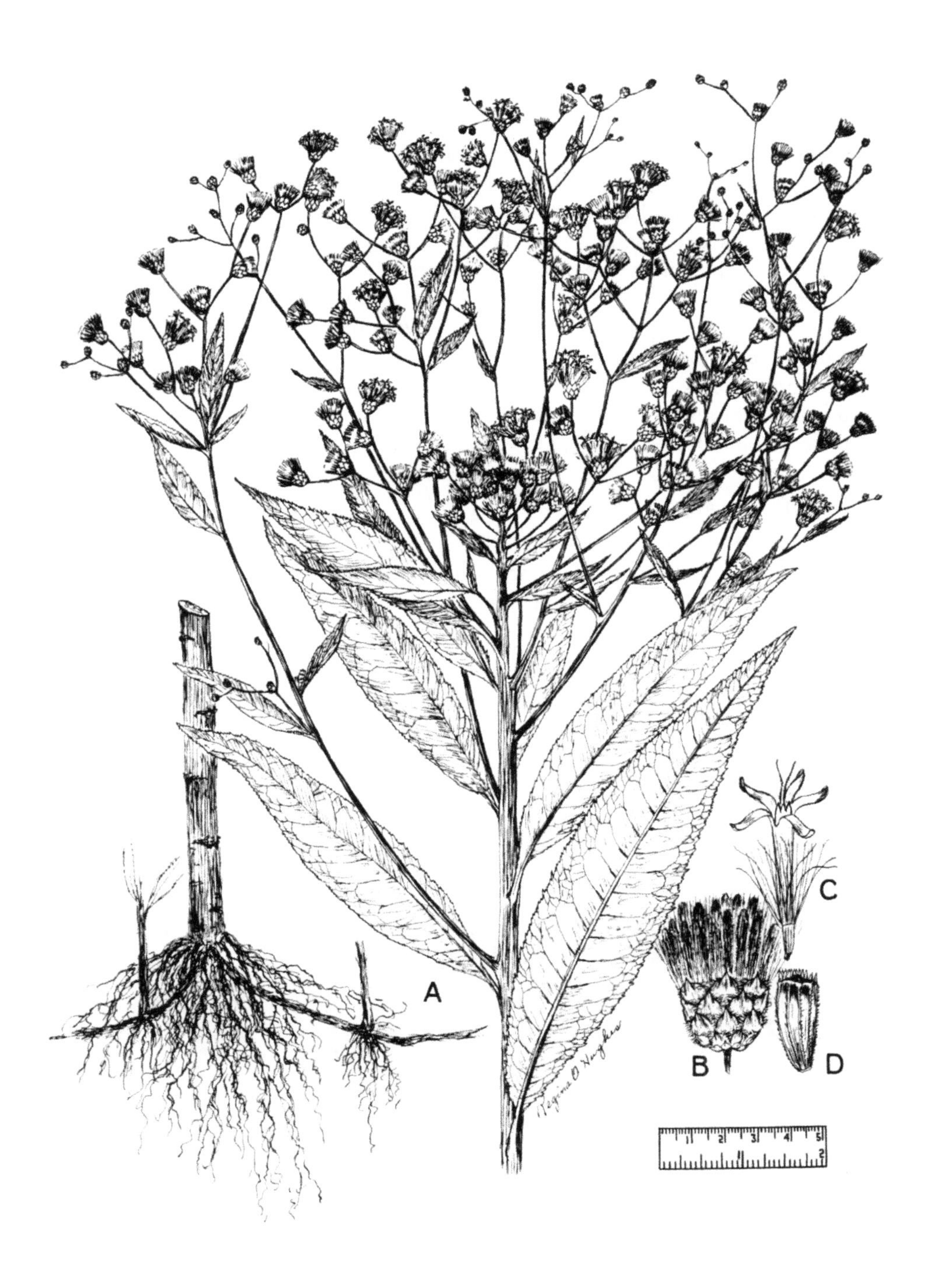

Vernonia gigantea, TALL IRONWEED. A, habit. B, flower head, showing the involucre. C, flower. D, achene.

Xanthium strumarium L. COMMON COCKLEBUR

Annual herb, robust, reproducing by seeds. Taproot rather woody, stout.

DESCRIPTION Stems erect, normally bushy, 2–9 dm tall, ridged, rough-hairy, often purple-spotted. **Leaves** alternate, simple, heart-shaped to truncate, triangular-ovate or broadly wedge-shaped, toothed or lobed, with petioles about as long as the blades, rough on both sides. Leaf axils without spines. **Flowers** small, male and female separate in clusters, the **staminate flowers** in short terminal spikes, dropping soon after shedding the pollen, the **pistillate flowers** in axillary clusters. Staminate involucre top-shaped, with 7–12 separate bracts. Pistillate involucre ovoid, tough, closed, hairy and spiny, bearing 2 pistillate flowers, developing into a hard, woody bur. **Bur** oval to oblong, light-brown, about 1–2 cm long, glabrous or nearly glabrous, with numerous crowded prickles 3–6 mm long, these hooked at the summit, glandular-pubescent and sometimes very sparsely spiny near the base. Beaks 2, stout, incurved. **Achene** 1–1.5 cm long, 2 in each bur, dark-brown, oblong, flattened, with pointed apex.

FLOWERING Late August–October.

SYNONYMS *Xanthium pensylvanicum* Wallr.

WHERE FOUND In cultivated fields, abandoned land, poor pastures, roadsides, bottomlands, waste places, and vacant lots.

ORIGIN Native of Eurasia.

NOTE A variable species, having several similar forms based on variations of the shape, hairiness, and spininess of the mature burs. The seedling is the most poisonous stage.

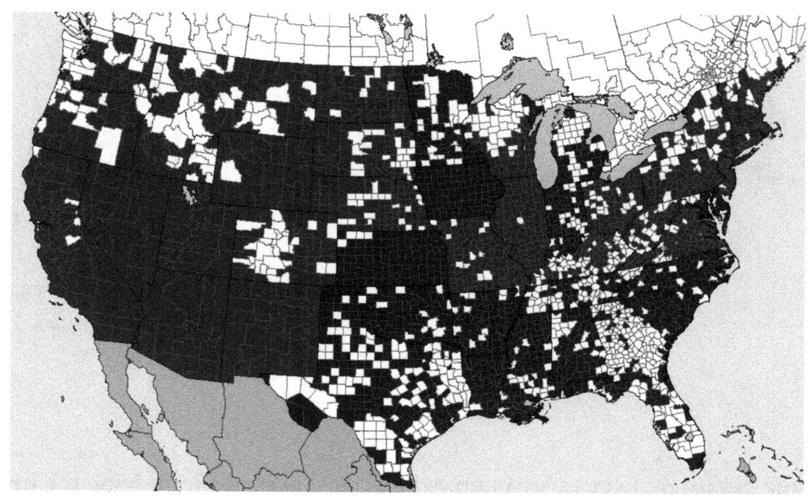

Distribution of **Xanthium pensylvanicum**, COMMON COCKLEBUR

Xanthium pensylvanicum, COMMON COCKLEBUR. A, habit. B, seedling. C, bur. D, seed.

Campsis radicans (L.) Seem. ex Bureau VIRGINIA CREEPER

Perennial vine, reproducing by seeds and vigorous running roots.

DESCRIPTION Stems glabrous, woody, vining, 6–12 m long, or on stems 4–30 dm long in cultivated fields. **Leaves** opposite, 20–40 cm long, pinnately compound, with 3–13 ovate to lanceolate leaflets, 4–8 cm long, acuminate with toothed margins, rounded at the base. **Flowers** in terminal short-stemmed corymbs. Calyx bell-shaped, 5-toothed, 10–15 mm long, the lobes about 5 mm long. Corolla funnel-shaped, orange to scarlet, rarely yellow, 58 cm long. 5-lobed, rarely regular. Stamens 4, included. **Capsule** 10–15 cm long, smooth, flattened, spindle-shaped, ridged at the edges of the 2 lengthwise halves of the pod, slightly curved. **Seed** broadly winged, about 15 mm long, in several rows in the capsule.

FLOWERING July–September.

WHERE FOUND In fields, yards, roadsides, waste places, and alluvial woods; on trees, posts, and old buildings; originally cultivated as an ornamental (and still is), but is an aggressive, widely distributed weed.

ORIGIN Native.

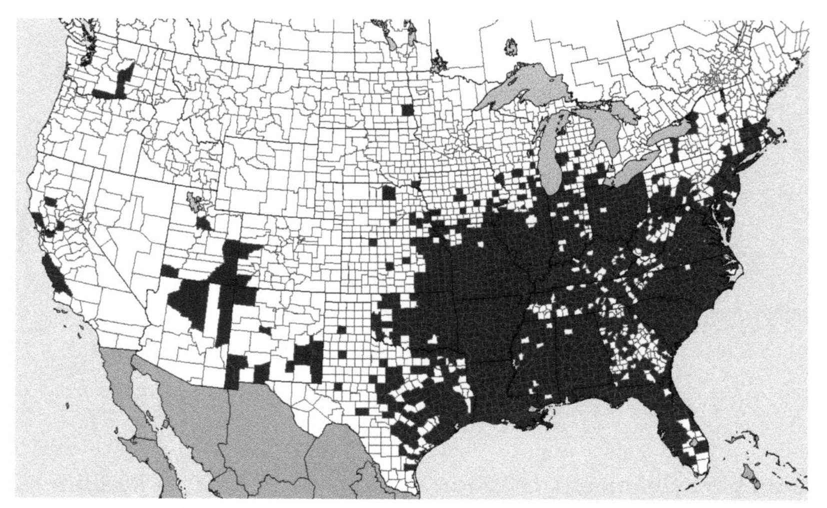

Distribution of **Campsis radicans,** VIRGINIA CREEPER

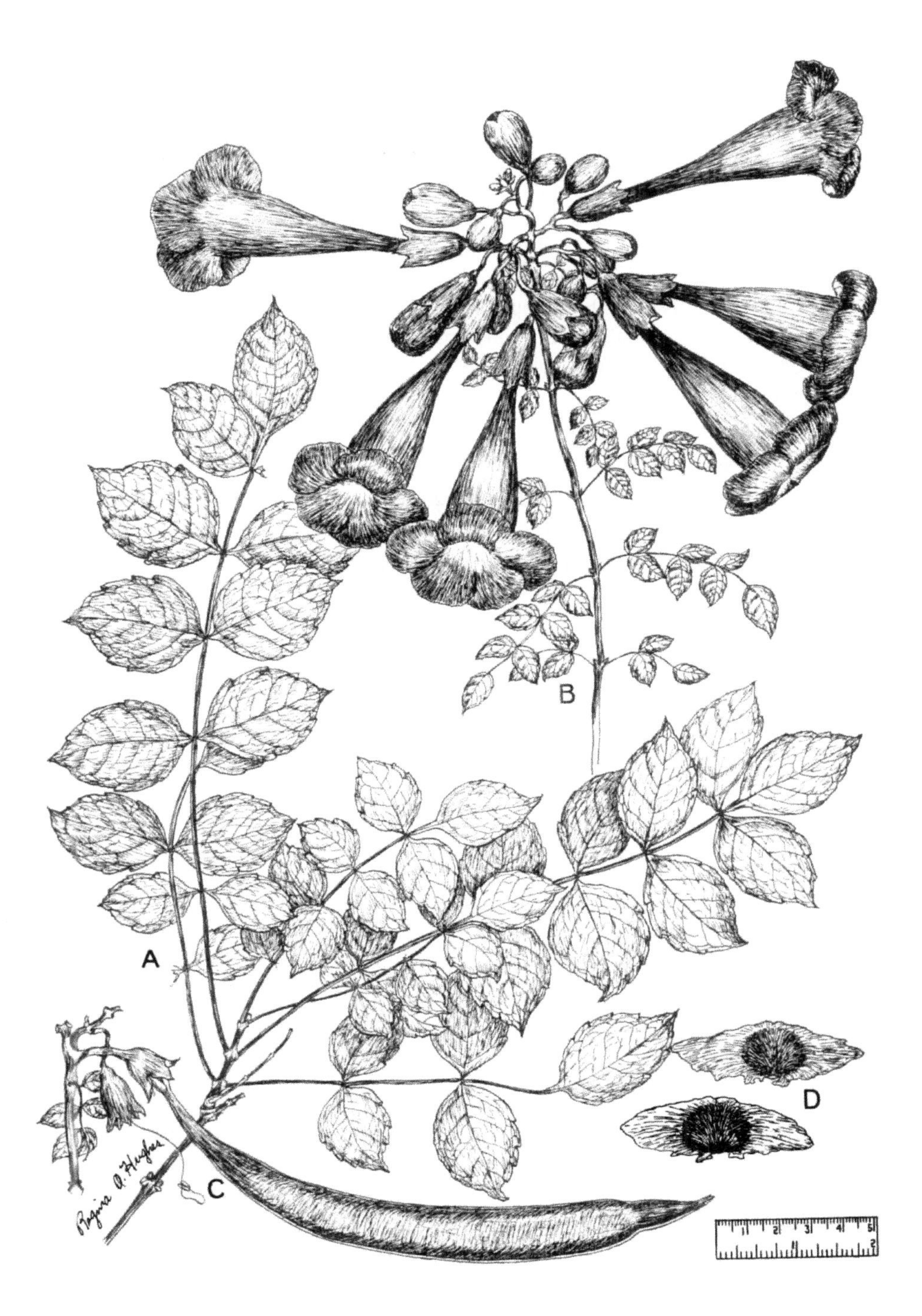

Campsis radicans, VIRGINIA CREEPER. A, Vegetative shoot. B, flower corymb. C, capsule. D, seed [back (below) and front views].

Lithospermum arvense L. CORN GROMWELL

Winter annual or biennial herb, reproducing by seeds.

DESCRIPTION Stems erect, slender, simple or branching at the base, 2-7 dm tall, minutely roughened and hairy, leafy at the top. **Leaves** alternate, simple, entire, sessile, lanceolate to linear, without lateral veins, 1-3 cm long, hairy on both sides. **Flowers** perfect, regular, nearly sessile, in the axils of leafy bracts of the terminal racemes. Calyx 5-pointed, 5-7 mm long, hairy. Corolla funnel-shaped, with 5 rounded spreading lobes, white to cream-colored, 5-7 mm long, with 5 short stamens inserted on its tube. Pistil with a deeply 4-lobed ovary and a simple style. **Fruit** of 4 small erect nutlets, conical, ovoid, about 3 mm long, one side angled, the other convex, the base with 2 small white tubercules and a scar, dull, prominently roughened, grayish-tan.

FLOWERING April–June.

SYNONYMS *Buglossoides arvensis* (L.) I. M. Johnst.

WHERE FOUND In grainfields, meadows, old gardens, and waste places; does not persist under cultivation; troublesome where winter wheat and rye are grown.

ORIGIN Introduced and naturalized from Europe.

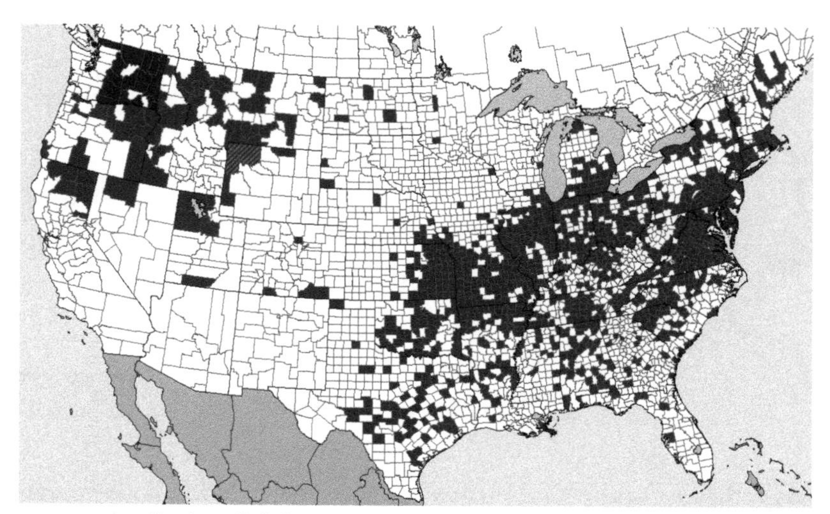

Distribution of **Lithospermum arvense**, CORN GROMWELL

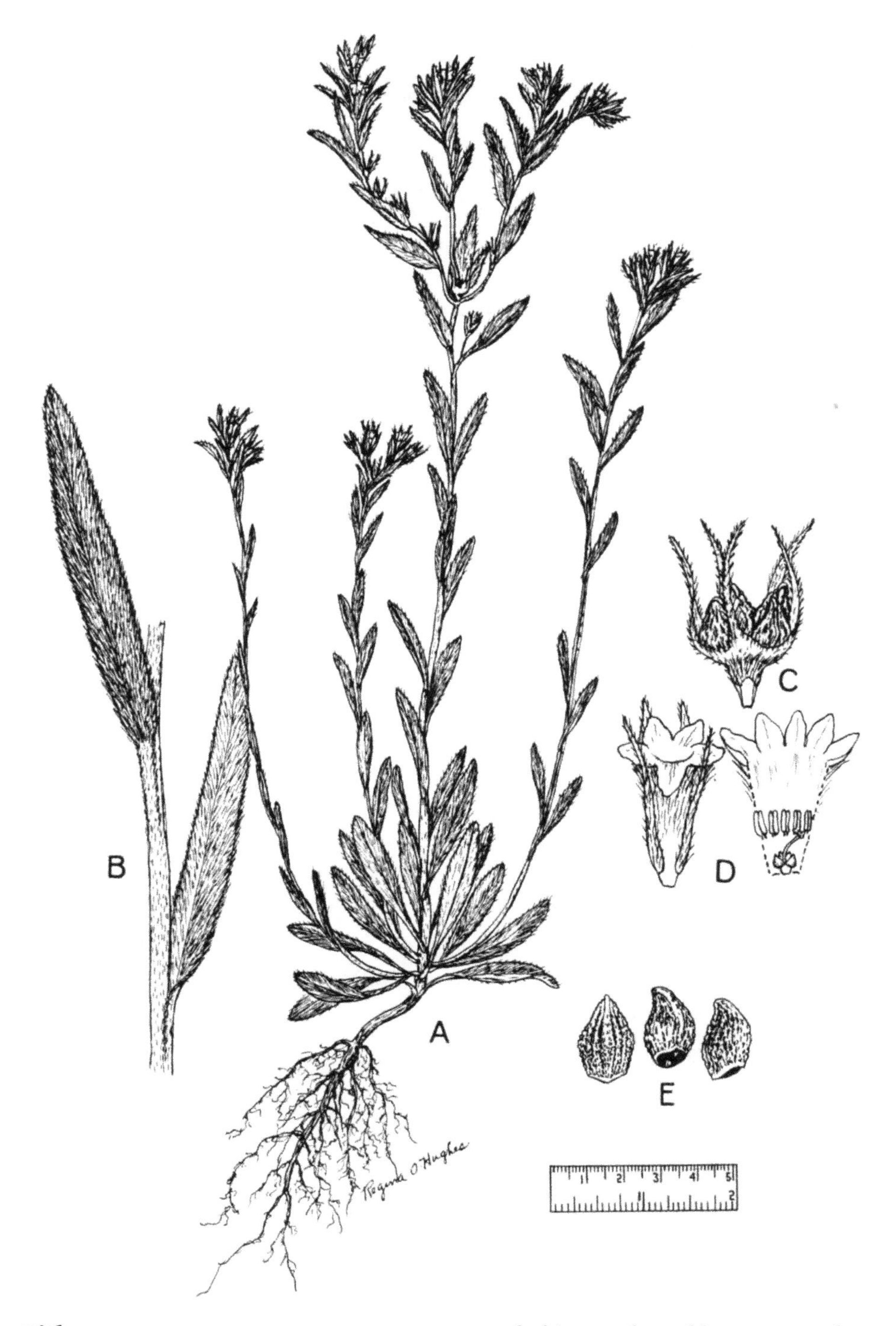

Lithospermum arvense, CORN GROMWELL. A, habit. B, enlarged leaves. C, nutlets in calyx. D, flowers. E, nutlets.

Barbarea vulgaris Ait. f. YELLOW ROCKET

Short-lived perennial herb, reproducing by seeds and sometimes by new shoots from the old crowns. Taprooted.

DESCRIPTION Stems numerous from a crown, erect, 3-6 dm tall, branched near the top, smooth, angular or ridged. **Leaves** 5-25 cm long, pinnately divided (rarely simple), the basal leaves with a large terminal lobe, forming a dense rosette, the stem leaves 1-4 pairs, becoming progressively shorter, with the uppermost leaves being about 2.5 cm long and less deeply lobed. **Flowers** in spike-like racemes at the end of each branch, bright lemon-yellow, 4-petaled, the petals narrowly obovate, 5.5-8 mm long, 2-3 mm broad. **Seed pod** (silique) slender-pedicelled, 2.5-5 cm long, 1.5 mm in diameter, nearly square in cross section, with a slender beak, 1.5-3 mm long. **Seed** light-yellow to yellowish-brown to grayish, about 1-1.5 mm long, wrinkled, short-oblong.

FLOWERING April-June.

WHERE FOUND New meadows and along roadsides; rich alluvial soil; clover and alfalfa fields; seeds live several years in the soil.

ORIGIN Introduced and naturalized from Eurasia.

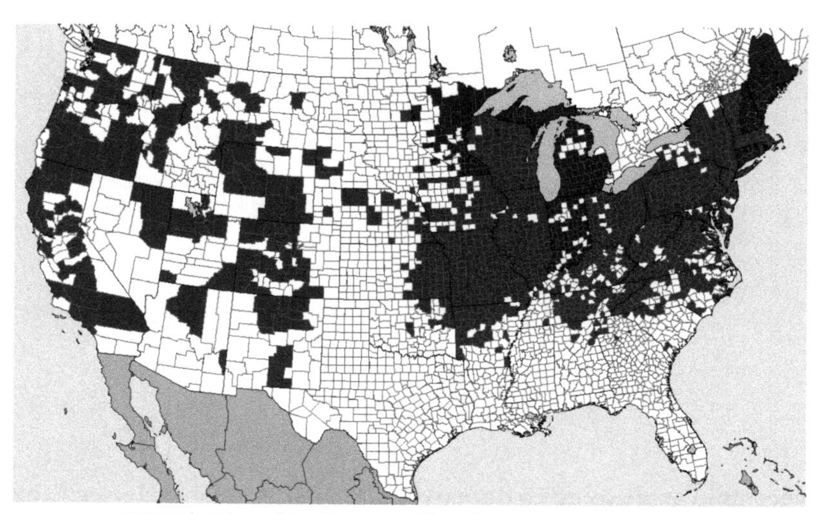

Distribution of **Barbarea vulgaris**, YELLOW ROCKET

Barbarea vulgaris, YELLOW ROCKET. A, habit. B, flower. C, raceme of fruits. D, silique. E, seeds.

Brassica nigra (L.) Koch — BLACK MUSTARD

Annual or winter annual herb, reproducing by seeds, with taproot.

DESCRIPTION Stems erect, green, simple or branched, up to 15 dm tall, usually bristly below, glabrous above. **Leaves** all slender-petioled, egg-shaped to obovate, the lower ones usually lobed with a large terminal lobe and a few small lateral lobes, the upper leaves merely toothed. **Flowers** 8–10 mm wide, the pedicels at maturity erect, 3–4 mm long. **Seed pod** (silique) erect, quadrangular, 1–2 cm long, smooth, the mid-vein nearly as strong as the sutures, the beak slender, 2.5–4 mm long. **Seed** reddish-brown to black, 1.5–2 mm long ellipsoidal, minutely roughly reticulate.

FLOWERING May–July; fruiting June–October.

WHERE FOUND Neglected fields and waste places, pastures, and ditches. Principal source of table mustard.

ORIGIN Naturalized from Eurasia.

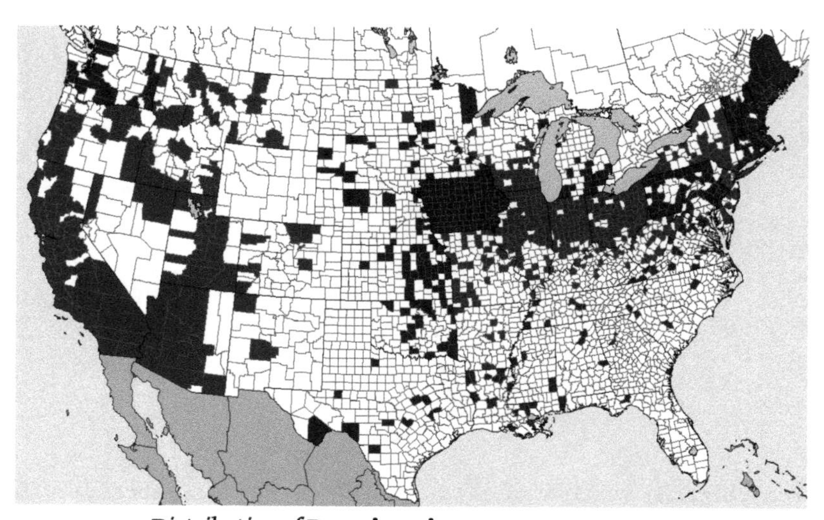

Distribution of **Brassica nigra**, BLACK MUSTARD

Brassica nigra, BLACK MUSTARD. A, Habit. B, basal leaf. C, siliques. D, seeds. E, reticulation pattern on seed.

Camelina microcarpa Andrz. ex DC. SMALL-SEED FALSE-FLAX

Annual or winter annual herb, reproducing by seeds.

DESCRIPTION Stem 30–80 cm tall, harsh with elongate, simple (1–2 mm long) and branching (stellate) hairs, simple or branched above. **Leaves**, the basal and lower ones narrowly spatulate, lanceolate, entire or nearly so, more or less hirsute, the leaves arising from stems linear to lanceolate, clasping by a sagittate auricle-like base. **Flowers** in elongated raceme, the fruiting raceme often over 20 cm long. Sepals erect, rounded at the end, the outer slightly sac-shaped at the base. Petals yellowish, spatulate, 3–4 mm long. **Silicles** (pods) 4–7 mm long, 4–5 mm thick, about twice as long as the style, walls thin, egg-shaped or pear-shaped, slightly flattened parallel to the partition, somewhat keeled along the seams, narrowed to the base and a short stipe. **Seed** dark-brown, oblong, mostly less than 1 mm long. Plants often show integration to *Camelina sativa.*

FLOWERING April–September.

WHERE FOUND Roadsides, waste places, and fields, often in flax-growing areas.

ORIGIN Naturalized from Europe.

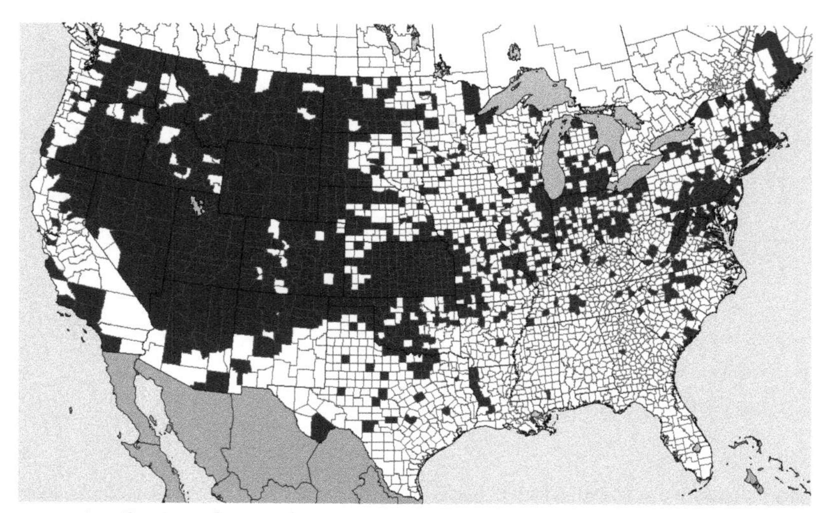

Distribution of **Camelina microcarpa**, SMALL-SEED FALSE-FLAX

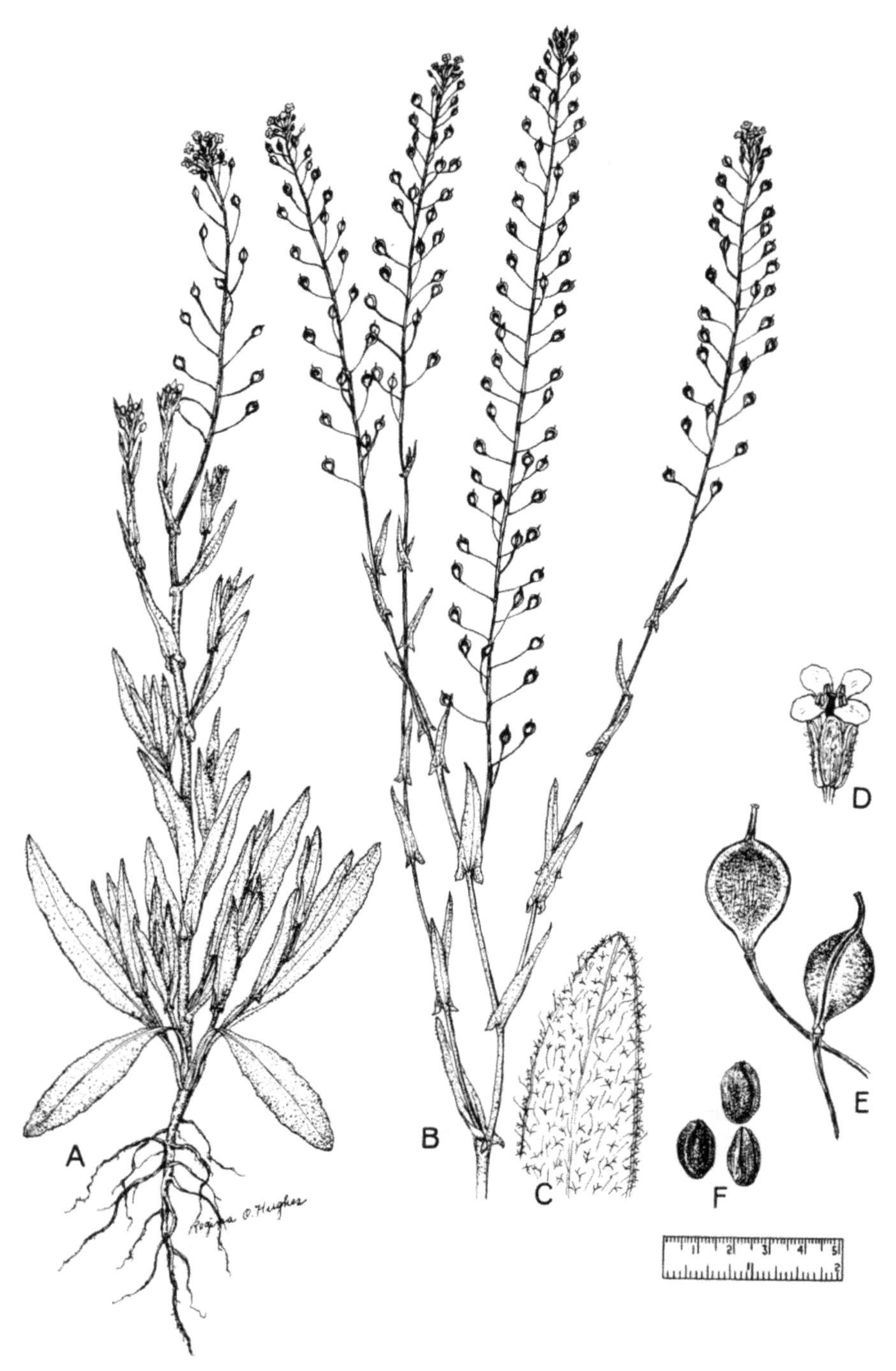

Camelina microcarpa, SMALL-SEED FALSE-FLAX. A, habit of young plant. B, elongated fruiting raceme of mature plant. C, enlarged portion of leaf, showing long simple and branched hairs. D, flower. E, silicles. F, seeds.

Capsella bursa-pastoris (L.) Medic. SHEPHERD'S-PURSE

Annual or winter annual herb, reproducing by seeds, with a branched thin taproot.

DESCRIPTION Stems erect, branched, 1–6 dm tall, covered with gray hairs. **Leaves** alternate, simple, variously toothed or lobed, in a rosette at base, coarsely lobed, clasping the stem with pointed lobes, coarsely serrate, 5–10 cm long, stem leaves arrow-shaped. **Flowers** small, white, 4-petaled, about 2 mm wide, borne in elongated racemes at the ends of the branches on slender pedicels. **Seed pod** (silique) triangular, 2-parted, 5–8 mm long, flattened at right angles to the partition, the values boat-shaped, each with several to many seeds. **Seed** small, about 1 mm long, yellow to orangish-brown, shiny, oblong, grooved on one side. Highly variable in leaf form.

FLOWERING March–December.

WHERE FOUND In practically all crops, gardens, and lawns, non-cultivated areas, and waste grounds. Seeds long-lived in soil.

ORIGIN Introduced and naturalized from southern Europe. A cosmopolitan weed. Common throughout North America.

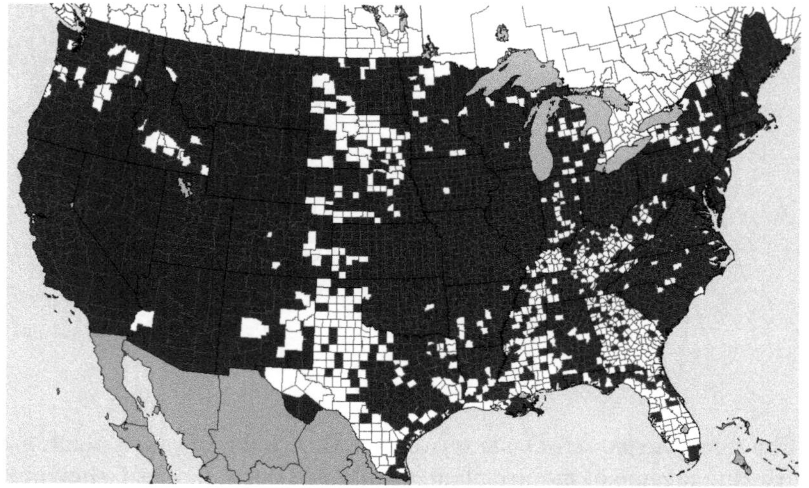

Distribution of **Capsella bursa-pastoris**, SHEPHERD'S-PURSE

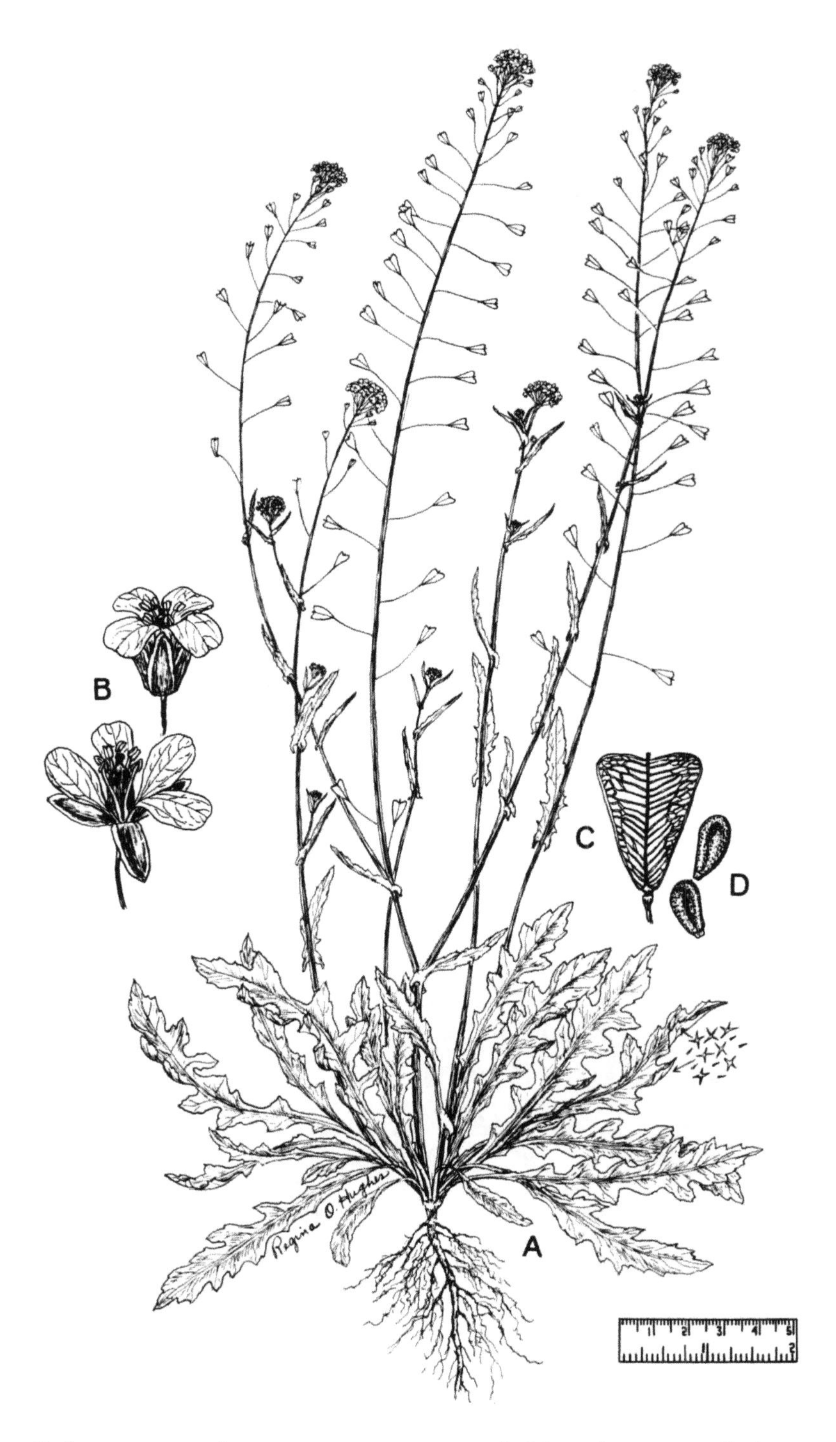

Capsella bursa-pastoris, SHEPHERD'S-PURSE. A, habit. B, flowers. C, silicle. D, seeds.

Conringia orientalis (L.) Dumort. HARE'S-EAR MUSTARD

Annual or winter annual herb, reproducing by seeds. Root system fibrous.

DESCRIPTION Stems glabrous, glaucous, 2.5–10 dm tall, erect, usually simple, leafy. **Leaves** alternate, sessile, entire, 4–13 cm long, obovate, oval to elliptical, deeply cordate-clasping, very smooth, whitish, somewhat fleshy. **Flowers** in elongate racemes, perfect. Sepals long and narrow, nearly equal, ascending. Petals pale-yellow, 8 mm long, narrow, tapering toward the point. Style rather long, the stigma entire or nearly so. **Pod** elongate-linear, a 2-celled capsule, 8–13 cm long, about 2 mm thick, ascending to erect, 4-angled, somewhat twisted or curved, the beak about 1.5 mm long. **Seed** numerous, 2–2.5 mm long, with 2 vertical grooves separated by a ridge, the seeds in 1 row in each cell, ellipsoid, oblong, thick, dark grayish-brown, minutely roughened, not winged.

FLOWERING May–July.

WHERE FOUND Grainfields, roadsides, waste places, gardens, and wasteland of the plains.

ORIGIN Native of Europe.

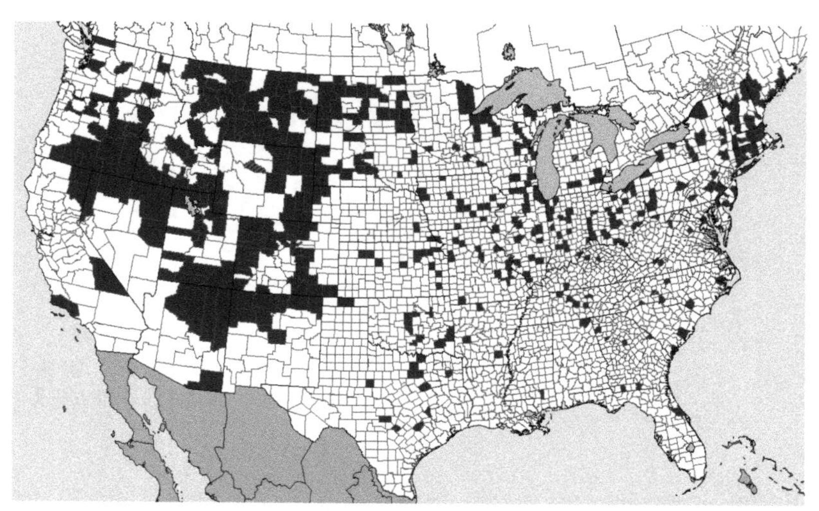

Distribution of **Conringia orientalis,** HARE'S-EAR MUSTARD

Conringia orientalis, HARE'S-EAR MUSTARD. A, habit. B, silique. C, seed.

Descurainia pinnata (Walt.) Britt. TANSY-MUSTARD

Annual herb, reproducing by seeds.

DESCRIPTION Stem erect, simple or branched, 1-8 dm tall, green, with densely hoary gray pubescence, glabrous to glandular. **Leaves** oblong, tapering toward the tip, alternate, pinnately dissected, often with grayish stellate pubescence. **Racemes** up to 3 dm long, glandular to glabrous. **Petals** yellow or yellowish-green to nearly white, 2-4 mm wide, nearly horizontally divergent. Siliques narrowly club-shaped, 5-16 mm long, 1-2 mm wide, on widely divergent to erect pedicels, the pedicels 0.6-1.5 cm long. **Seed** 2-ranked in each locule, less than 1 mm long. A variable species.

FLOWERING March-August.

WHERE FOUND Waste places, prairies, dry or sandy soils, and open woods; especially abundant in arid and semiarid regions.

ORIGIN Native.

NOTE ssp. *pinnata*—Leaves with hoary gray pubescence, not glandular. Raceme with hoary gray pubescence, occasionally also glandular. Siliques spreading or ascending, 5-10 mm long. On dry or sandy soils. North Carolina to Florida and Texas.

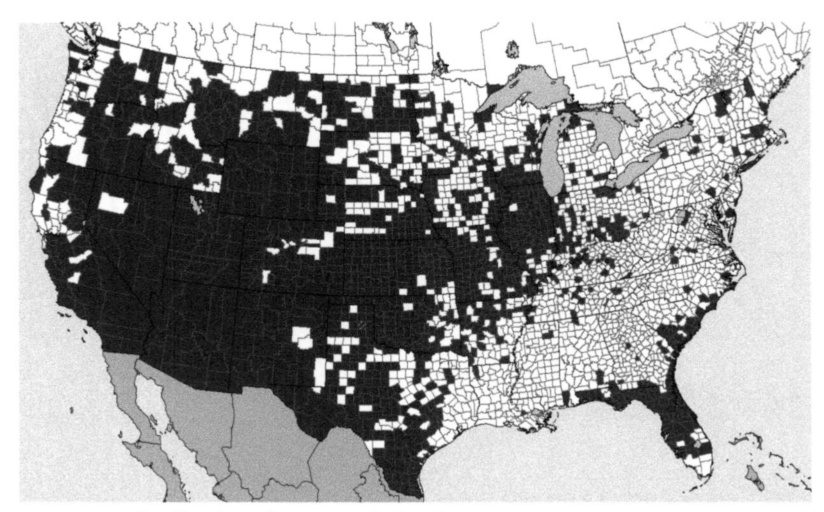

Distribution of **Descurainia pinnata**, TANSY-MUSTARD

Descurainia pinnata, TANSY-MUSTARD. A, habit. B, enlarged cauline leaf. C, flower. D, silique. E, seeds.

Lepidium campestre (L.) Ait. f. FIELD PEPPERWEED

Winter-annual or annual herb, reproducing by seeds.

DESCRIPTION Stems erect, often clustered with stiff ascending branches above, hoary-pubescent, or rarely glabrous, 1.5–7 dm tall, very leafy. **Leaves** on stem alternate, covered with soft hairs, arrow-shaped, the bases clasping the stem; the basal leaves arrow-shaped, spathulate to lanceolate, often pinnately lobed or toothed. **Flowers** inconspicuous, borne in dense racemes at the top of the plant. Petals 4, white or greenish, slightly exceeding the sepals. Stamens 6. **Seed pods** (silicles) oblong-ovate, 5–6 mm long, 4 mm wide, slightly winged, on short spreading pedicels, flattened at right angles to the partition, curved upward and the upper surface concave, often with small projections, dehiscent, each valve with 1 seed. **Seed** dark-brown, 2–2.5 mm long, obovoid, 1 side flattened, coarsely granular, pointed at the tip.

FLOWERING May–September.

WHERE FOUND Clover, alfalfa, and winter wheat fields, other grainfields, first-year meadows, and wastelands; locally abundant.

ORIGIN Introduced and naturalized from Europe.

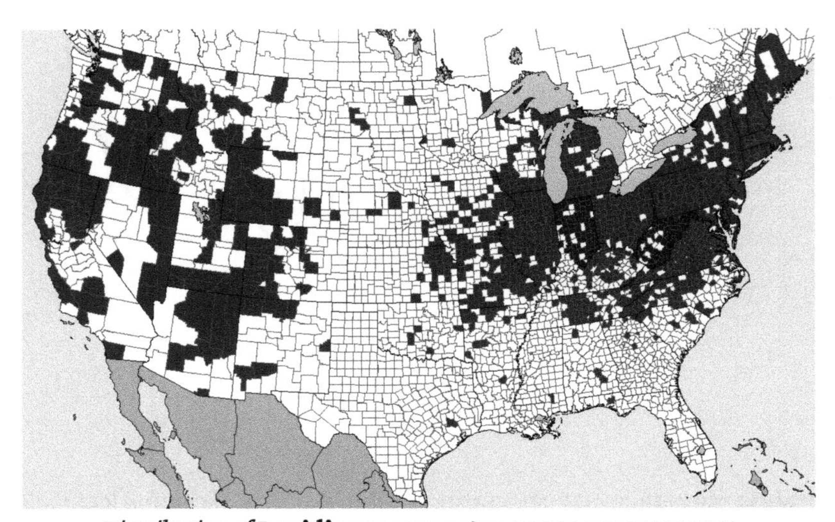

Distribution of **Lepidium campestre**, FIELD PEPPERWEED

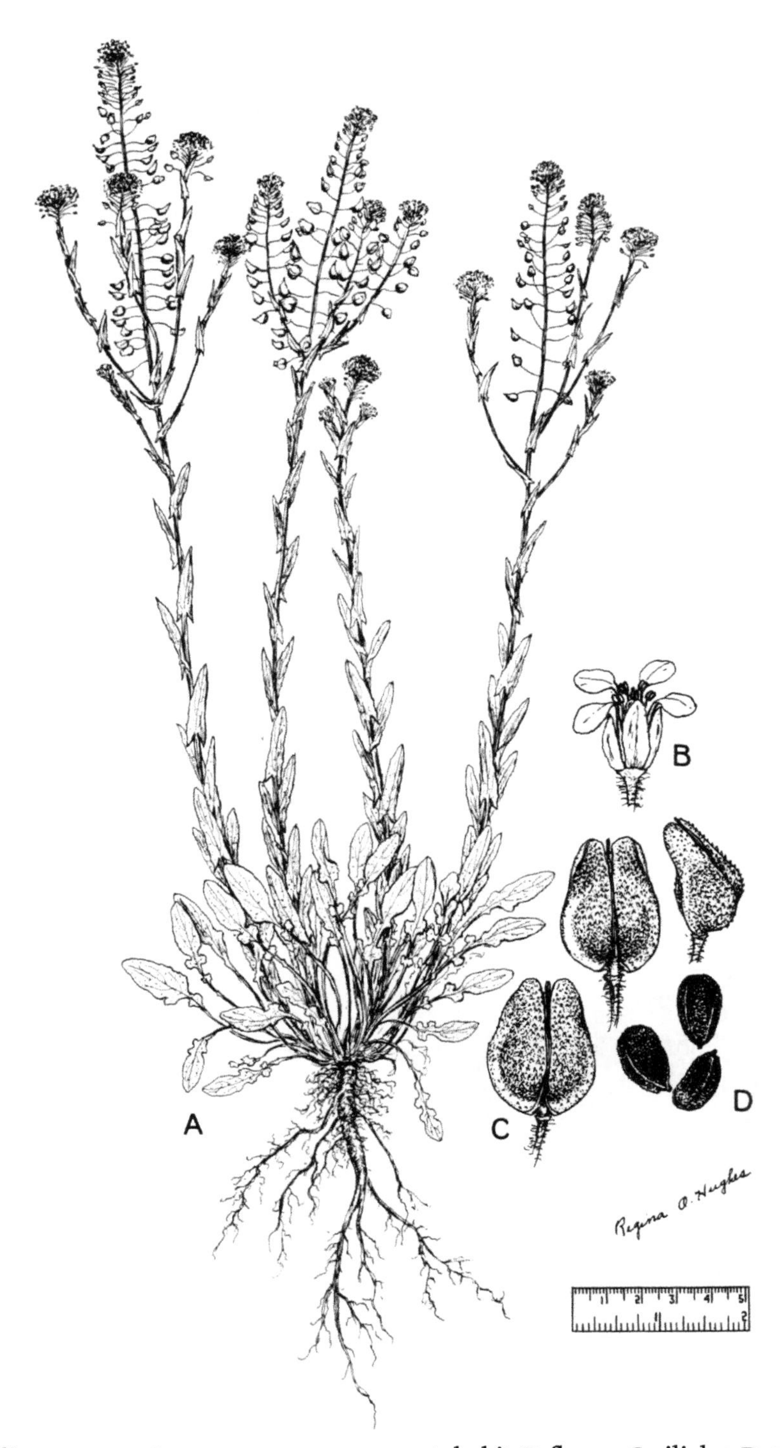

Lepidium campestre, FIELD PEPPERWEED A, habit. B, flower. C, silicles. D, seeds.

Lepidium draba L. HOARY CRESS

Perennial herb, reproducing by seeds and by horizontal creeping roots.

DESCRIPTION Stems stoutish, erect or spreading, 1–8 dm tall, branched, sparsely pubescent to ash-colored hoary. **Leaves** alternate, simple, mostly toothed, the basal leaves 4–10 cm long and petioled, oblanceolate to obovate, the cauline leaves sessile, 2–6.5 cm long, oblong or tapering to the point, entire, with broad clasping bases. **Flowers** in terminal corymb-like racemes. Petals white, clawed, 3–5 mm long, exceeding the sepals (1.5–2 mm long). Ovary glabrous. **Seed pods** (silicles) kidney-shaped to heart-shaped, on divergent slender pedicels (6–15 mm long), glabrous, 2–5 mm long, often oblique and inflated, obovate to triangular-egg-shaped to egg-shaped, netted-veined, with a prominent slender persisting style (1 mm long), barely if at all notched, dehiscent, each valve with 1 seed. **Seed** 2–4, wingless, about 2 mm long, obovate, slightly flattened, granular, reddish-brown.

FLOWERING April–June; fruiting late June–August.

SYNONYMS *Cardaria draba* (L.) Desv.

WHERE FOUND Cultivated fields, grainfields, grasslands, waste places, meadows, and roadsides.

ORIGIN Naturalized from Europe.

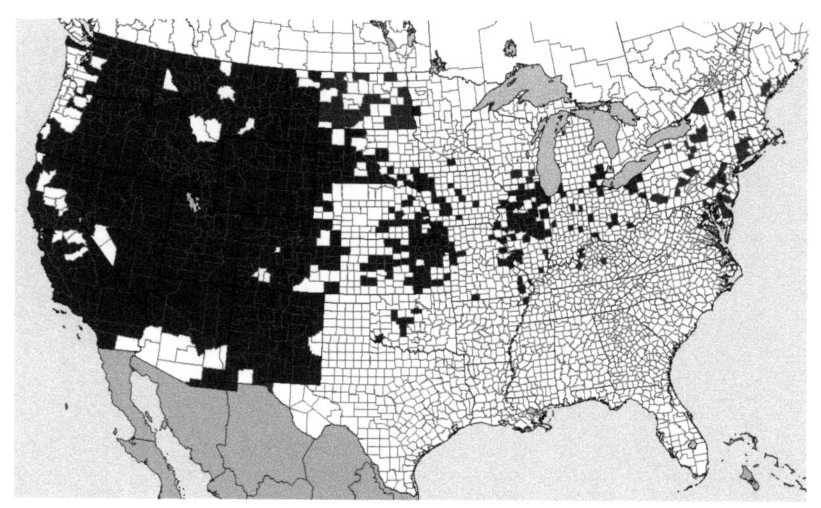

Distribution of **Lepidium draba**, HOARY CRESS

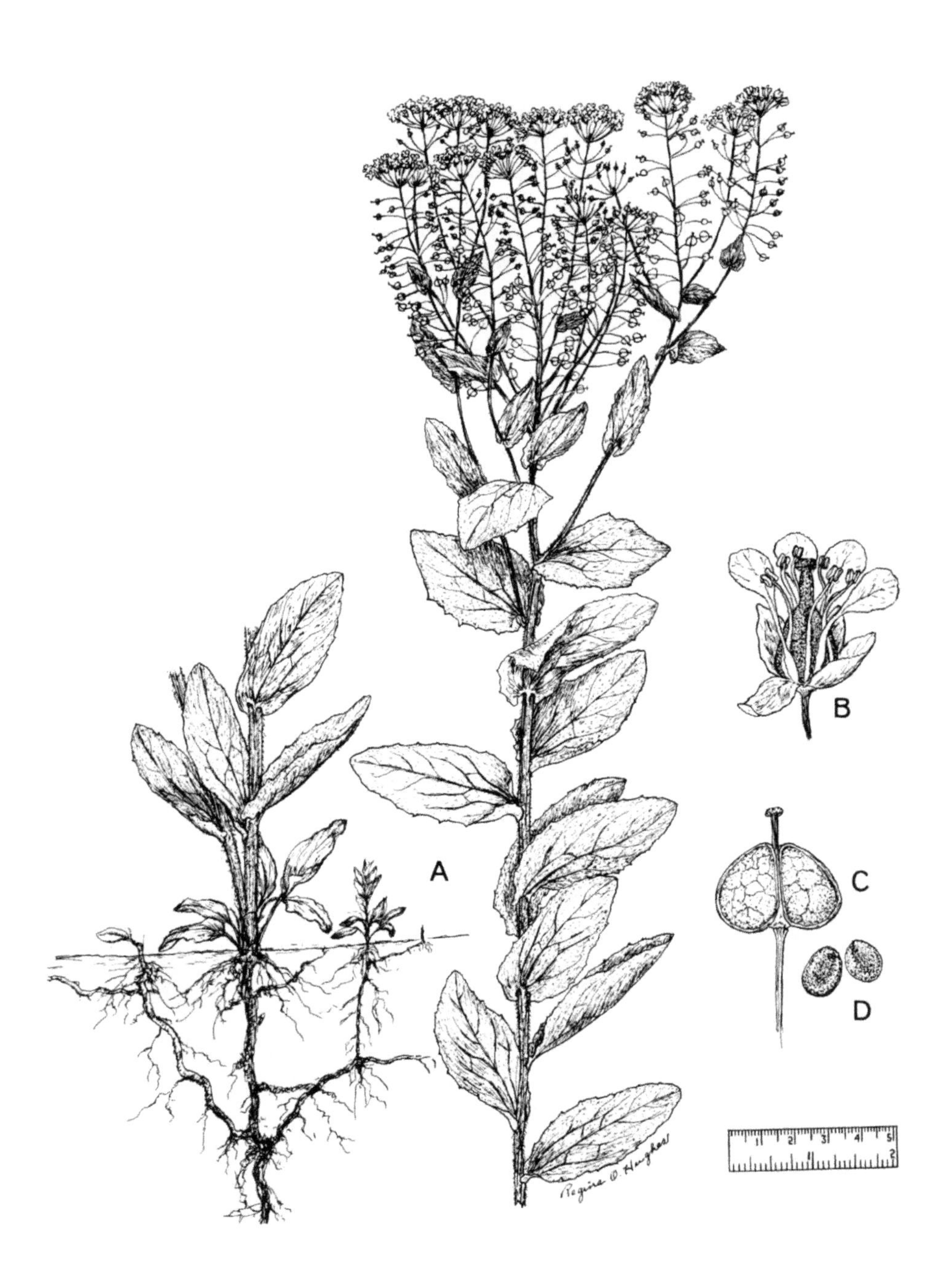

Lepidium draba, HOARY CRESS. A, habit. B, flower. C, silicle. D, seeds.

Lepidium virginicum L. VIRGINIA PEPPERWEED

Annual or winter-annual herb, reproducing by seeds.

DESCRIPTION Stems erect, muchbranched, 2-9 dm tall, smooth to minutely pubescent, green. **Leaves** on stem lanceolate to linear, coarsely toothed, irregularly cut or entire, usually sessile, the basal leaves obovate, irregularly cut, pinnatifid or pinnate, with 1 large terminal lobe and several smaller dentate lateral ones, glabrous. **Inflorescence** in elongate racemes, that grow and flower for long periods, often with seed pods below and flowers above. **Flowers** small, with 4 white or greenish petals, the petals equaling or exceeding the sepals. Stamens 2 (rarely 4). **Seed pods** (silicles) nearly orbicular, 2.5-4 mm wide, containing 2 seeds, shallow-notched at the summit. **Seed** 1.5-2 mm long, obovate, with 1 edge straight, the other rounded, slightly winged at the apex, granular, chestnut-brown.

FLOWERING May-November.

WHERE FOUND Fields, meadows, roadsides, and waste places; dry soils. A common weed.

ORIGIN Native.

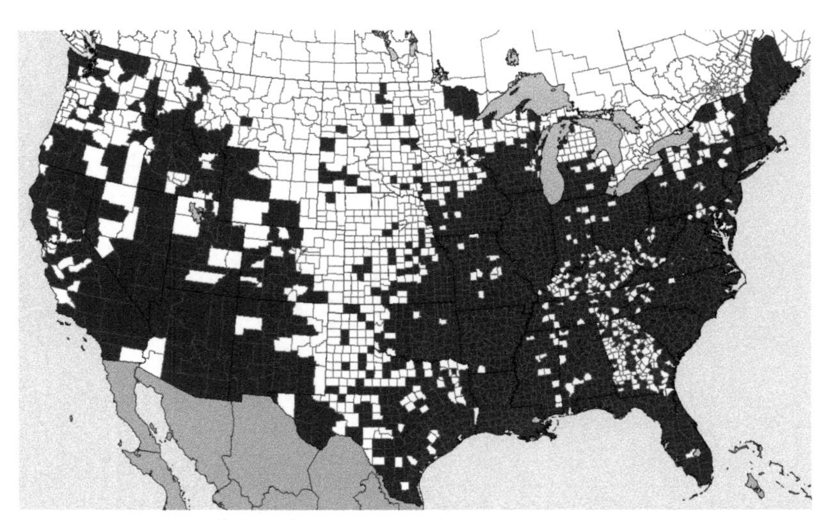

Distribution of **Lepidium virginicum**, VIRGINIA PEPPERWEED

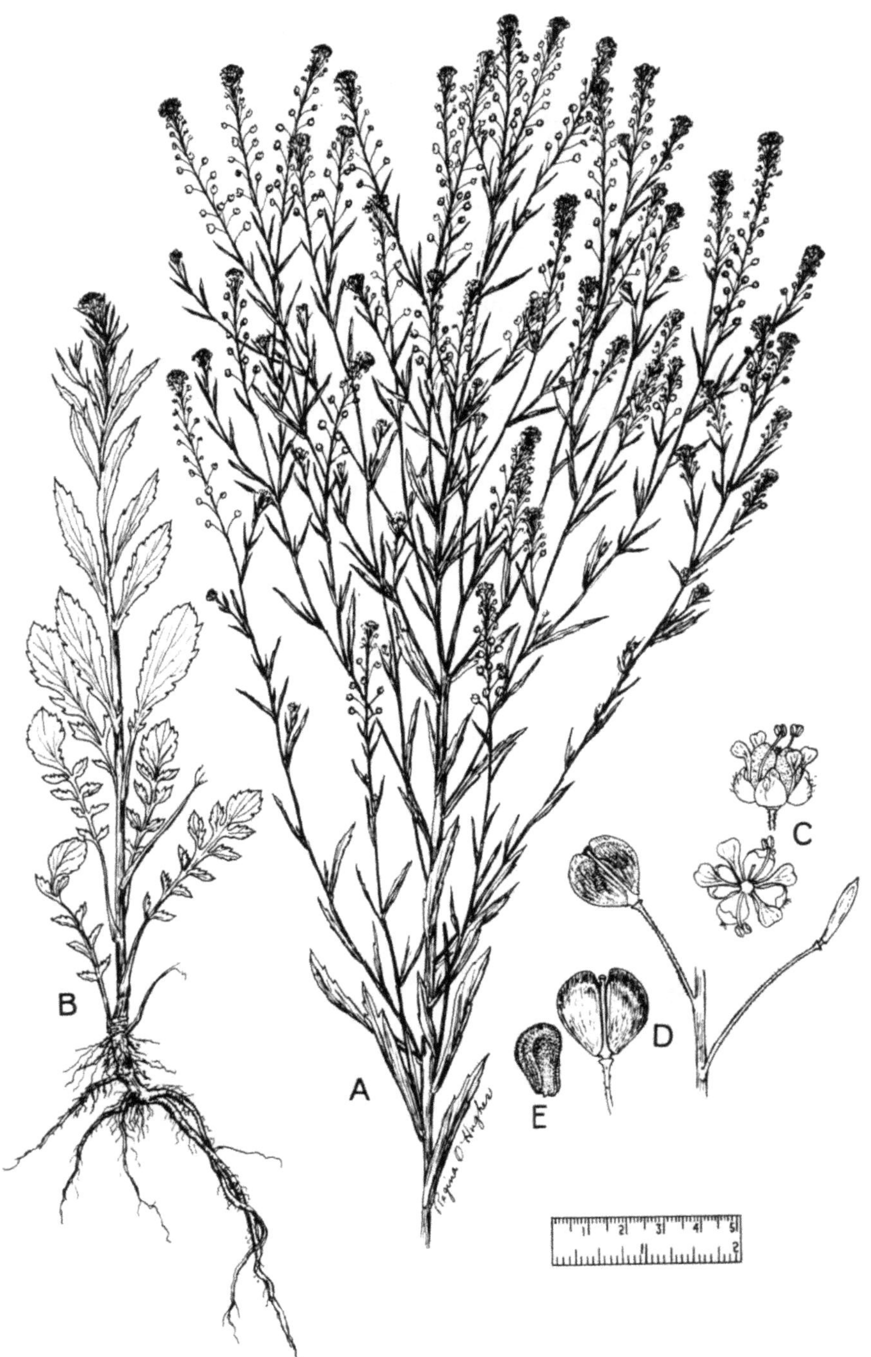

Lepidium virginicum, VIRGINIA PEPPERWEED. A, habit, upper part of plant. B, young plant, showing basal and cauline leaf forms. C, flowers. D, silicles, mature and immature. E, seed.

Nasturtium officinale Ait. f. — WATERCRESS

Perennial aquatic or marsh herb, reproducing by seeds and rooted stems.

DESCRIPTION Stems succulent, smooth, creeping or floating, freely rooting. Leaves pinnate, the 3-11 leaflets roundish to oblong or elongate, somewhat fleshy, nearly entire, the terminal leaflet roundish, oval, or oblong. **Flowers** perfect, in racemes. Sepals equal at the base, spreading during anthesis. Petals white, twice the length of the calyx, without nectaries. **Seed pods** (siliques) linear-cylindric, 1-2.7 cm long, 3 mm thick, somewhat curved, ascending on divergent pedicels, the convex valves nerveless or 1-nerved, the beak about 1 mm long. **Seed** in 2 rows in each locule, small, turgid, marginless.

FLOWERING April–October.

SYNONYMS *Rorippa nasturtium-aquaticum* (L.) Hayek

WHERE FOUND Brooks, rills, and springheads; often cultivated in cool waters.

ORIGIN Native of Eurasia; throughout most of the United States.

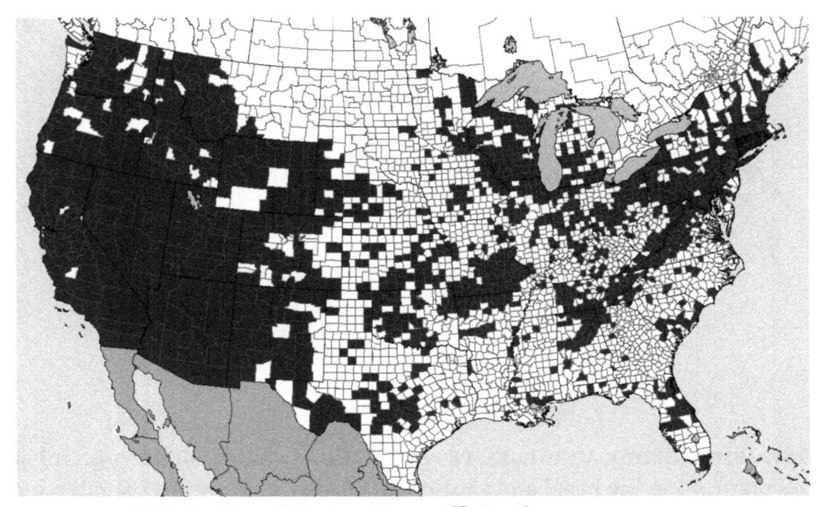

Distribution of **Nasturtium officinale,** WATERCRESS

Nasturtium officinale, WATERCRESS. A, habit. B, elongated mature plant. C, flower. D, silique. E, seeds.

Sinapis arvensis L. WILD MUSTARD

Annual or winter-annual, reproducing by seeds.

DESCRIPTION Stems erect, 2-8 dm tall, branched near the top, with a few bristly hairs. **Leaves** obovate in general outline, the lower ones sometimes lobed, more often merely coarsely toothed, the upper ones progressively smaller, coarsely toothed, roughly pubescent to nearly glabrous. **Flowers** in clusters at the ends of the branches, with 4 yellow petals, about 15 mm wide, the pedicels at maturity spreading or ascending, 5-7 mm long. **Seed pod** (silique) linear, nearly circular in cross section, ascending, glabrous or nearly so, 1-2 cm long, 1.5-2.5 mm wide, the beak angular, shorter than the valves, often containing 1 seed. **Seed** globular, 1-1.5 mm in diameter, black or dark purplish-brown, smooth or minutely reticulate.

FLOWERING May-August.

SYNONYMS *Brassica arvensis* (L.) Kuntze, *Brassica kaber* (DC.) L. C. Wheeler

WHERE FOUND Gardens, spring grainfields (especially oats), cultivated land, and waste places; commonest mustard of grainfields in northeastern states; seeds live in soil for many years; a serious weed in many areas.

ORIGIN Introduced and naturalized from Europe.

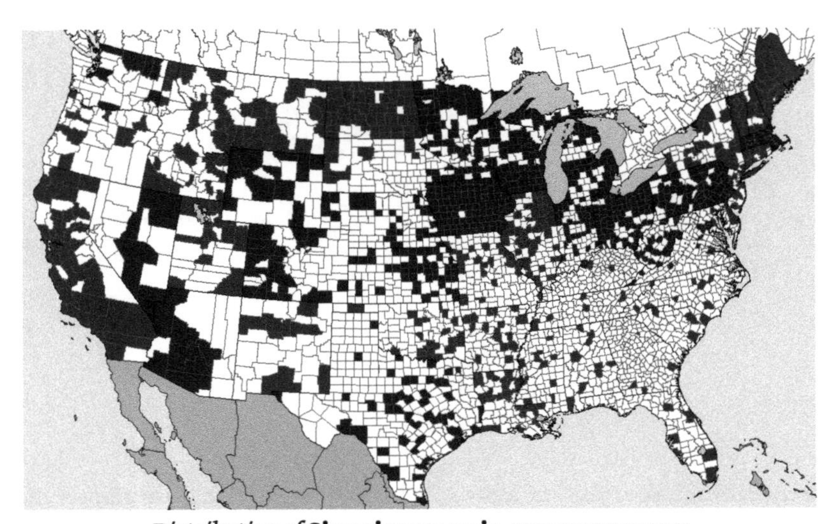

Distribution of **Sinapis arvensis**, WILD MUSTARD

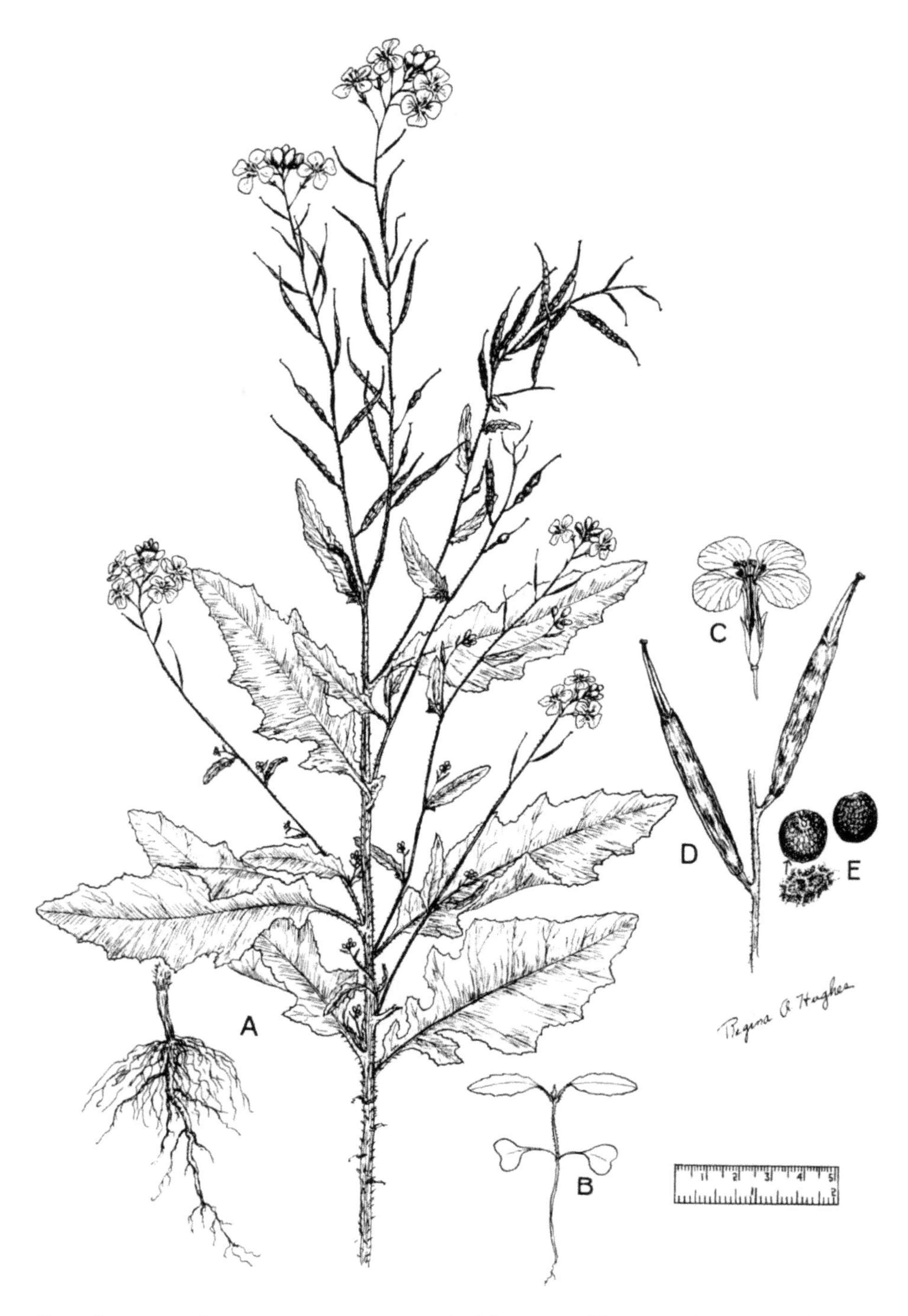

Sinapis arvensis, WILD MUSTARD. A, habit. B, seedling. C, flower. D, siliques. E, seeds.

Sisymbrium altissimum L. TUMBLE-MUSTARD

Annual or winter-annual herb, reproducing by seeds, with a taproot. The stem often breaks off at maturity causing the plant to be blown about by the wind.

DESCRIPTION Stem erect, up to 6 dm tall, bushy, branching, the stem and branches smooth above, but somewhat hairy below, pithy. **Leaves** alternate, pale-green, the lower ones large and pinnately lobed with broad, irregular, toothed lateral segments and a large terminal lobe, the upper ones with smaller narrow segments, arrowhead-shaped to lanceolate or entire. **Flowers** small, yellowish-white, 0.9–1.4 cm in diameter, 4-petaled, in numerous but short spike-like racemes. **Seed pods** (siliques) stiff, pubescent to finely dense pubescent, awl-shaped, 1–2 cm long, 1–1.5 mm wide at the base, divided into 2 parts on very short pedicels, thus causing the pods to resemble stems rather than seed pods. **Seed** oblong, about 1.5 mm long, dark reddish-brown, quite smooth.

FLOWERING May–September.

WHERE FOUND Fields of small grain, gardens, and waste places.

ORIGIN Naturalized from Europe.

NOTE Also common in the eastern United States is the weedy *Sisymbrium officinale* (L.) Scop., with siliques essentially glabrous and the plant greener than plants of related species.

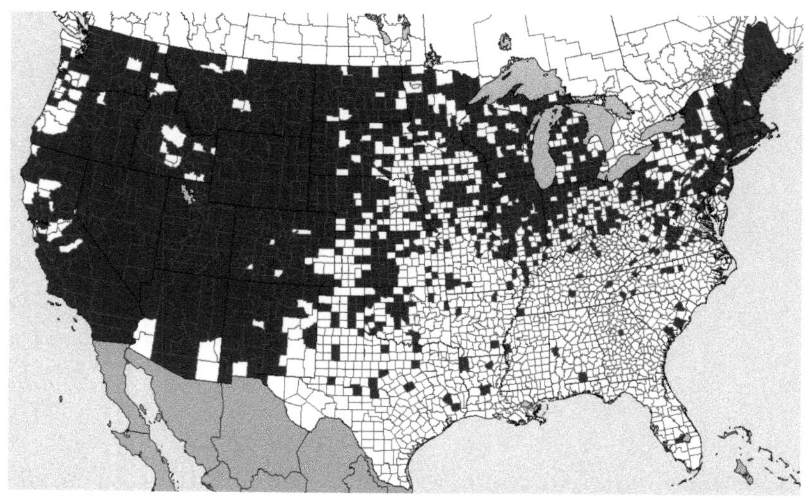

Distribution of **Sisymbrium altissimum**, TUMBLE-MUSTARD

Sisymbrium altissimum, TUMBLE-MUSTARD. A, habit. B, leaves, basal and cauline. C, silique. D, seeds.

Thlaspi arvense L. FIELD PENNYCRESS

Annual or winter-annual herb, reproducing by seeds.

DESCRIPTION Stem simple to much-branched above, erect, smooth, up to 8 dm tall. **Leaves** alternate, the lowest ones petioled, narrowly obovate, soon drooping; the middle and upper leaves membranaceous, oblong, toothed or entire, sessile, with two narrow, divergent, acute auricles 1-5 mm long. **Flowers** in racemes. Petals white, 3-4 mm long, equal, twice as long as the sepals. **Seed capsules** (silicles) orbicular to rounded-oblong, 10-18 mm long, notched (2-3 mm) at the top, dehiscent by 2 winged values, each with 2-8 seeds per locule. **Seed** 2-2.3 mm long, compressed, ovoid, with 10-14 concentric granular ridges on each side, dark reddish-brown to blackish.

FLOWERING April-August.

WHERE FOUND Waste places, fields, roadsides, grainfields, grasslands, and gardens. A troublesome weed in grainfields in the northern and northwestern states.

ORIGIN Naturalized from Eurasia.

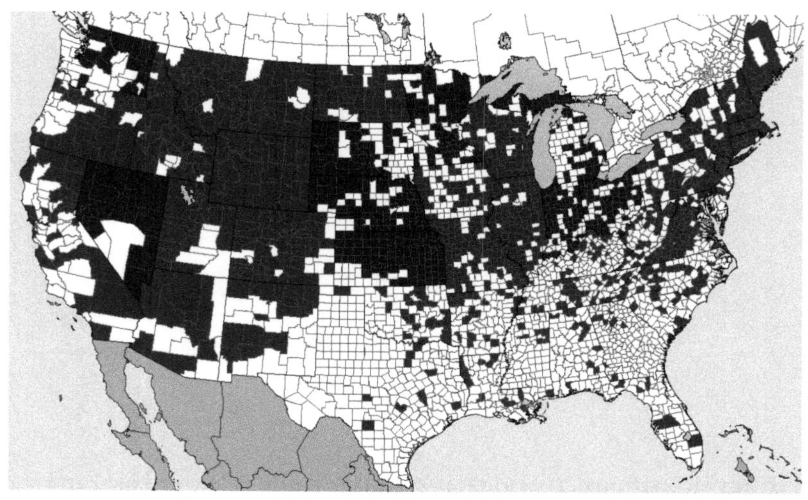

Distribution of **Thlaspi arvense**, FIELD PENNYCRESS

Thlaspi arvense, FIELD PENNYCRESS. A, habit. B, silicle. C, seeds.

Cannabis sativa L. HEMP, MARIJUANA

Annual herb, reproducing by seed. Taproot much-branched. Plants bushy unless crowded.

DESCRIPTION Stems simple or sparingly branched, 6–30 dm tall, coarse, somewhat grooved, rough and hairy, the hairs on the upper parts exuding a sticky resin with a characteristic odor, the inner bark of very tough fibers. **Leaves** opposite, but alternate above, palmately divided, with 5 to 9 hairy leaflets with notched edges. **Flowers** dioecious, small and green, the male and female flowers borne on separate plants: **male plants** turning yellow and dying after shedding pollen, male flowers in axillary compound racemes or panicles in the axils of the upper leaves, 5 perianth segments and 5 stamens; **female plants** vigorous and dark-green until frost, female flowers in spike-like clusters, without petals, in the axils of the leaves. **Fruit** (achene) about 4 mm long, ovoid to nearly round with obtuse edges, yellow to olive-brown. **Seed** oval, mottled-brown, about 3 mm long.

FLOWERING June–October.

WHERE FOUND Neglected fields, farmyards, ditches, roadsides, moist fertile soil, on wastelands, and fence rows; cultivated varieties grown as fiber crop; chiefly sporadic and derived from foreign packing, ballast lots.

ORIGIN Adventive from Asia.

NOTE Plants prolific seed and pollen producers; source of marijuana and of hemp fibers.

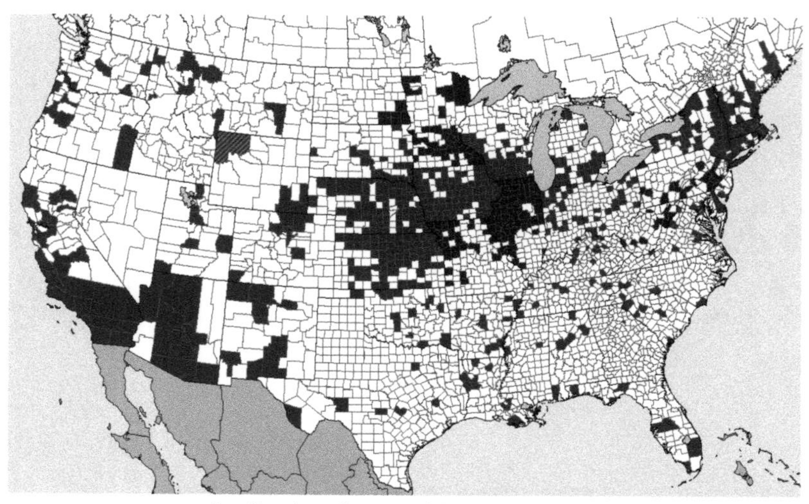

Distribution of **Cannabis sativa**, HEMP

Cannabis sativa, HEMP. A, male plant; a, habit; b, enlarged staminate flower panicle; c, flower. B, female plant; a, habit. B, enlarged pistillate flower cluster; c, flower; d, achene; e, seeds.

Lonicera japonica Thunb. JAPANESE HONEYSUCKLE

Perennial high-twining, climbing or trailing shrub, reproducing by seeds and underground rhizomes.

DESCRIPTION **Stems** woody, pubescent to glabrous, long and twining, green to purplish, freely rooting at the nodes. **Leaves** opposite, simple pubescent, mostly persistent throughout the winter especially southward, ovate or oblong, thickish, entire to variously lobed, short-petioled, green or purplish. Bracts leaf-like, becoming smaller. **Flowers** in pairs on the summit of solitary, axillary peduncles. Corolla 3–4 cm long, with 5 fused petals white to white tinged with purple, becoming yellow, sometimes carmine on the outside, very fragrant, with a pubescent tube. Sepals 5, fused with the ovary. Stamens 5, attached to the corolla. **Berries** black to purplish-black, with 2–3 seeds. **Seed** ovate to oblong, 2–3 mm long, flattened, 3-ridged on the back, flat or concave on the inner face, dark-brown or gray-brown.

FLOWERING Late April–November; fruiting September–November.

WHERE FOUND Orchards, gardens, fence rows, thickets, borders of woods, and roadsides, a pernicious and dangerous weed, often overwhelming and strangling the native flora.

ORIGIN Naturalized from Asia.

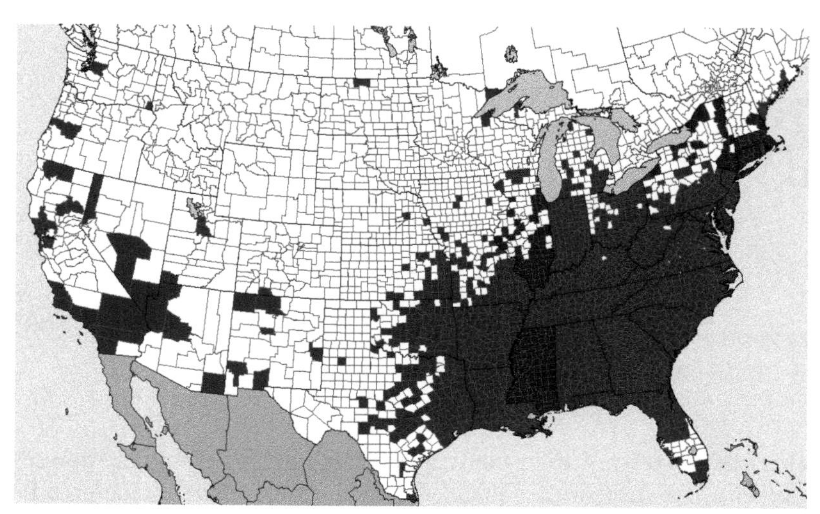

Distribution of **Lonicera japonica**, JAPANESE HONEYSUCKLE

Lonicera japonica, JAPANESE HONEYSUCKLE. A, habit. B, flowers. C, fruits. D, seeds.

Agrostemma githago L. CORN COCKLE

Winter-annual, reproducing by seeds. Taproot shallow.

DESCRIPTION Stems rough, silky, hairy, erect, linear 6–10 dm tall, swollen at the nodes, simple or branching slightly above. **Leaves** linear or lanceolate, acute, hairy, 8–12 cm long, 5–10 mm wide, opposite and jointed at the base. **Flowers** in cymes or solitary, on long peduncles up to 2 dm long, large, 2.5–4 cm in diameter, purplish-red, with narrow green sepals longer than the colored petals. **Capsule** 1-locular, 14–18 mm long, ovoid-oblong, enclosed in a 10-ribbed urn-like calyx tube (12–18 mm long). **Seed** black, 2–3 mm in diameter, triangular, rounded on the back, covered with rows of sharp tubercules.

FLOWERING May–September.

WHERE FOUND Cultivated land, in association with fall-sown grain crops, especially winter wheat and rye. Seeds poisonous, highly objectionable in grains used for milling or feed, dangerous to stock and poultry.

ORIGIN Naturalized from Eurasia; widely distributed throughout the United States.

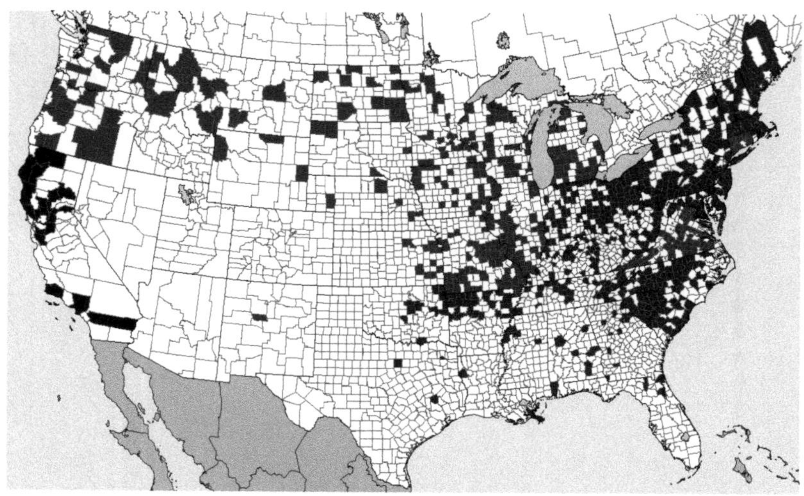

Distribution of **Agrostemma githago**, CORN COCKLE

Agrostemma githago, CORN COCKLE. A, habit. B, flower. C, calyx. D, capsule. E, seed.

Cerastium arvense L. FIELD CHICKWEED

Perennial herb, forming mats or tufts, reproducing by seeds and creeping stems, rooting at the lower nodes.

DESCRIPTION Stems slender, erect or ascending, glabrous to densely villous, glandular or non-glandular, 1.5-6 dm tall. **Leaves** opposite, simple, linear to lanceolate or narrowly ovate, acute to obtuse, 2-7 cm long, 115 mm wide, glabrous to pubescent, glandular or non-glandular; axillary fascicles or short sterile shoots often subtended by the primary leaves. **Inflorescences** with few to many flowers, the pedicles subcapillary. Sepals 5 (rarely 4), lanceolate acute, mostly 5-8 mm long, 23 times exceeded by the conspicuous white petals. Petals 5 (rarely 4), about 1 cm long. **Capsule** 1-celled, about 1 cm long, many-seeded, cylindric, equaling or much exceeding the sepals. **Seed** 0.6-1 mm long, obovoid, flat on the sides, strongly papillate-tuberculate, especially dorsally, reddish-brown to chestnut-brown. Has many different forms.

FLOWERING April-August.

WHERE FOUND Lawns, wastes, barrens, grasslands, abandoned fields, and meadows; rocky, gravelly, and sandy areas, chiefly on calcareous and magnesian soils; most introduced where it appears in lawns; does not persist under cultivation.

ORIGIN Native of North America and Eurasia.

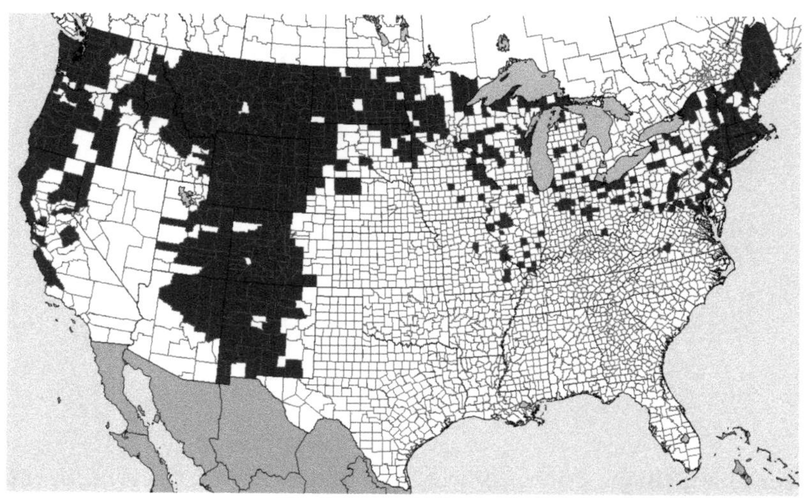

Distribution of **Cerastium arvense**, FIELD CHICKWEED

Cerastium arvense, FIELD CHICKWEED. A, habit. B, enlarged leaves. C, flower. D, capsule. E, seeds.

Cerastium fontanum Baumg. MOUSE-EAR CHICKWEED

Perennial herb, reproducing by seeds, occasionally by root development on the creeping stems.

DESCRIPTION Roots shallow, branched, fibrous. **Stems** sticky-pubescent, slender, 1.5-5 dm long, partly spreading to erect, often rooting at the lower nodes, forming mats. **Leaves** opposite, small, 1-2 cm long, 3-12 mm wide, 1-nerved, very hairy, attached directly at the stem, mostly oblong to ovate-lanceolate, or oblanceolate, or spathulate. **Inflorescence** rather open, the mature pedicels 5-12 mm long. Sepals 4-6 mm long, oblong-lanceolate, acute or obtuse, pubescent, 1-nerved toward the base, scarious-margined. Petals (sometimes absent) about the same length as the sepals, white, notched at the tips. **Capsules** very small, 8-10 mm long, 2-3 mm wide, cylindrical, sometimes curved, containing many seeds. **Seed** chestnut-brown, very small, 0.75 mm long, circular to angular-obovate, flattened with a rounded back bearing irregular knobs; seed long-lived in the soil.

FLOWERING April-October.

SYNONYMS *Cerastium vulgatum* L.

WHERE FOUND Woods, lawns, pastures, abandoned cultivated land, and meadows; often a troublesome weed.

ORIGIN Introduced and naturalized from Eurasia.

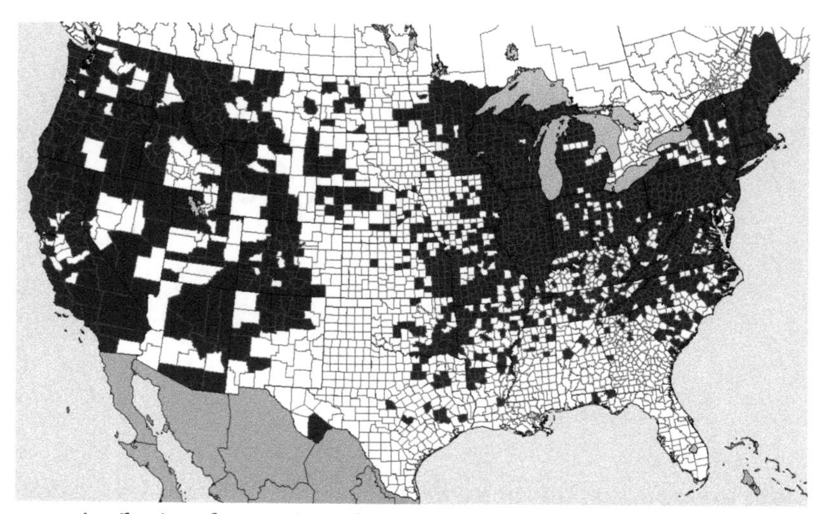

Distribution of **Cerastium fontanum**, MOUSE-EAR CHICKWEED

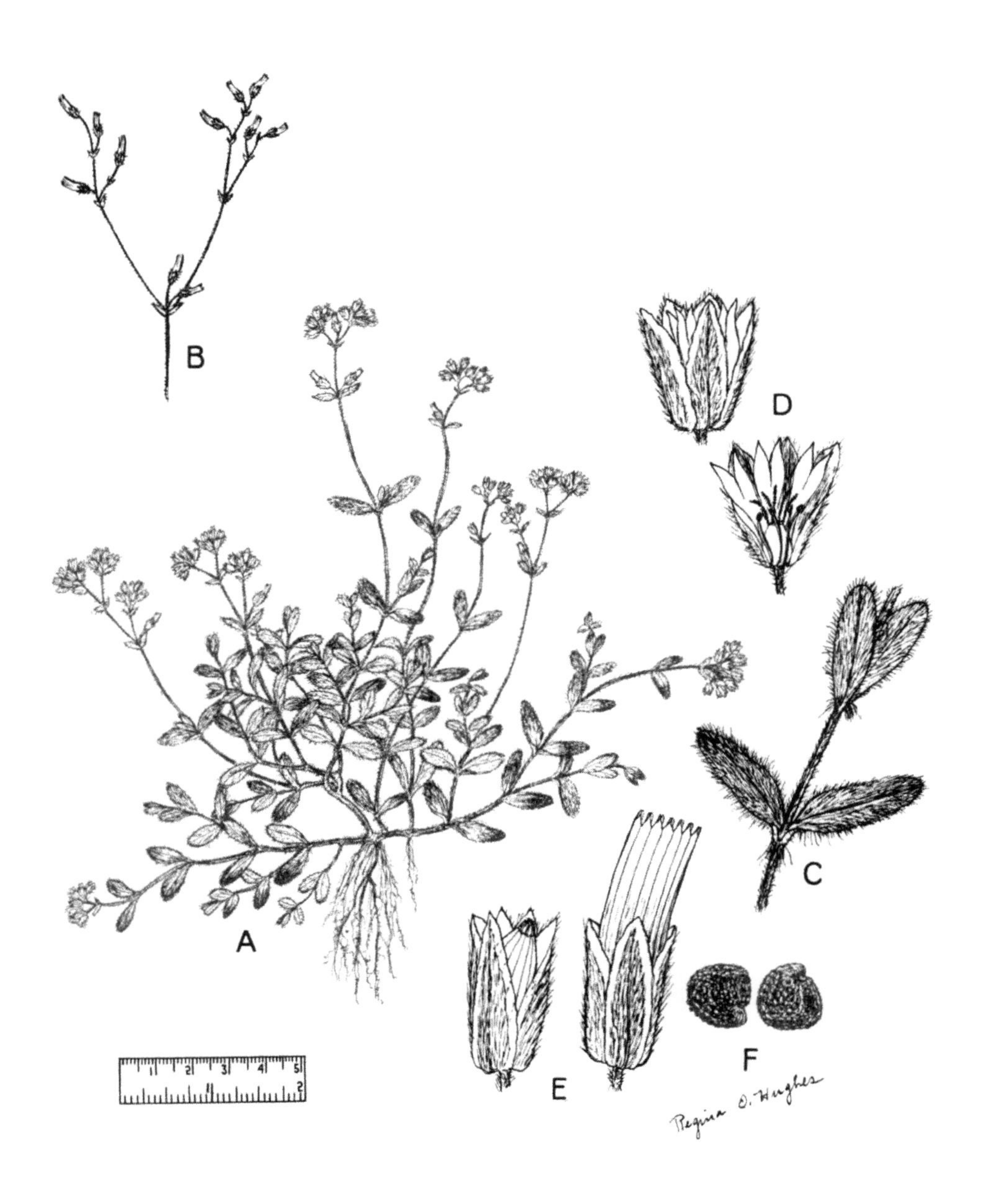

Cerastium fontanum, MOUSE-EAR CHICKWEED. A, habit. B, mature dichotomous cymes. C, enlarged leaves. D, flowers. E, capsules. F, seeds.

Scleranthus annuus L. ANNUAL KNAWEL

Annual or winter-annual herb, reproducing by seeds.

DESCRIPTION Stems much-branched, low, spreading, 5–15 cm high, glabrous or puberulent. **Leaves** opposite, simple, awl-shaped, without stipules. **Flowers** small, green, sessile to nearly sessile, in axillary clusters. Sepals 5, fused below into a cup that surrounds the 1-seeded fruit, the calyx lobes scarcely margined. **Fruit** a utricle, enclosed in the 5-toothed, 10-angled, straw-colored calyx, about 3–4 mm long. **Seed** ovoid, beaked at the micropylar end, straw-colored, 1–1.3 mm long.

FLOWERING May–October.

WHERE FOUND A common weed in gardens, fields, waste places, and lawns; on dry, sandy, or gravelly soils.

ORIGIN Introduced from Europe.

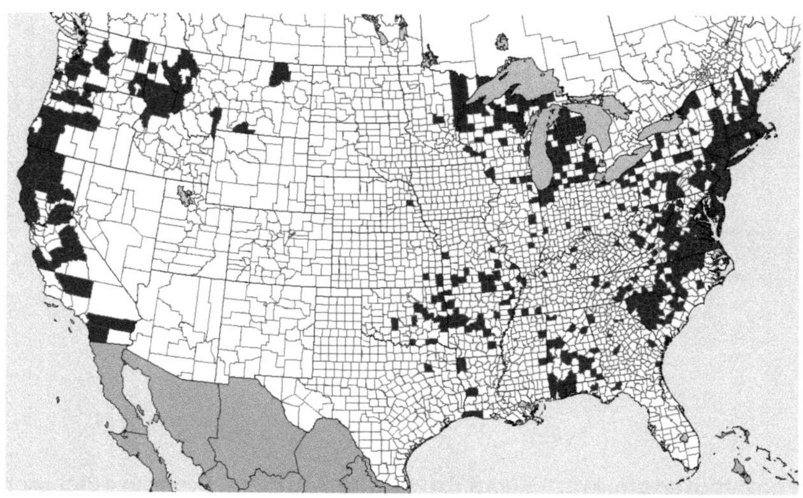

Distribution of **Scleranthus annuus**, ANNUAL KNAWEL

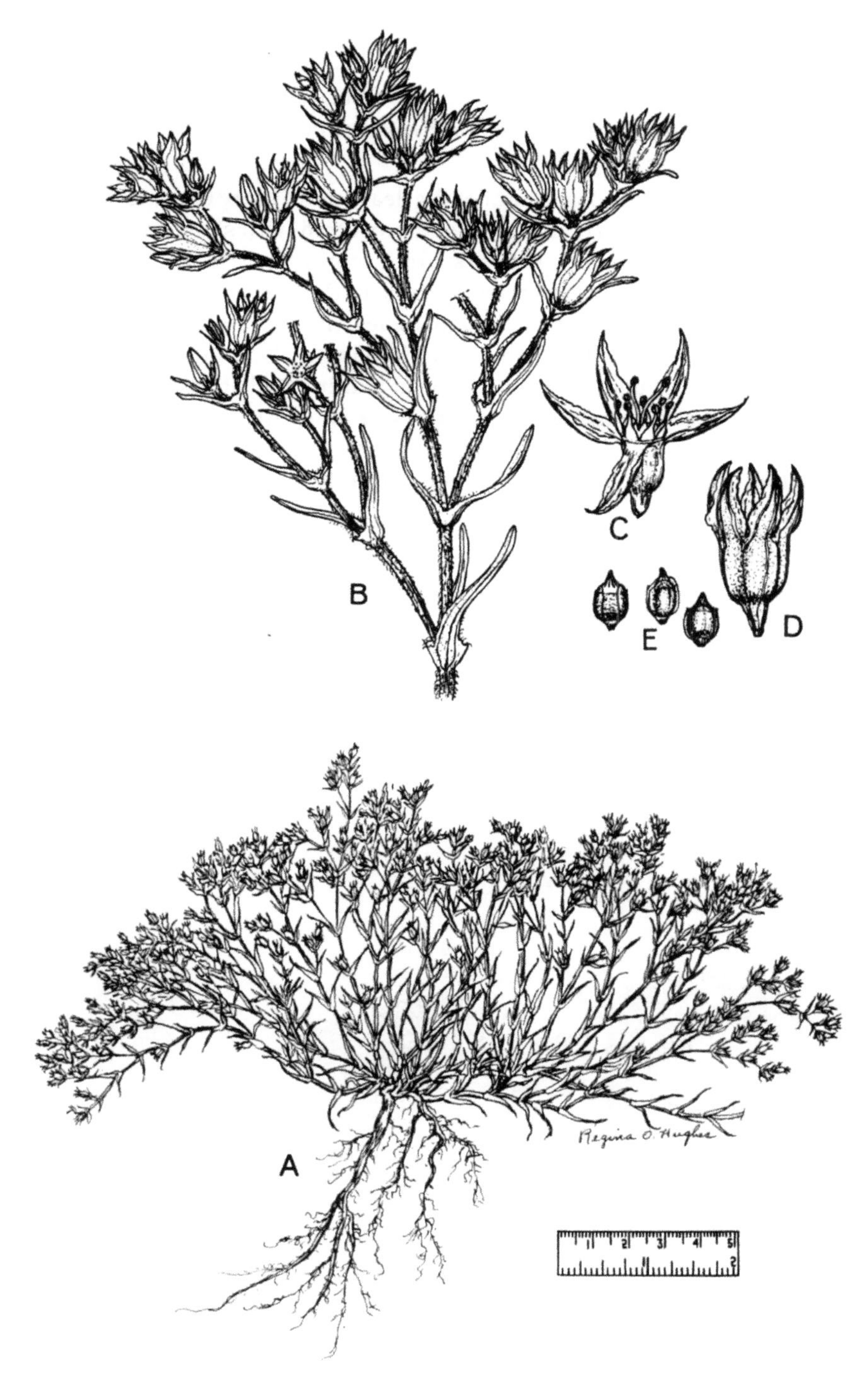

Scleranthus annuus, ANNUAL KNAWEL. A, habit. B, enlarged branch. C, flower. D, fruit (utricle, 1-seeded). E, seeds.

Silene latifolia Poir. WHITE COCKLE

Biennial herb or short-lived perennial, reproducing by seeds, plants green, glandular, minutely pubescent above. Lateral roots thick, sending up a few short barren shoots and long, branching flowering stems.

DESCRIPTION **Stems** erect, loosely branching, leafy, quite hairy and sticky, 3-7.5 dm tall. **Leaves** opposite, long and narrow, sessile, covered with short hair, pointed at the tip and rather light-green. **Flowers** white to pink, several to many, with 5-notched petals, up to 2 cm in diameter, opening in the evening, fragrant, borne on erect stems in the leaf axils or in loose panicles, the flowers at the forks longer pediceled. **Male and female flowers** on separate plants. Calyces of staminate flowers ellipsoid, those of the pistillate flowers ovoid and inflated at maturity, the teeth elongate, lance-linear, gradually tapering. **Capsule** conic-ovoid, with a narrow opening, with 10 teeth at the top, erect or only slightly spreading. **Seed** numerous, flat, nearly round, pale-gray, covered with small knobs, about 1 mm in diameter.

FLOWERING Late May–September.

SYNONYMS *Lychnis alba* Mill., *Melandrium album* Garcke.

WHERE FOUND Borders of fields, waste places, and roadsides; a troublesome weed in fields of grains and legumes.

ORIGIN Naturalized from Eurasia.

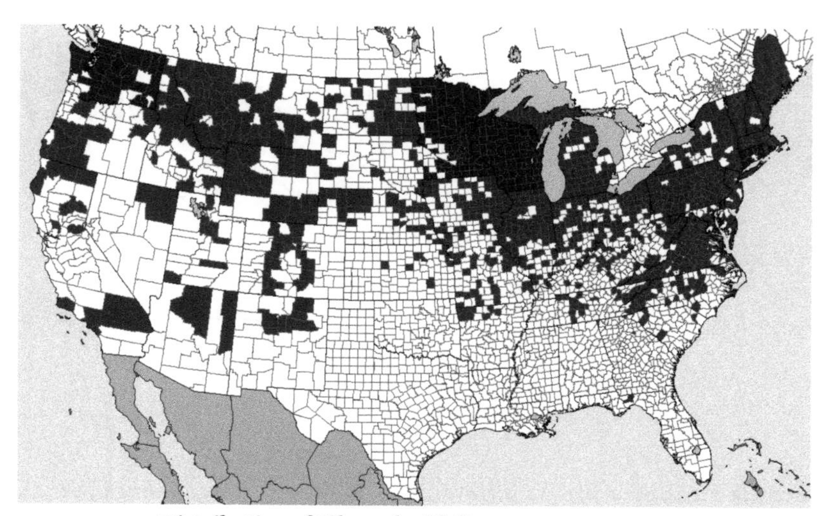

Distribution of **Silene latifolia**, WHITE COCKLE

Silene latifolia, WHITE COCKLE. A, habit, branchlet of staminate plant; a, staminate flower. B, habit, branchlet of pistillate plant; a, pistillate flower. B, capsule; c, seeds.

Silene noctiflora L. NIGHT-FLOWERING CATCHFLY

Annual herb, reproducing by seeds.

DESCRIPTION Stems erect, stout, up to 1 m tall, simple or branched, covered with sticky hairs, at least above. **Leaves** lanceolate or ovate-lanceolate, 5–12 cm long, 2–4 cm wide, the lower leaves obovate to oblanceolate, blunt, narrowed to a broad petiole, the upper leaves ovate-lanceolate, acute or tapering to a protracted point and sessile. **Inflorescence** an open, loosely branched cyme or panicle. **Flowers** fragrant, opening at night, erect, rather few, often unisexual. Calyx 1.5–2.3 cm long, with 10 prominent green ribs, glandular and freely cross-veined, the 5 awl-shaped teeth enclosing the capsule, the lobes linear-lanceolate, cylindrical at first, in fruit becoming inflated-ovoid. Petals 5, white, shading into pink, usually pinkish at base, the deeply cleft limb creamy-white, 7–10 mm long. Styles normally 3, sometimes 4, rarely 5. **Capsule** 3-celled, sessile, ellipsoid. **Seed** rounded, 0.8–1 mm in diameter, gray, uniformly wrinkled with minute projections.

FLOWERING Late May–September.

WHERE FOUND Cultivated ground and waste places; often a troublesome weed.

ORIGIN Introduced.

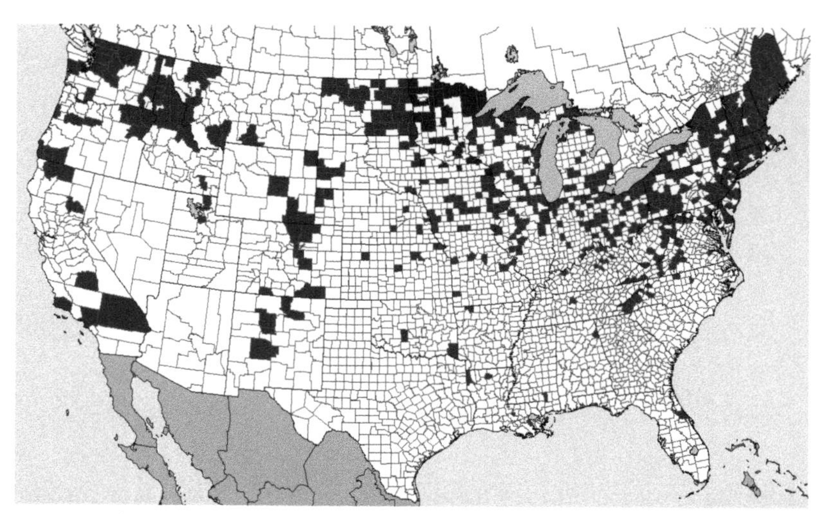

Distribution of **Silene noctiflora**, NIGHT-FLOWERING CATCHFLY

Silene noctiflora, NIGHT-FLOWERING CATCHFLY. A, habit of young plant. B, upper part of mature plant. C, capsule. D, seeds.

Spergula arvensis L. CORN SPURRY

Annual herb, reproducing by seeds.

DESCRIPTION Stems erect or spreading, much-branched, very slender, jointed, 15–45 cm high, glabrous or only slightly hairy or sticky. **Leaves** bright-green, narrowly linear or awl-shaped, 2–5 cm long, in whorls of 6–8 at each node, appearing disposed in a whorl, channel at base of leaf, stipules minute. **Flowers** small, perfect, in terminal clusters, or open cymes. Sepals nearly separate. Petals 5, white. Stamens 10 (occasionally 5). Pistil with 5 styles, 5-valved. **Capsule** 1-celled, many-seeded, broadly ovoid, splitting into 5 sections. **Seed** lens-shaped, 1–1.5 mm wide, dull-black with minute whitish projections, with a very narrow whitish wing margin.

FLOWERING March–October.

WHERE FOUND Cultivated ground and waste places, especially on sandy and gravelly soils; locally in small grains and other annual crops in the Midwest; abundant on the Piedmont and Coastal Plain.

ORIGIN Naturalized from Europe (sown formerly in Europe as forage for cattle and sheep).

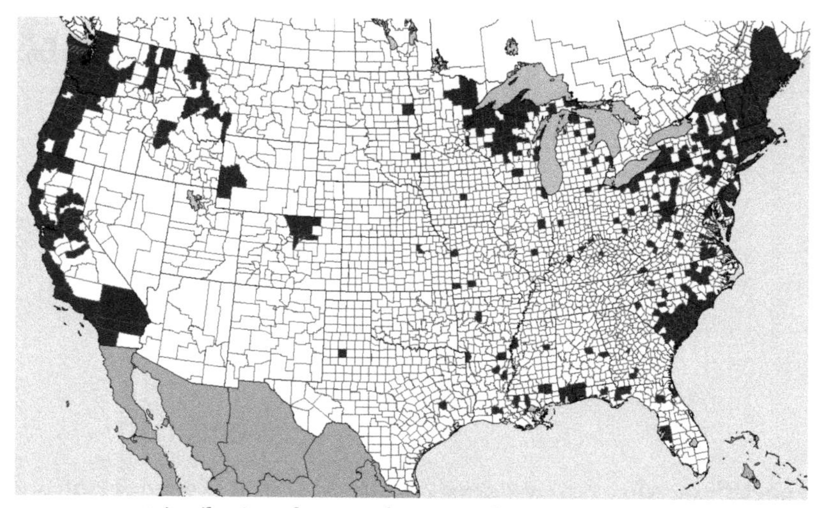

Distribution of **Spergula arvensis,** CORN SPURRY

Spergula arvensis, CORN SPURRY. A, habit. B, enlarged branch. C, flowers. D, seeds.

Stellaria media (L.) Cyrillo COMMON CHICKWEED

Annual or winter-annual, weakly tufted, reproducing by seeds and creeping stems. Root system fibrous, shallow.

DESCRIPTION Stems much-branched, often trailing, matted or loosely ascending, up to 8 dm long, minutely pubescent in lines, rooting at the nodes. **Leaves** opposite, simple, usually 1–3 cm long, ovate, elliptic, oblong-ovate or obovate, pointed at the tip, glabrous, the upper ones sessile, the lower ones petiolate, ovate, often hairy toward the base or on the petioles, 0.5–4 cm long. **Flowers** solitary or in few-flowered terminal, leafy cymes; pedicels nearly capillary, ascending, reflexed or recurved, frequently pubescent. Sepals lanceolate-oblong, 3.5–6 mm long, blunt to acute, usually with long soft hairs. Petals 5, white, small, shorter than the sepals, 2-parted or absent. Stamens 3–10. **Capsule** ovoid, usually a little longer than the sepals, breaking into 5 segments at maturity, many-seeded. **Seed** 1–1.2 mm long, nearly circular, dull reddish-brown, conspicuous projections, the marginal projections more prominent. Has many different forms.

FLOWERING February–December, but may flower throughout the year.

WHERE FOUND Dooryards, lawns, waste places, cultivated areas, woodlands, thickets, and meadows; especially bad in gardens, alfalfa, strawberry beds, and nurseries; a cosmopolitan weed.

ORIGIN Introduced from the Old World, but often appears to be native.

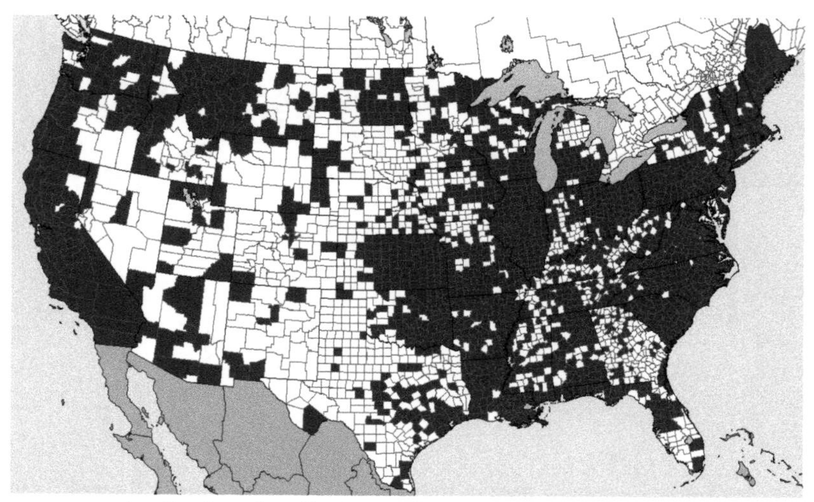

Distribution of **Stellaria media**, COMMON CHICKWEED

Stellaria media, COMMON CHICKWEED. A, habit. B, flower. C, capsule. D, seeds.

Ceratophyllum demersum L. COON'S-TAIL

Submersed aquatic plants, usually without roots.

DESCRIPTION Stems elongate, freely branched, forming large masses, varying from brittle and stiffly branched to more cord-like and flexuous. **Leaves** in whorls of 5-12, 2 or 3 times palmately divided, 1-3 cm long, the ultimate divisions linear, flat, about 0.5 mm wide, conspicuously serrate on one side, often more crowded toward the end of the branches. **Flowers** unisexual, minute, solitary and sessile in the axils, each subtended by a calyx-like involucre of 8-14 bracts. Perianth none. Stamens 12-16, filaments very short, anthers sessile, oblong, terminating in two short sharp points. Pistil 1, simple, 1-celled, narrowed above into a filiform style (4.5-6 mm long), with 1 suspended ovule. **Achene** ellipsoid, wingless, smoothish, 4-6 mm long, with 2 basal spines 2-5 mm long.

FLOWERING July-October.

WHERE FOUND Quiet waters and ponds. Achenes important food for aquatic birds.

ORIGIN Native.

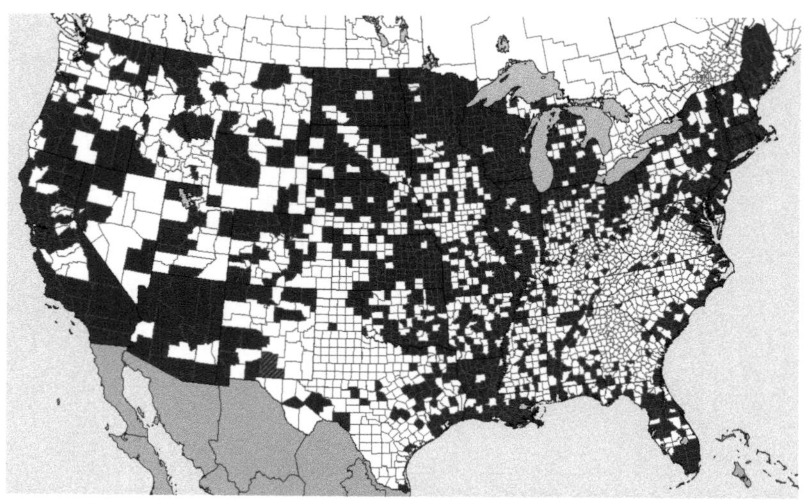

Distribution of **Ceratophyllum demersum**, COON'S-TAIL

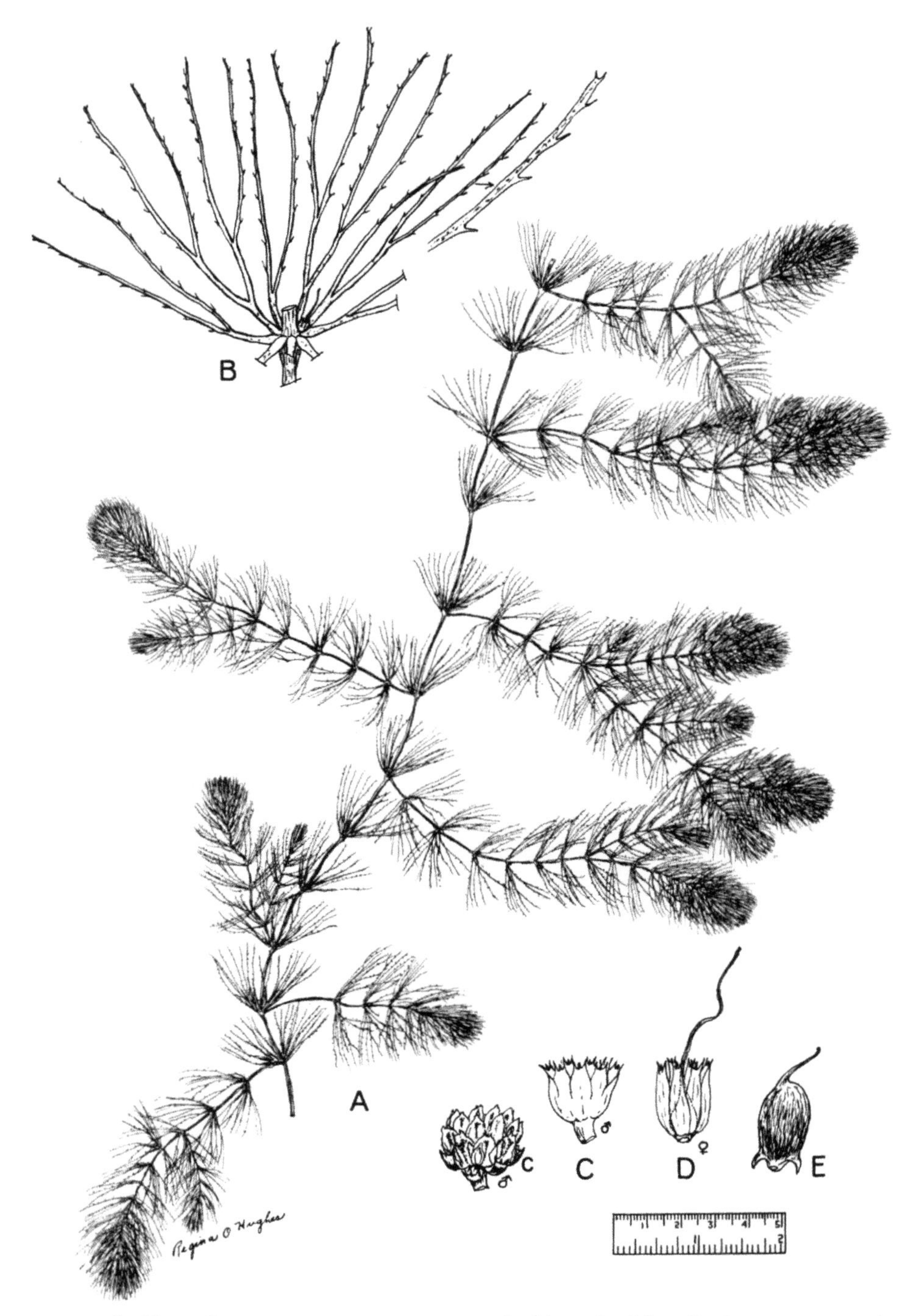

Ceratophyllum demersum, COON'S-TAIL. A, habit. B, leaf detail, showing a pistillate flower. C, staminate flower; c, staminate flower showing developed stamens. D, pistillate flower. E, achene.

Chara vulgaris L. CHARA

Perennial aquatic green algae.

DESCRIPTION Plants monoecious, moderately to heavily encrusted, rarely unencrusted. **Stems** 5-60 (occasionally 100) cm long, moderately slender to stout, 0.2-1.1 mm in diameter, the internodes 1-3 times as long as the branchlets, up to 4 (rarely 10) cm long. **Cortex** essentially 2-layered, furrowed, equal-sided or lumpy. **Spine cells** solitary, rarely in pairs, variable, rudimentary and globose to well-developed (up to 1.3 mm long) and spreading, ovoid and blunt. **Stipulodes** in 2 tiers, 2 sets per branchlets, typically contiguous but often separated, variable in size from obscure up to 1.6 mm long, exceeding the branchlet segment but characteristically ovoid and regular, the upper tier frequently longer than the lower tier. **Branchlets** 6-10 per whorl, 0.8-5 cm long, straight, incurved or rarely reflexed; the segments 4-8 of which 3-6 have two covering layers and 1-2 are naked; the end segment 2- to 3-celled, the basal 1 or 2 occasionally enlarged, the end cell sausage-shaped or conical, blunt to pointed, the nodes rarely swollen. **Bract cells** 2-6, one-sided, occasionally whorled, rarely forked; the anterior bract cells variable, 0.3-2.4 mm long, one-half as long as the branchlet diameter to three times as long as the oogonium, tapering, blunt or pointed; the posterior bract cells usually rudimentary, globular or ovoid, but occasionally conical, rarely similar to the anterior bract cells. **Bracteoles** 2, larger (rarely smaller) than the anterior bract cells, shorter to 4 times longer than the oogonium, 0.3-3 mm long, tapering to obtuse, occasionally inflated, rarely forked. Bractlet rare, shorter than the oogonium. **Gametangia** (male and female sex organs) on different branches of the same plant, at the lowest 1-3 or all the branchlet nodes, solitary or paired, commonly encrusted. **Odgonia** (female sex organ) 0.5-0.8 mm long (excluding the coronula), 0.3-0.5 mm wide, the convolutions 12-16. Coronula 0.07-0.17 mm high, 0.2-0.3 mm wide, the cells spreading or rarely erect or meeting, blunt or with divergent tips, occasionally deciduous. Oospores dark-red, brown (rarely golden) to black, 0.4-0.8 mm long, 0.2-0.25 mm wide, often encrusted; the ridges of 10-15 low to prominent, often prolonged into basal claws or cage; the fossa 0.041-0.053 mm across; the membrane obscurely to clearly granulate, granular or papillate or diffusely spotted. **Antheridia** (male sex organ) peltate, 0.2-0.5 mm in diameter, in 8 planes. **Bulbils** rare, spheroid.

WHERE FOUND Frequent to common in fresh water habitats throughout the United States. Since this plant belongs to a group of plants with a reproductive system different from the other plants considered in this book, much of its terminology is unique to it. Most of the structures are identified on the illustration opposite.

ORIGIN Native.

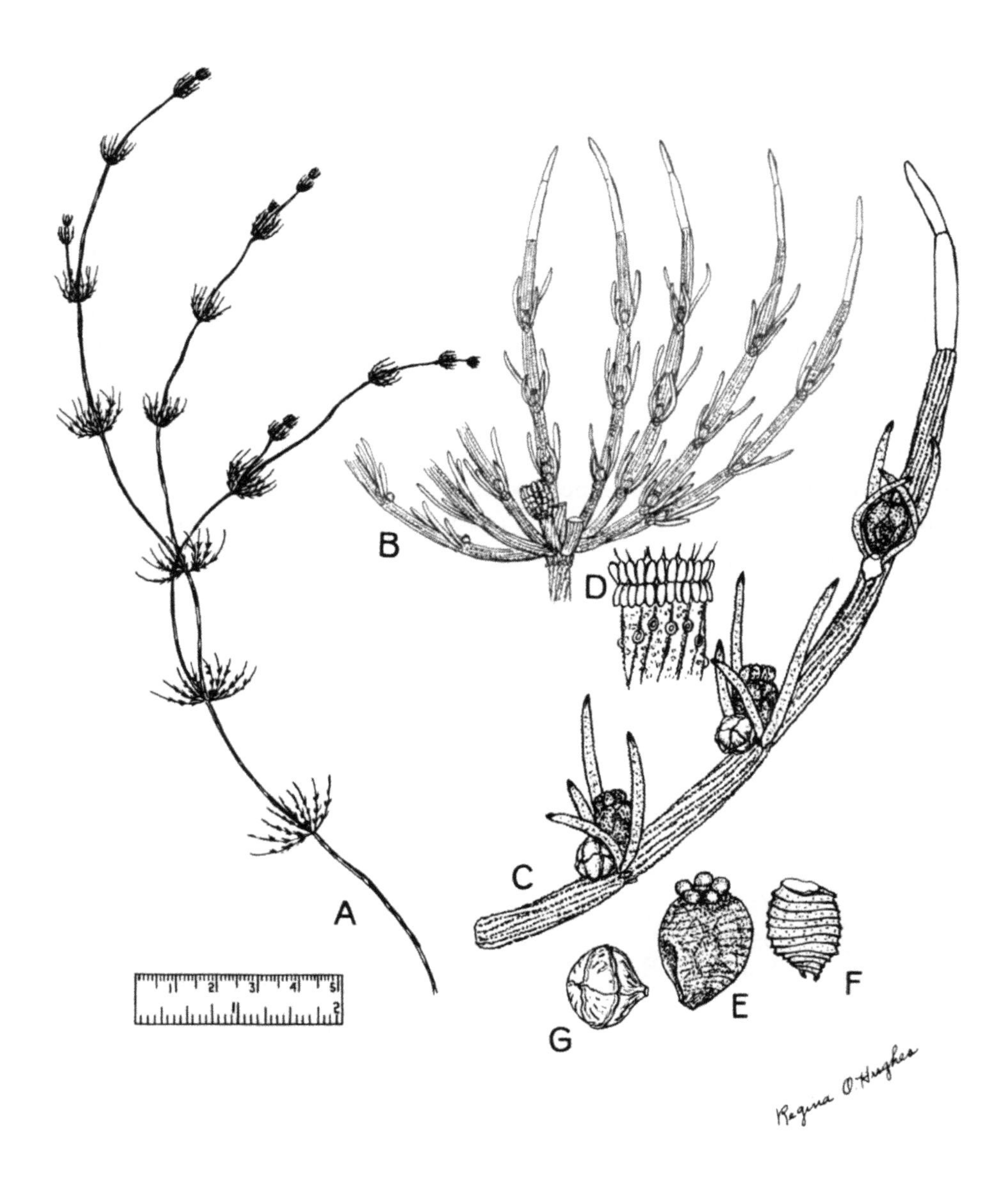

Chara vulgaris, CHARA. A, habit. B, whorl of branchlets. C, branchlet. D, stipulodes. E, oogonium. F, oospore. G, antheridium.

Calystegia sepium (L.) R. Br. HEDGE BINDWEED

Perennial vining herb, reproducing by seeds and fleshy creeping rootstocks. Roots extensive but relatively shallow.

DESCRIPTION **Stems** coarse, glabrous to pubescent, 9-30 dm long, twining on plants or trailing on the surface of the ground. **Leaves** alternate, simple, large, about 1 dm long, elongated, heart-shaped, or triangular-ovate to narrowly lanceolate, usually sharp-pointed at the tip, the basal lobes large, long-petioled. **Flowers** axillary, on 4-angled, short to prolonged spreading or ascending stalks. Bracts paired, heart-shaped, ovate, 1.5-3.5 cm long. Calyx of 5 overlapping sepals. Corolla bell-shaped to funnelform, white to pinkish, 4-8 cm high, nearly or quite as broad. Stamens 5, 2-3.3 cm long, attached to the corolla. **Capsule** globose, about 8-10 mm in diameter, 2-celled, containing from 2-4 seeds, usually covered by the bracts and the calyx. **Seed** ovoid, 4-5 mm long, slate-colored to black, dull, usually with 1 rounded and 2 flattened sides. A variable species.

FLOWERING Mid-May-September.

SYNONYMS *Convolvulus sepium* L.

WHERE FOUND Cultivated fields, fence rows, shores, thickets, waste places, and bottomlands; a serious weed, often densely covering banks and roadside shrubbery.

ORIGIN Native of Eurasia.

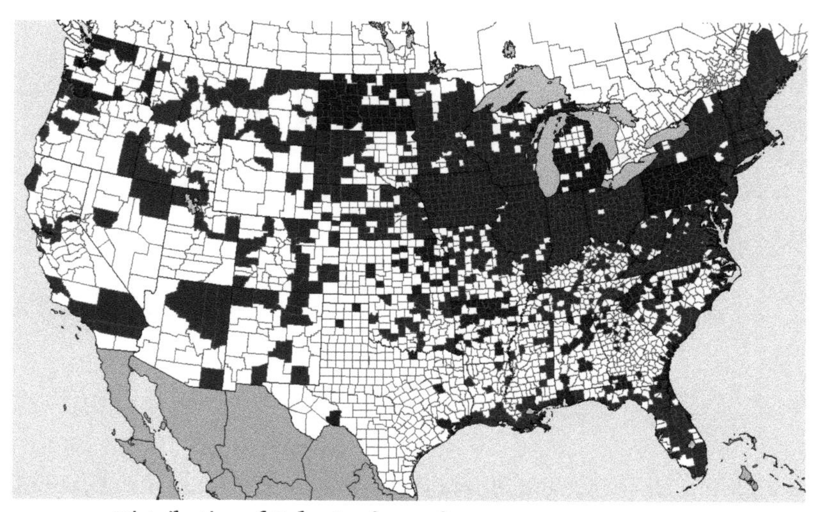

Distribution of **Calystegia sepium**, HEDGE BINDWEED

Calystegia sepium, HEDGE BINDWEED. A, habit. B, rootstock. C, diagram of flower, showing structure. D, seeds.

Convolvulus arvensis L. FIELD BINDWEED

Perennial herb, reproducing by seeds and creeping roots. Root system extensive, may extend to depth of 6–9 m below ground surface.

DESCRIPTION Stems glabrous to pubescent, slender, 1–3 m long, twining or spreading over the surface of the ground. **Leaves** alternate, simple, very variable, long-petioled, entire, ovate-oblong, with hastate, cordate or sagittate bases, glabrous, up to 5 cm long. **Flowers** perfect, regular, 1.5–2 cm wide, usually borne singly in the axils of the leaves. Flower stalk with 2 bracts, 1.2–2.5 cm below the flower. Calyx small, bell-shaped. Corolla white or pink, funnel-shaped. Stamens 5, attached to the corolla. Pistil compound with 2 threadlike stigmas, the ovary 2-celled. **Capsule** ovate, usually 2-celled, containing 4 seeds. **Seed** ovoid, dull, dark brownish-gray, coarsely roughened, about 3–5 mm long, with 1 rounded and 2 (or only 1) flattened sides.

FLOWERING June–September.

WHERE FOUND All uncultivated areas and waste places, grainfields, and gardens; able to grow under most cultivated conditions; a most serious weed wherever it grows.

ORIGIN Introduced and naturalized from Eurasia.

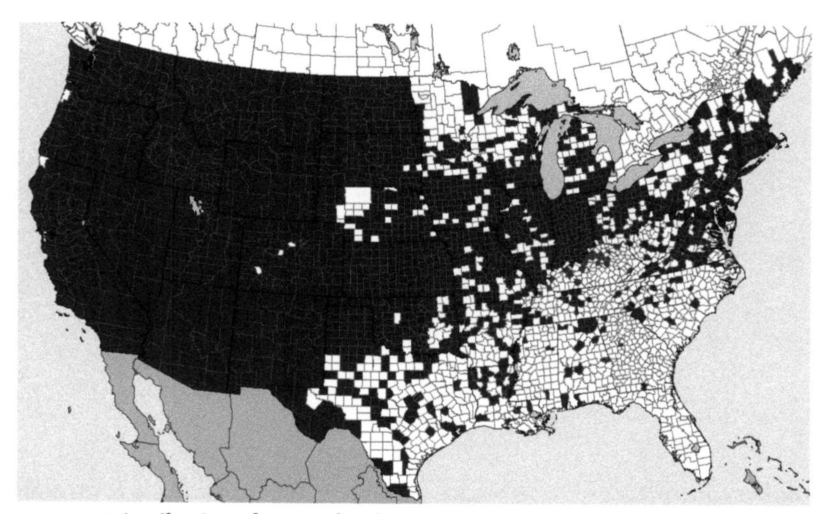

Distribution of **Convolvulus arvensis**, FIELD BINDWEED

Convolvulus arvensis, FIELD BINDWEED. A, habit. B, rootstock. C, leaf variation. D, flower, showing 5 stamens of unequal length. E, capsule. F, seeds.

Cuscuta approximata Bab. ALFALFA DODDER

Annual parasitic herb, reproducing by seeds.

DESCRIPTION Stems yellow or yellowish, slender, leafless, climbing or twining on its host, with small suckers adhering to the host. **Scales** few, very minute. **Flowers** sessile, whitish, perfect, regular, 3–4 mm long, in fewto several-flowered clusters 5–10 mm in diameter. Calyx inverted cone-shaped, of 4–5 fused sepals, the lobes equaling the corolla, enclosing the corolla tube, broadly ovate to triangular-ovate, overlapping at the base, usually broader than long, slightly keeled, the tips fleshy and turgid, drying golden-yellow and shining with large prominent cells. Corolla of 4–5 petals, bell-shaped, soon becoming round, about the developing capsule, the lobes ovate-rounded, obtuse and bent backward, fused below, shorter than the narrow cylindrical tube. Scales below the stamens oblong, mostly scarcely reaching the filaments, fringed about the top. Styles separate. Stigmas filiform, about as long as the styles, recurved, reddish. **Capsule** depressed-globose, opening in a definite line near the base, capped by the persistent corolla. **Seed** usually 4, about 1 mm long, oval to oblong, angular, finely granular and more or less scurfy, light-brown. **Hilum** short, oblong.

FLOWERING June–October.

WHERE FOUND Commonly parasitizing cultivated legumes, clover, alfalfa, and other hosts, often doing considerable damage.

ORIGIN Native of Eurasia and Africa, introduced from southern Europe.

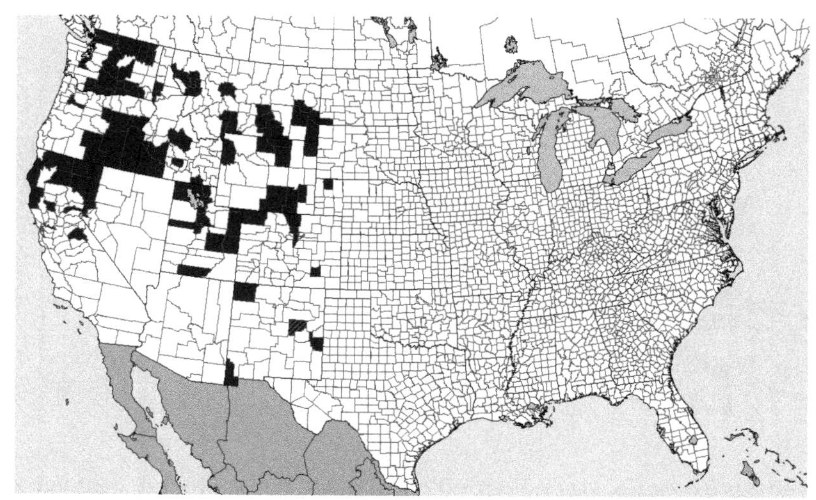

Distribution of **Cuscuta approximata**, ALFALFA DODDER

Cuscuta approximata, ALFALFA DODDER. A, habit. B, flower cluster and haustoria. C, corolla. D, calyx enclosing the corolla. E, corolla opened, showing fringed scales and ovary. F, seeds.

Cuscuta indecora Choisy — LARGE-SEED DODDER

Annual parasitic herb, reproducing by seeds.

DESCRIPTION Stems relatively coarse, loosely matted, yellow, leafless, upon rising from the ground after germination soon adhering to a host by means of suckers developed on the surface in contact, hence climbing or twining on its host. **Flowers** whitish, parts in 5's, 2–5 mm long, fleshy, papillose, on stalks longer than the flowers, forming rather open panicles or cymose clusters. Calyx much shorter than the corolla tube, with narrow triangular-ovate or acute lobes. Corolla 3–5 mm long, bell-shaped, with spreading-ascending to erect triangular lobes, with inflexed acute tips. Stamens about half as long as the lobes of the corolla. Scales oblong, equal in length to the corolla tube, deeply and regularly fringed. Styles equal to or longer than the depressed-globose ovary. **Capsule** thickened at the summit, wrapped in the withering but persistent corolla. **Seed** 1–1.7 mm long, globose or broader than long, somewhat scurfy.

FLOWERING July–September.

WHERE FOUND Parasitic on a wide variety of herbs and shrubs, mainly Asteraceae and Fabaceae, in damp pinelands, wet bottomlands, sandy springs, and alfalfa fields.

ORIGIN Native.

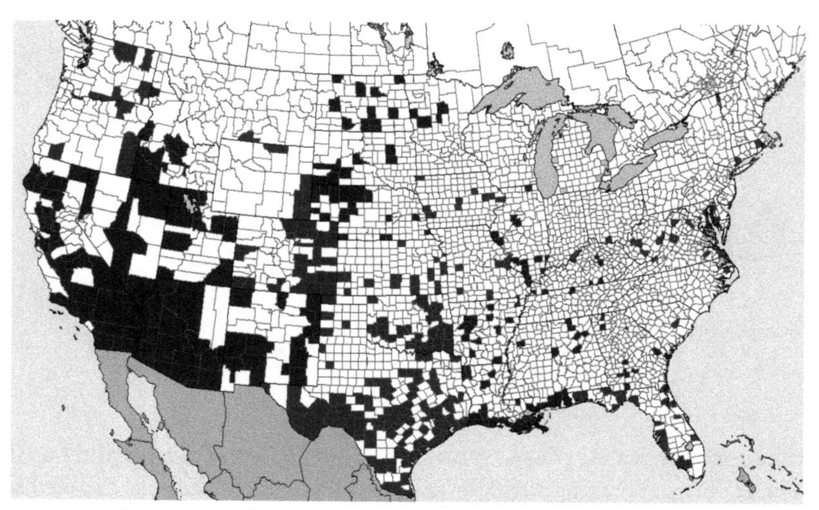

Distribution of **Cuscuta indecora,** LARGE-SEED DODDER

Cuscuta indecora, LARGE-SEED DODDER. A, Habit. B, enlarged habit. C, flower. D, capsule. E, seeds.

Cuscuta pentagona Engelm. BUSH-CLOVER DODDER

Annual parasitic herb, reproducing by seeds.

DESCRIPTION Stems pale, very slender and low. Glomerules globular, scattered or slightly confluent. **Flowers** whitish, parts in 5's, 1.5-2 mm long, somewhat glandular, short-pediceled, in loose clusters. Calyx nearly inclosing the corolla tube, the calyx lobes obtuse to very broadly ovate or depressed, often broader than long, overlapping at the grooves. Corolla lobes 4-5, lance-acuminate, about equaling the broad tube, usually wide-spreading, the acute elongate tips inflexed. Stamens arising below the grooves, exserted in flower. Scales oblong, reaching slightly above the middle of the corolla tube, their longer fringe about one-fifth as long as the blades. Stigma with a head. **Capsule** depressed-globose, no longer than wide, not opening, protruding from the withering corolla. **Seed** 1-1.5 mm long, one side rounded, the other flattened and often with an obtuse ridge, minutely pitted, yellow to reddish-brown. **Hilum** whitish within a smooth circular area on the flattened side.

FLOWERING April-October.

WHERE FOUND Dry open soils, on many herbaceous and slightly ligneous hosts.

ORIGIN Native. Widely distributed throughout all the United States.

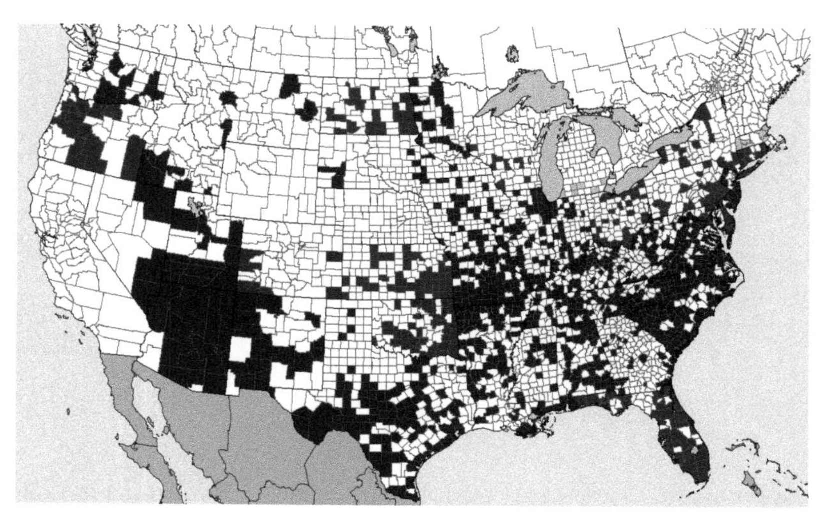

Distribution of **Cuscuta pentagona**, FIELD DODDER

Cuscuta pentagona, FIELD DODDER. A, habit. B, enlarged habit. C, flowers. D, capsule. E, seeds.

Ipomoea hederacea Jacq. IVY-LEAF MORNING-GLORY

Annual herb, reproducing by seed.

DESCRIPTION Stems hairy, twining or spreading on the ground, 1–2 m long. **Leaves** 3-lobed, occasionally 5-lobed, deeply heart-shaped with rounded basal lobes, 5–12 cm wide and long. Peduncles shorter than or equaling the petioles below, bearing 1–3 flowers. **Flowers** funnel-shaped, sessile or short-pediceled, the corolla 3–5 cm long, pale to sky-blue when fresh, quickly changing to rose-purple, with a white tube. Sepals lanceolate, 15–25 mm long, narrowed from below the middle into a slender, linear, recurved tip, densely hairy or bristly. **Capsule** egg-shaped, partly covered by the calyx, usually with 4–6 seeds. **Seed** about 6 mm long, dark-brown to black, with 1 rounded and 2 flattened sides.

FLOWERING July–October.

WHERE FOUND Gardens, fields, and waste places; a troublesome weed in cultivated fields, especially in corn and soybean, where it ties the plants together before harvest.

ORIGIN Naturalized from tropical America.

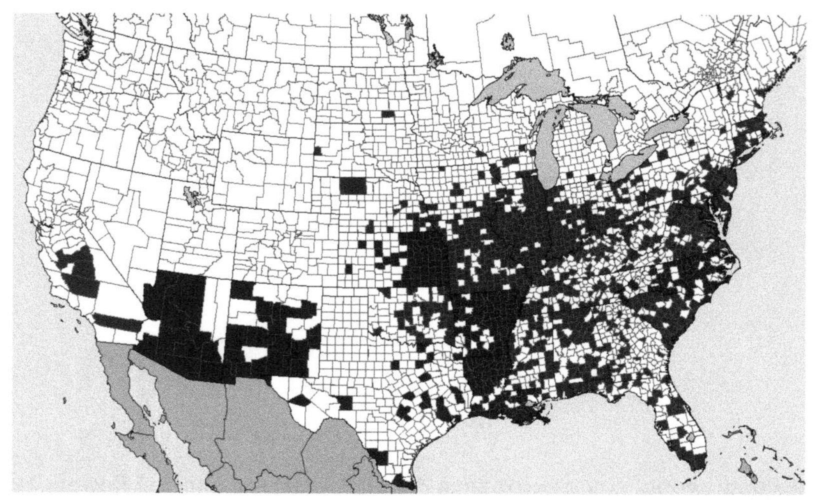

Distribution of **Ipomoea hederacea**, IVY-LEAF MORNING-GLORY

Ipomoea hederacea, IVY-LEAF MORNING-GLORY. A, habit. B, sepals. C, capsule. D, seeds.

Ipomoea purpurea (L.) Roth — COMMON MORNING-GLORY

Annual herb, reproducing by seeds.

DESCRIPTION Stems pubescent to glabrous, twining, up to 5 m long. **Leaves** rounded-cordate to ovatecordate, entire, or rarely 3-lobed, short, pointed, glabrous to sparsely pubescent. Peduncles about equaling the subtending leaves, 1- to 5-flowered. Sepals ovate-lanceolate to oblong, acute or acuminate, hairy below the base, 10–15 mm long. Corolla blue, purple, red, white, or variegated, 4–7 cm long. **Capsule** 10–12 mm in diameter, globular, pointed, 4- to 6-seeded, 2to 4-valved. **Seed** granular, pubescent, 4–5 mm long, 3- to 4-angled, brownish-black.

FLOWERING July–September.

WHERE FOUND Often a noxious weed, originally spread from cultivation; fields, waste places, gardens, and roadsides.

ORIGIN Introduced and naturalized from tropical America.

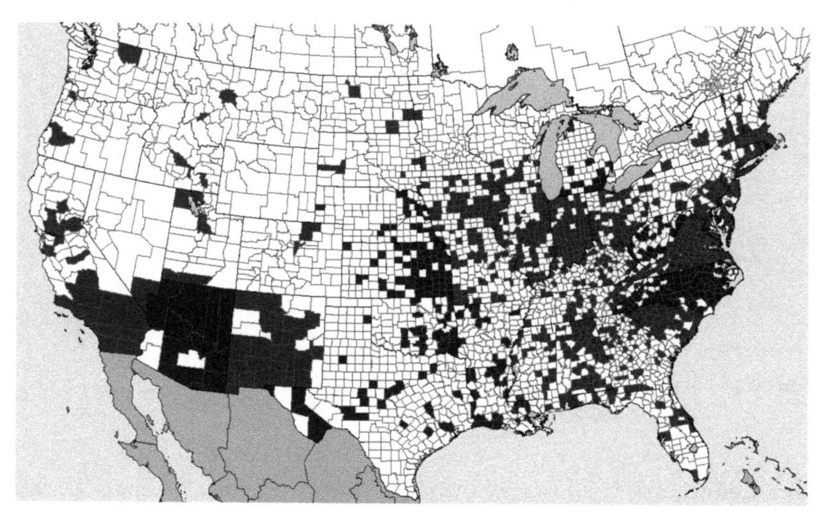

Distribution of **Ipomoea purpurea,** TALL MORNING-GLORY

Ipomoea purpurea, TALL MORNING-GLORY. A, habit. B, flower diagram. C, capsule. D, seeds.

Carex nebrascensis Dewey — NEBRASKA SEDGE

Perennial sedge. Rhizome long-creeping and stoloniferous.

DESCRIPTION Stems (culms) 2.5–12 dm tall. **Leaves** 3–12 mm wide, flat or channeled, more or less divided into partitions and swollen at the nodes. Sheaths smooth. **Staminate spike** usually solitary or with smaller ones at base, terminal. **Pistillate spike** 1.5–6 cm long, 5–9 mm wide, oblong to cylindrical, erect, upper nearly sessile, the leaves short- to long-peduncled, rather contiguous, 2–5, with many ascending perigynia. Lowest bract leaf-like, sheathless, usually longer than the inflorescence. **Scales** narrower and from longer to shorter than the perigynia, lanceolate, rounded at the end to tapering to a gradual point, purplish, brown or brownish-black, often with narrow hyaline margins, lighter centers, 1- to 3-nerved. **Perigynia** ascending, 3–3.5 mm long, about 2 mm wide, oblong-ovate, flat-convex or unequally double convex, flattened, straw-colored, leathery in texture, strongly many-ribbed, abruptly contracted to a 2-toothed beak 0.4–1 mm long. Styles jointed. Stigmas 3.

FLOWERING May–July.

WHERE FOUND Wet meadows and swamps.

ORIGIN Native.

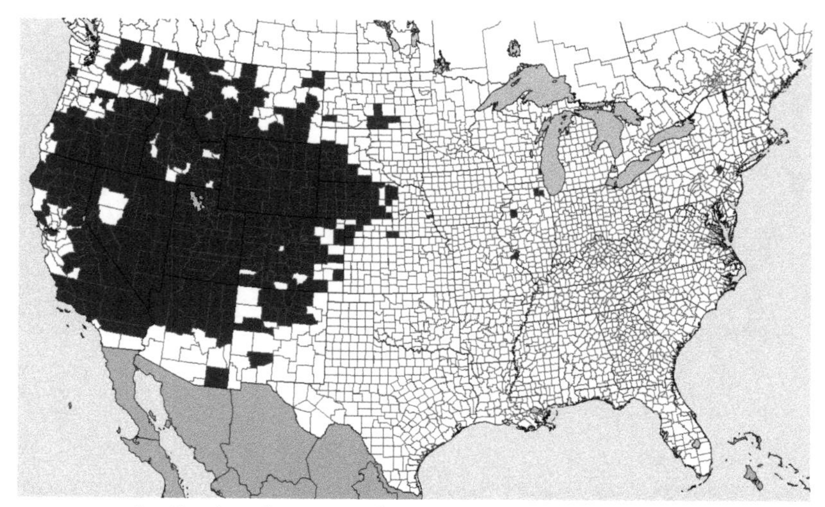

Distribution of **Carex nebrascensis**, NEBRASKA SEDGE

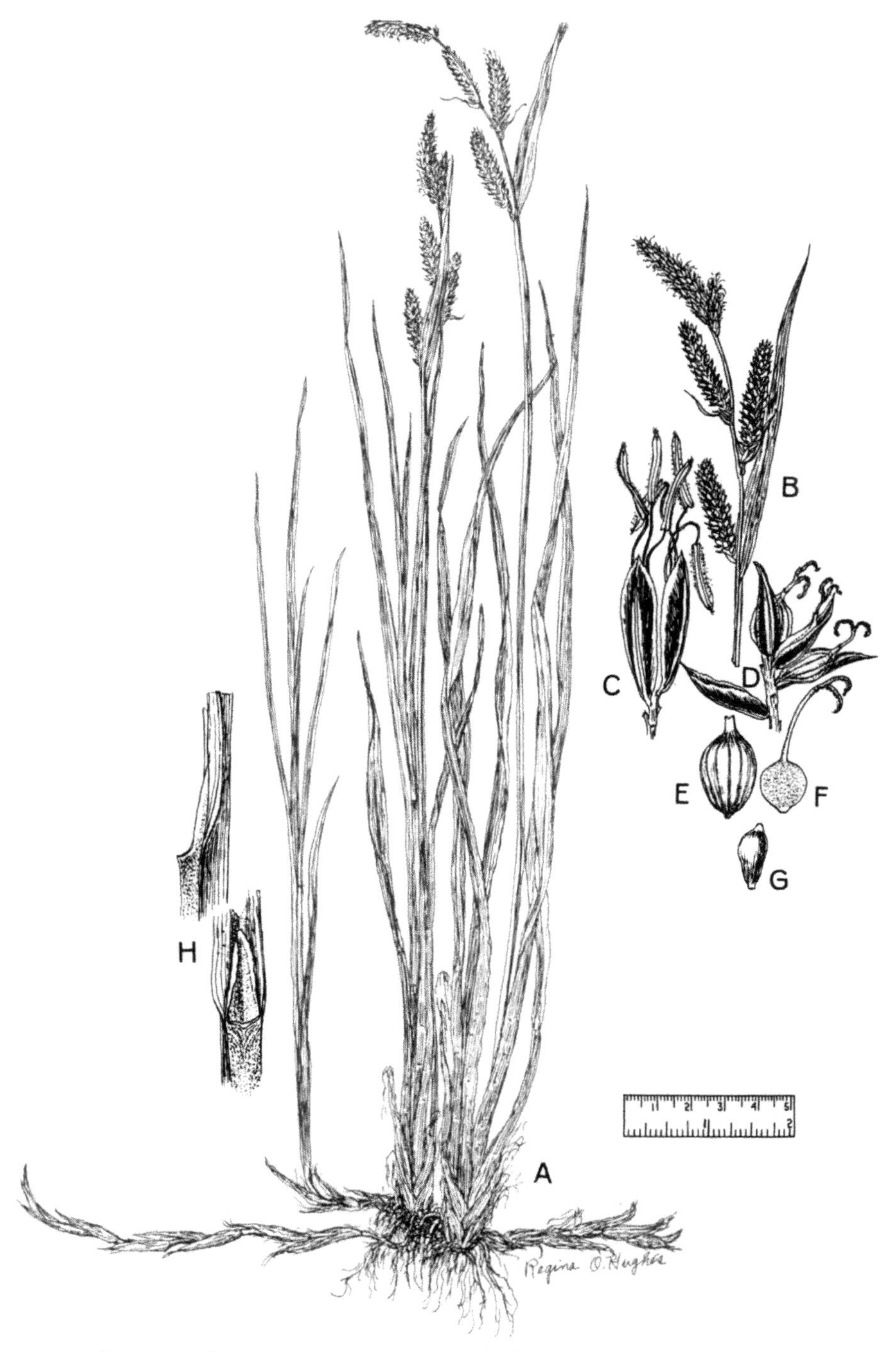

Carex nebrascensis, NEBRASKA SEDGE. A, habit. B, enlarged flowering spike. C, staminate flower. D, pistillate flowers, showing scales. E, perigynium. F, pistillate flower without perigynium. G, achene. H, ligule.

Carex pellita Muhl. ex Willd. WOOLLY SEDGE

Perennial rhizomatous sedge, forming large colonies.

DESCRIPTION Stems (culms) 5–10 dm tall, 3-sided, acutely angled and rough to the touch in the upper regions. Leaves very long and slenderly tapering, the leaf blades rough to the touch, 2–5 mm wide, flat, the margins rolled backward. **Staminate spikes** usually 2, 2–6 cm long. **Pistillate spikes** 1–3, widely spaced, cylindric, 1–3 cm long, rarely longer, sessile or nearly so. Lowest bract over-topping the culm. **Pistillate scales** ovate-lanceolate, from shorter than to longer than the perigynia, purplish-brown, acute to abruptly tipped or gradually tapering to the end. **Perigynia** oblong egg-shaped, 2.5–5 mm long, 1.2–3 mm thick, densely pubescent, the teeth 0.3–0.8 long. **Achene** angled, egg-shaped. Variable in the leaf characters.

FLOWERING May–August.

SYNONYMS *Carex lanuginosa* Michx., *Carex lasiocarpa* Ehrh. var. *latifolia* (Boeckl.) Gilly

WHERE FOUND Marshes, swamps, swales, rich meadows, and shores.

ORIGIN Native.

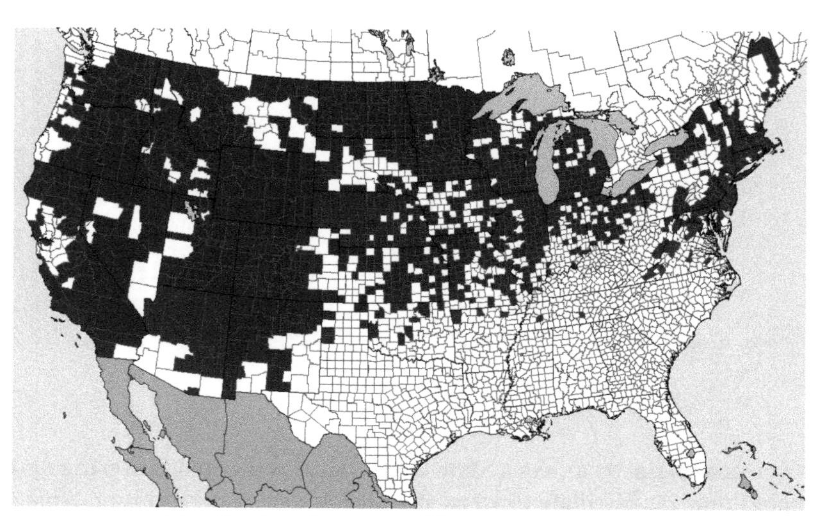

Distribution of **Carex pellita**, WOOLLY SEDGE

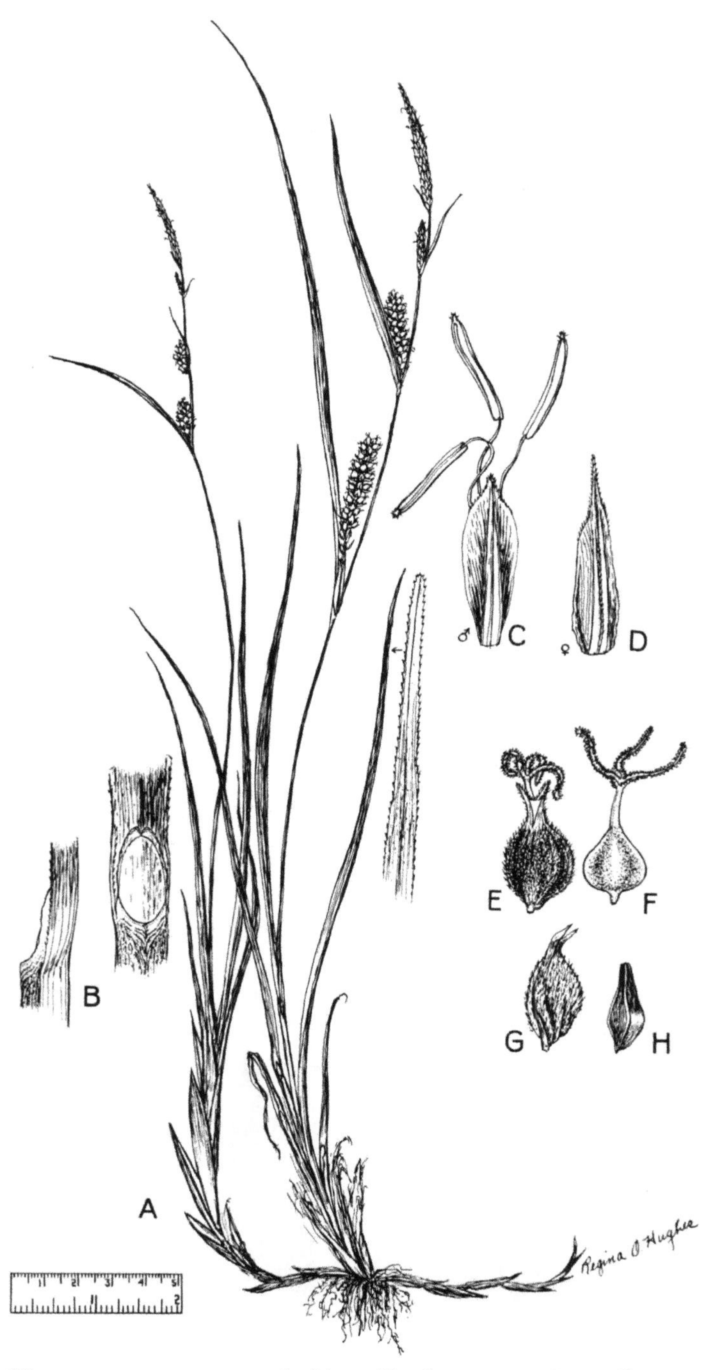

Carex pellita, WOOLLY SEDGE. A, habit. B, ligules. C, staminate flower (scale subtending 3 stamens). D, pistillate scale. E, pistillate flower with perigynium. F, pistillate flower without perigynium. G, perigynium enclosing the achene. H, achene.

Cyperus esculentus L. CHUFA, YELLOW NUTSEDGE

Perennial grass-like herb, reproducing by seeds and weak thread like stolons, terminated by hard tubers 1–2 cm long. Roots fibrous.

DESCRIPTION Stems (culms) erect, 2–9 dm tall, simple, triangular. **Leaves** 3-ranked, pale-green, 4–9 mm wide, about as long as the stem, with closed sheaths mostly basal. **Umbel** terminal, simple to compound, the longest involucral leaf much exceeding the umbel. **Spikelets** 0.5–3 cm long, 1.5–3 mm broad, yellowish to golden-brown, strongly flattened, mostly 4-ranked (occasionally 2-ranked) along the wing-angled rachis, blunt, the tip acute to round. **Scales** thin, oblong, obtuse, distinctly nerved, thin and dry at the tip, 2.3–3 mm long. **Achene** yellowish-brown, 3-angled, lustrous, ellipsoid or linear to oblong-cylindric, rounded at the summit, 1.2–1.5 mm long, granular-streaked.

FLOWERING July–September.

WHERE FOUND Cultivated fields, gardens, and grainfields; rich or sandy soils; often limited to low, poorly drained areas in fields; a serious weed.

ORIGIN Native of North America.

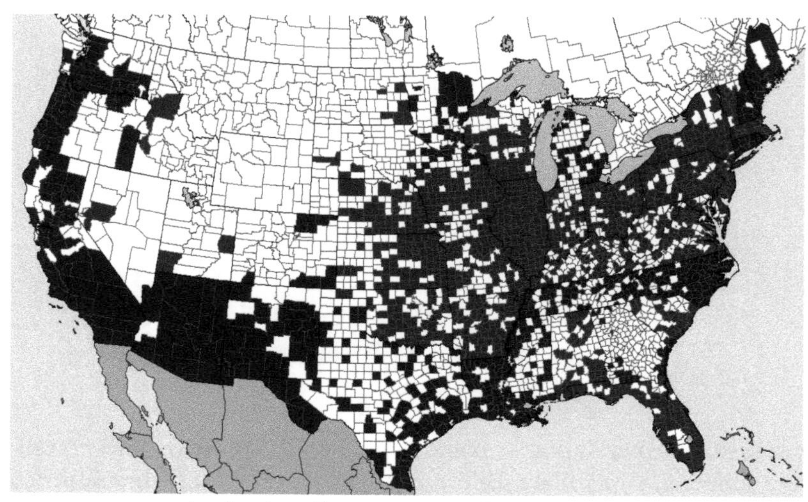

Distribution of **Cyperus esculentus**, CHUFA

Cyperus esculentus, CHUFA. A, habit. B, spikelet. C, achene.

Cyperus rotundus L. PURPLE NUTSEDGE

Perennial grass-like herb, reproducing by seeds and tuber-bearing rhizomes.

DESCRIPTION Stems (culms) erect, 1-6 dm tall, simple, triangular, longer than the leaves. **Leaves** 3-6 mm wide. **Umbel** simple or slightly compound, about equaling the involucre. **Spikelets** 0.8-2.5 cm long, chestnut-brown to chestnut-purple, acute, 12- to 40-flowered. **Scales** ovate, closely appressed, nerveless except on the keel, 2-3.5 mm long, bluntish. **Achene** linear-oblong, 1.5 mm long, 3-angled, the base and apex obtuse, granular, dull, olive-gray to brown, covered with a network of gray lines.

FLOWERING July-October.

WHERE FOUND Cultivated sandy fields and gardens, especially in the Cotton Belt; often a troublesome weed.

ORIGIN Introduced and naturalized from Eurasia.

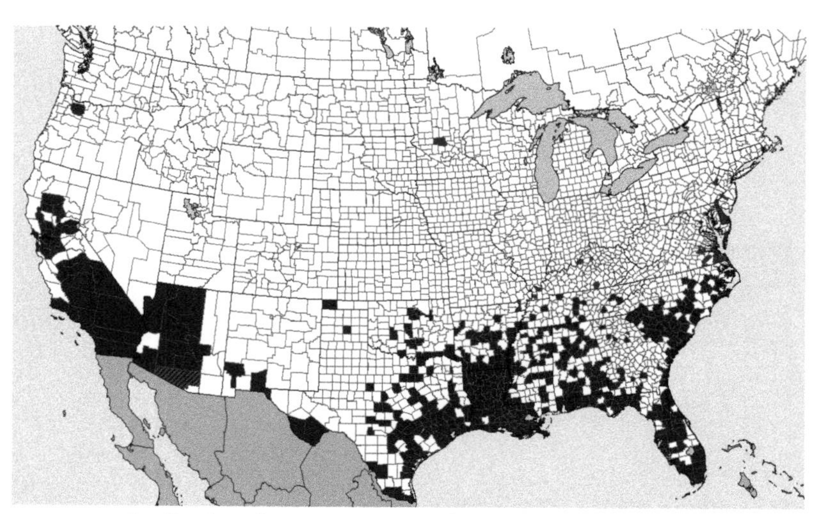

Distribution of **Cyperus rotundus**, PURPLE NUTSEDGE

Cyperus rotundus, PURPLE NUTSEDGE. A, habit. B, flowering spikelet. C, achenes, showing type of reticulation and cross section in detail.

Schoenoplectus acutus (Muhl. ex J.M.Bigelow) Á. & D.Löve

HARDSTEM BULRUSH

Perennial herb. Rhizome stout, often drab and brown.

DESCRIPTION Culms circular in cross section, erect, up to 3 m tall, olive-green. **Inflorescence** of a few nearly sessile spikelets and of several flattened peduncles 1–8 cm long, each having 1–5 spikelets. **Bract** erect, usually 1–4 cm (occasionally up to 10 cm). **Spikelets** mostly in compact clusters, several, ovate to cylindric, acute, 7–20 mm long, reddish to grayish-brown. **Scales** red-spotted, sticky, broadly ovate, about 4 mm long, conspicuously exceeding the achene, usually short-pubescent, the margin thin and hyaline and often fringed, the apex notched and abrupt tipped. **Achene** becoming black and lustrous, thickly flat-convex, 2–2.5 mm long, abrupt short-pointed, bristles variable in length.

FLOWERING July–August.

SYNONYMS *Scirpus acutus* Muhl.

WHERE FOUND Swamps, shores, and shallow water.

ORIGIN Native.

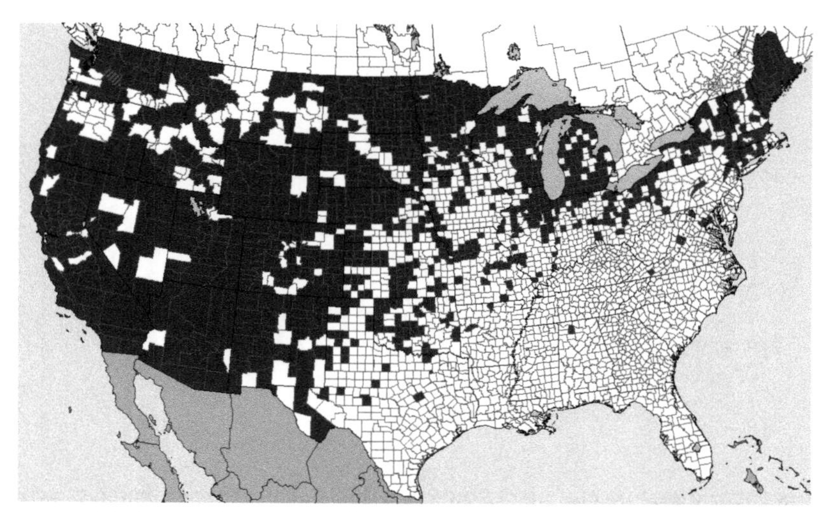

Distribution of **Schoenoplectus acutus**, HARDSTEM BULRUSH

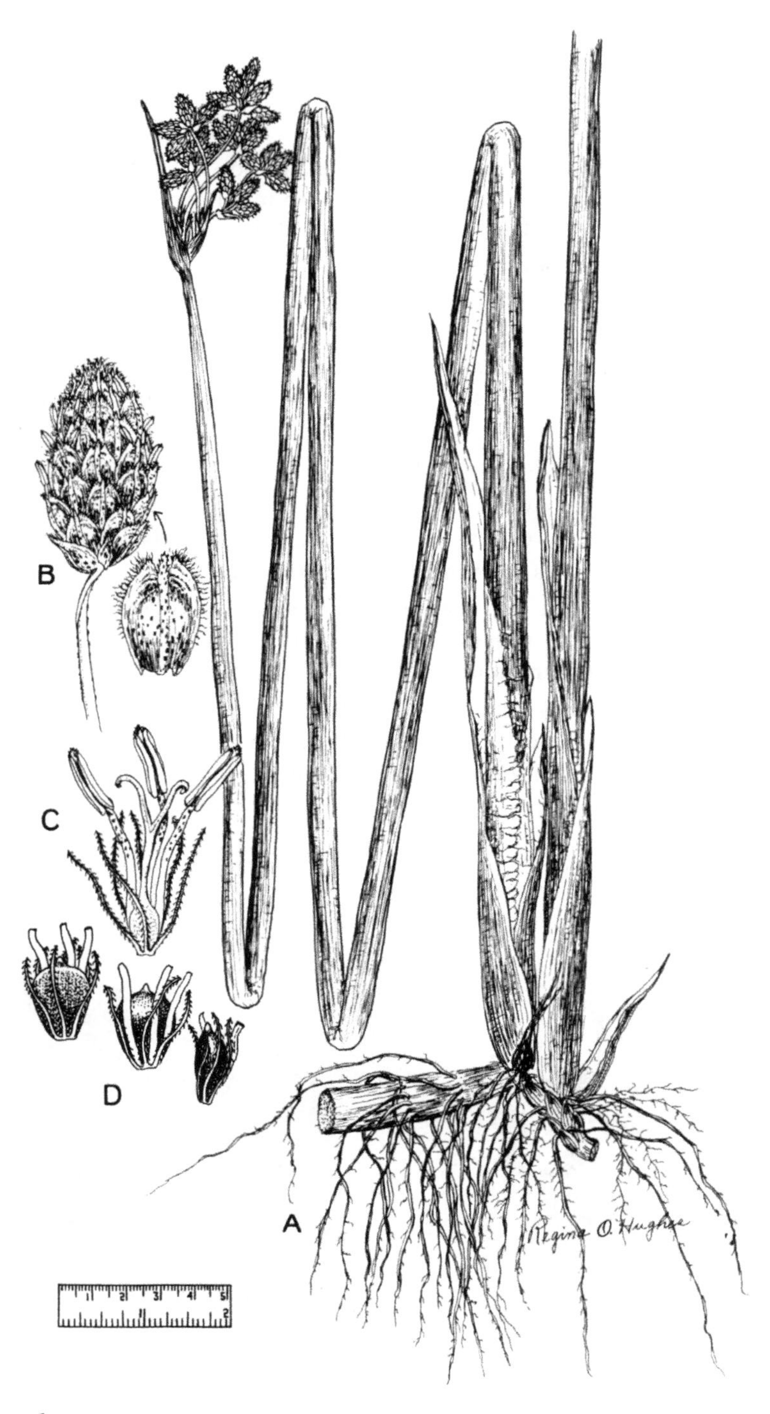

Schoenoplectus acutus, HARDSTEM BULRUSH. A, habit. B, spikelet. C, flower. D, achenes.

Pteridium aquilinum (L.) Kuhn BRACKEN FERN

Perennial fern, reproducing by spores, or by shoots from a thick, black scaly rhizome that may grow to 6 m long.

DESCRIPTION Fronds coarse, deciduous, 3–15 dm (up to 5 m) tall, arising from the rhizome, broad, triangular, and divided into three main parts, each branch being compounded by numerous leaflets (pinnae and pinnules), which in turn may be segmented, glabrous or pubescent. **Sporangia** on the underside of the pinnules of the fronds, along the margin as a narrow brown band, covered by a thin membranaceous indusium. **Spores** tetrahedral, with a perispore.

WHERE FOUND Open pastures, woodlands, pinewoods, barrens, and hillsides, mainly on acid soils. Fronds and petioles poisonous to cattle and horses if consumed over a period of time.

ORIGIN Native throughout North America; also in Europe and eastern Asia.

NOTE Several varieties defined:

var. *pubescens* Underwood—WESTERN BRACKEN. Southern Alaska to Ontario and Quebec, southwest to Texas and northwestern Mexico.

var. *latiusculum* (Desv.) Underwood ex Heller—EASTERN BRACKEN. Newfoundland and Quebec to Minnesota, south to Virginia and upland North Carolina, Mississippi, Oklahoma, Wyoming, Colorado, and northeastern Mexico.

var. *pseudocaudatum* (Clute) Heller—SOUTHERN BRACKEN. On or near the Coastal Plain from southeastern Massachusetts to Florida and inland to southern Illinois, Indiana, southern Missouri, and eastern Oklahoma.

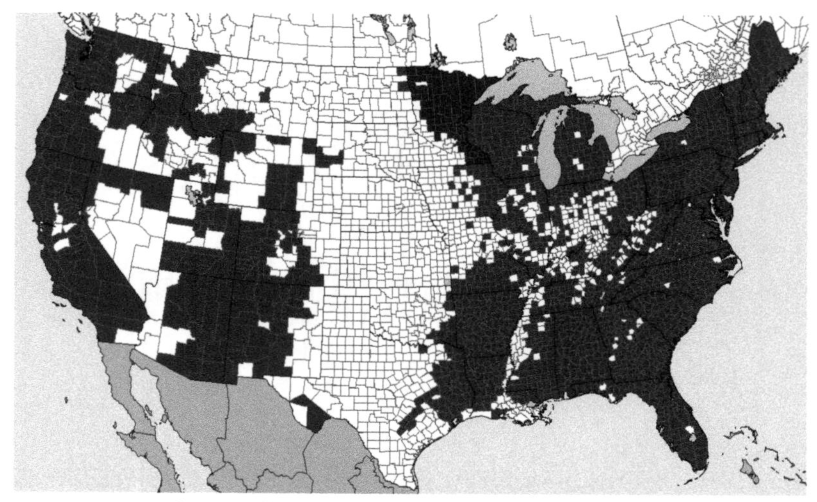

Distribution of **Pteridium aquilinum**, BRACKEN FERN

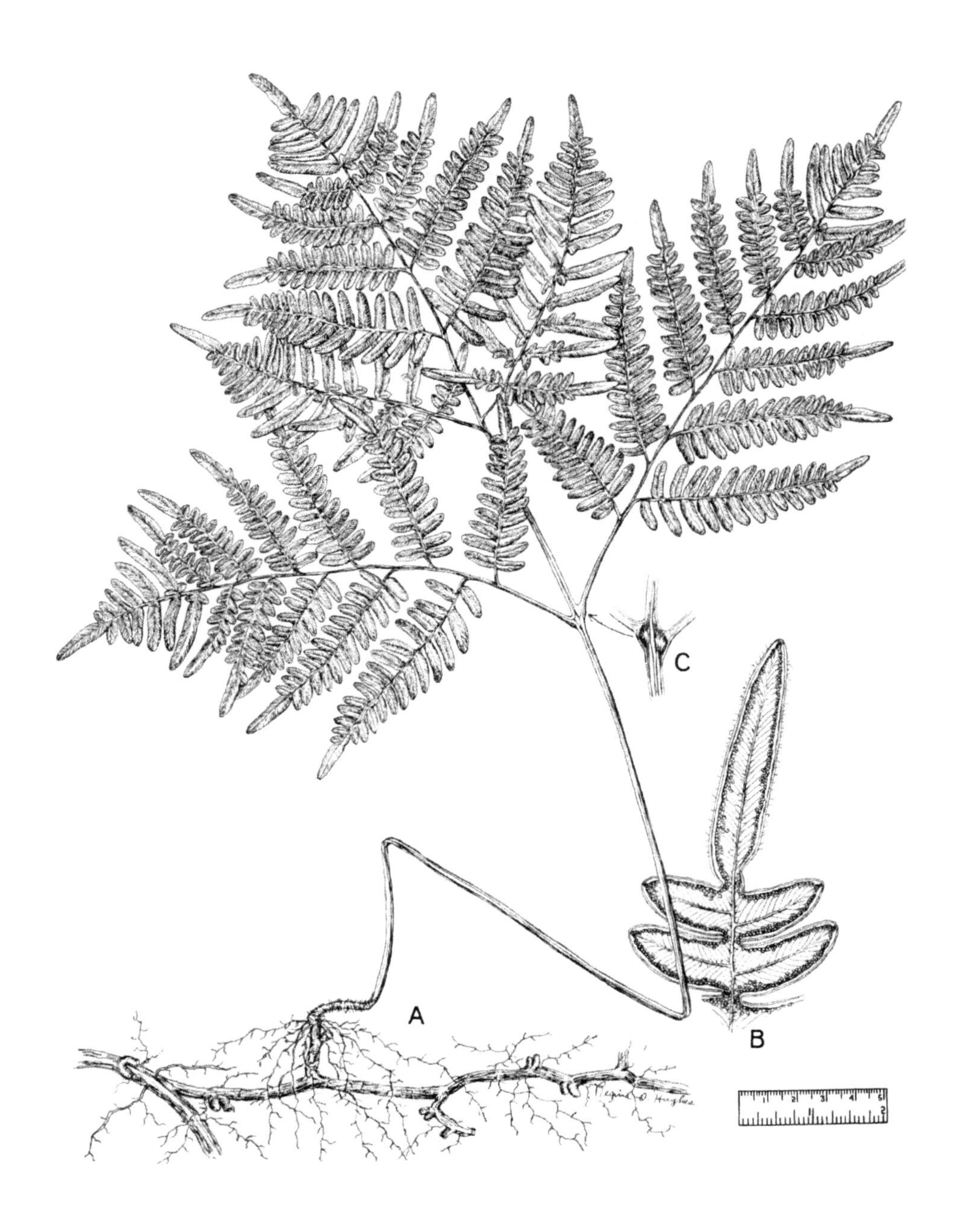

Pteridium aquilinum, BRACKEN FERN. A, habit. B, pinnules showing marginal sori. C, nectary glands.

Dipsacus fullonum L. TEASEL

Biennial herb, reproducing by seeds.

DESCRIPTION Stems the second year erect, 5-20 dm tall, stout and coarse, prickly at the angles. **Leaves** a rosette the 1st year, the basal leaves oblanceolate and scallop-margined, 20-60 cm long, sessile or the upper ones clasping the stem lance-oblong, toothed and often prickly on the midveins below and on the margins; the stem leaves opposite the 2nd year, lanceolate, entire, sessile. **Inflorescence** in dense ovoid-ellipsoid heads, 3-10 cm long, terminating long naked peduncles. Involucral bracts numerous, some of them surpassing the head, becoming stiff-hooked prickles at maturity. **Flowers** with a 4-leaved calyx-like involucel, investing the ovary and fruit (achene). Calyx silky, 1 mm long. Corolla slender, nearly regular, 10-15 mm long, 4-cleft, the 4 petals (tube) lilac or white, the short (1 mm) lobes pale-purple, pubescent. Stamens 4, inserted on the corolla, distinct. **Achene** 2-3 mm (rarely up to 8 mm) long, 4-angled, ridged, hairy, grayish-brown.

FLOWERING July-October.

SYNONYMS *Dipsacus sylvestris* Huds.

WHERE FOUND Roadsides, waste grounds, old fields, and pastures and along ditches and edges of forests.

ORIGIN Naturalized from Europe.

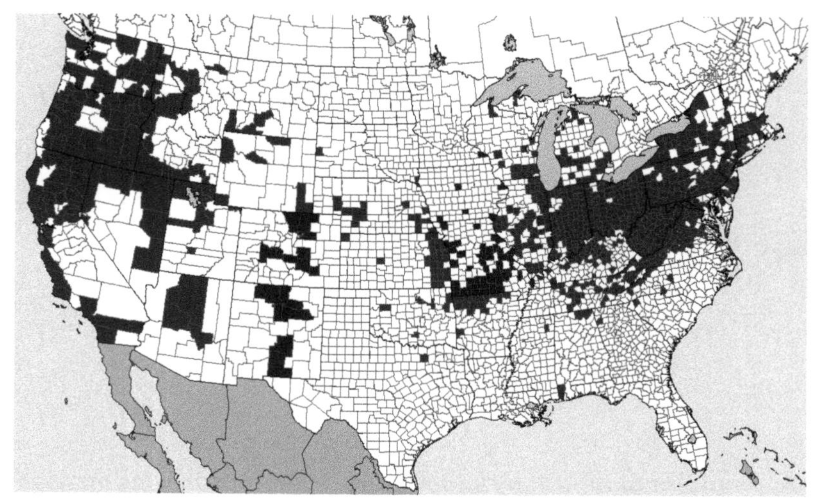

Distribution of **Dipsacus sylvestris,** TEASEL

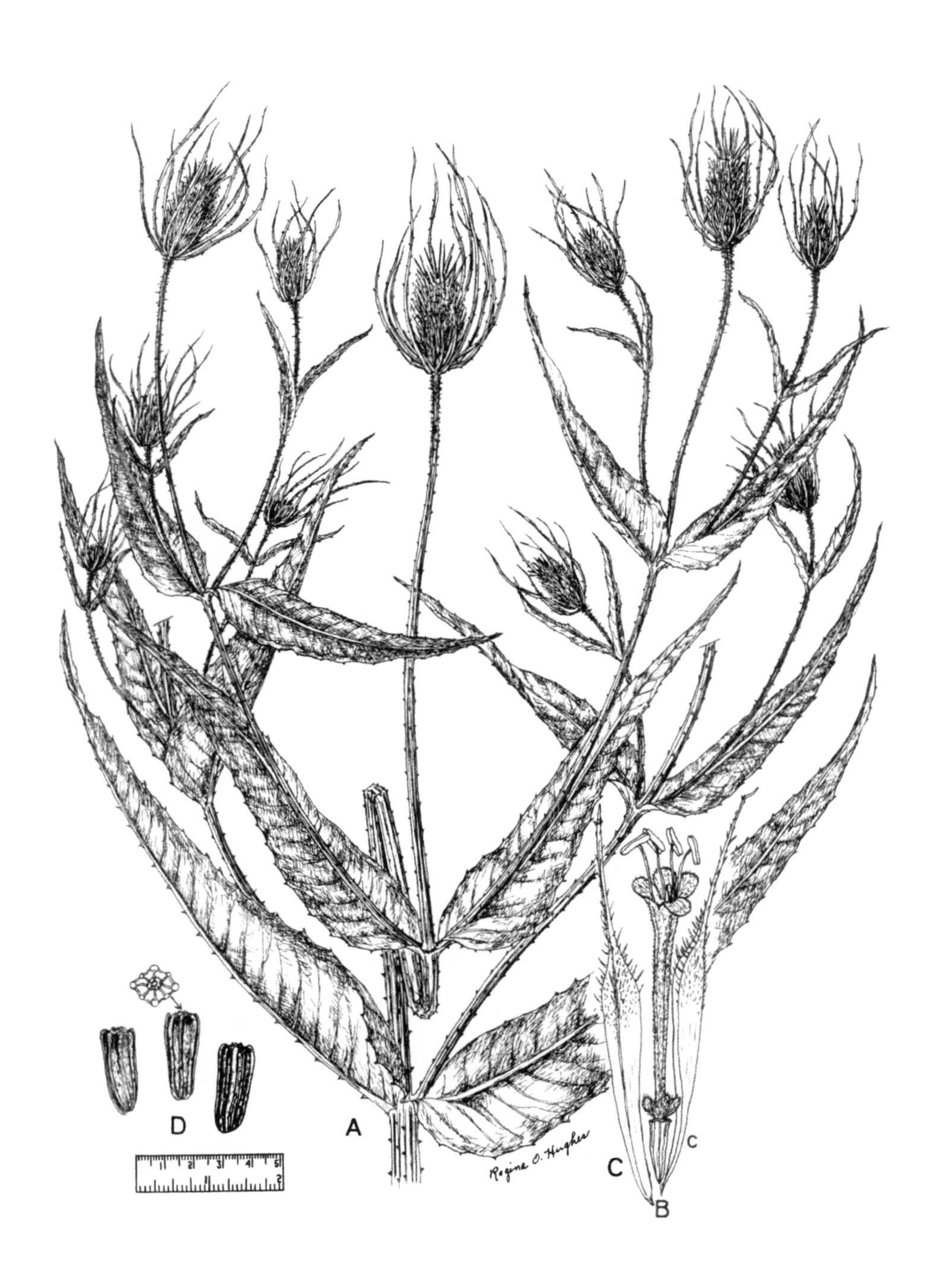

Dipsacus sylvestris, TEASEL. A, Habit. B, flower. C, and c, bracts (chaff). D, achenes.

Equisetum arvense L. FIELD HORSETAIL

Perennial fern ally, reproducing by spores and by rhizomes attached to small tubers. Rhizome deep-seated and long-running.

DESCRIPTION Stems erect or decumbent, 1.5-6 dm tall or long, hollow, jointed. **Fertile stems** stout, producing a terminal cone, with large easily separated joints, up to 8 mm thick, the sheaths 14-20 mm long, the teeth large, partly united, 5-9 mm long; unbranched; flesh-colored, yellowish or brownish. **Sterile or vegetative stems** tough, wiry, with much smaller joints, the lateral branches in whorls around the main stem, green, often rough with silicaceous crystals along it (hence the name scouring rush), solitary or in clusters, normally 10- to 12-ridged (dwarf plants only 4-ridged), internodes 1.5-6 cm long, 1.5-5 mm thick. **Branches** solid, numerous, regularly whorled, 3- or 4-angled, the sheaths with 3 or 4 teeth, the basal sheath pale-brown, the first internode longer than the subtending stem sheath. **Leaves** green, on the sterile stems only, forming cup-shaped sheaths at the joints, the sheaths 5-10 mm long, gradually widening upward, the teeth dark-brown to light-tan, persistent, free or partly united, 1.5-2 mm long, thin, dry, membranaceous, margined. **Fruiting heads**, or cones, terminal, containing masses of pale-greenish to yellow spores, usually on the fertile stems only, long-peduncled, not ending in an abrupt sharp-pointed tip. **Spores** globose, with a pair of elators for dissemination. Fertile shoots appear in early spring and soon disappear; sterile shoots coming up in late spring and persisting until frost.

WHERE FOUND Mostly on wet sandy or gravelly soils; railroad embankments, stream banks, woods, and roadsides.

ORIGIN Cosmopolitan, but native.

NOTE Poisonous to livestock, especially to horses and cattle, causing equisetosis.

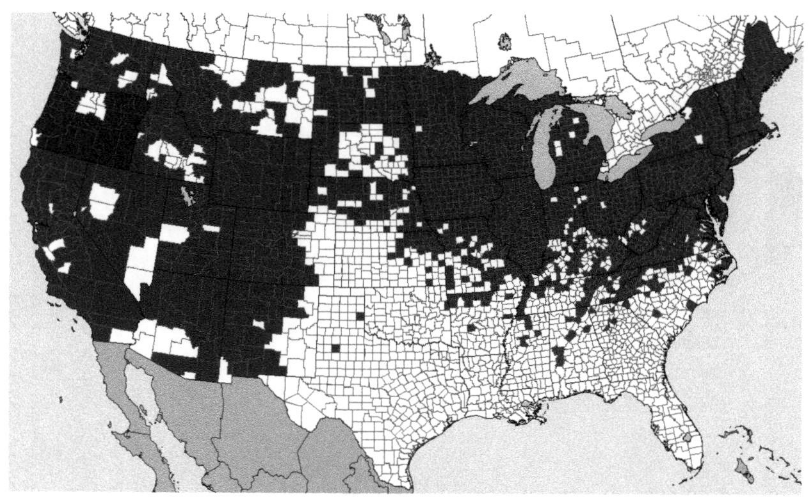

Distribution of **Equisetum arvense**, FIELD HORSETAIL

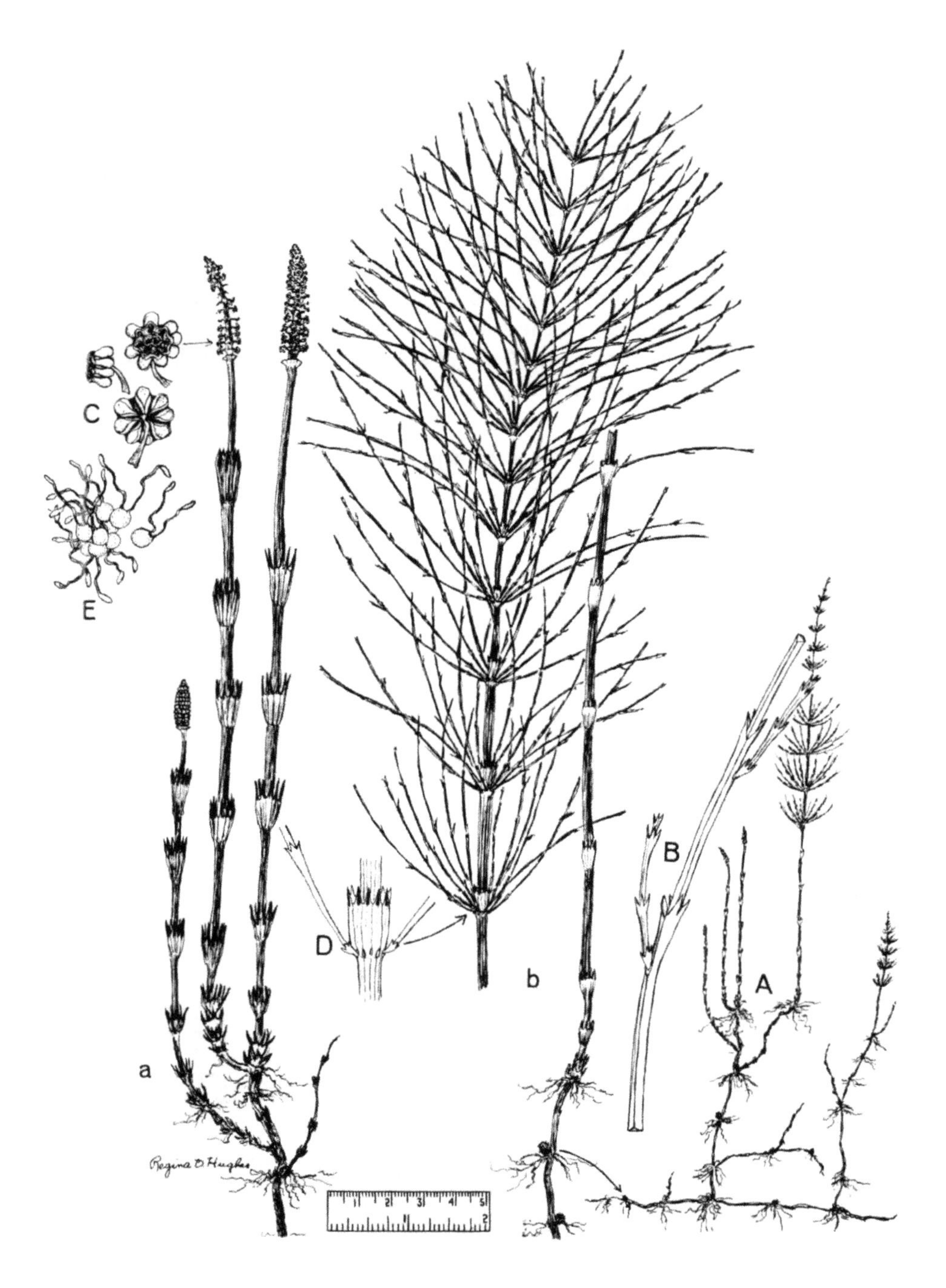

Equisetum arvense, FIELD HORSETAIL. A, habit; a, early, sporophyll-bearing plants. 6, later, vegetative stems. B, enlarged branch, habit. C, sporangiophores. D, sheath. E, spores, showing elaters.

Croton capitatus Michx. WOOLLY CROTON

Annual herb, reproducing by seeds. Taproot shallow.

DESCRIPTION Stems stout, erect, up to 2 m tall, extensively branched, densely soft-woolly with star-shaped hairs, somewhat glandular. **Leaves** usually alternate, simple, with long petioles, lanceolate-oblong to oblong or oval, entire, rounded or nearly heart-shaped at the base, 4–10 cm long, woolly with short hairs. Trichomes of inflorescence brownish. **Flowers** in terminal clusters, 1–3 cm long, surrounded by a leafy bract. **Pistillate flowers** in head-like groups, crowded at base of staminate spike. **Staminate flowers** with 5 sepals, 5 petals, 7–14 stamens; pistillate flowers without petals, calyx 6–12 parted, the 3 styles two or three times 2-parted. **Capsule** globose, 7–9 mm in diameter, densely hairy, 3-sided, containing 3 seeds. **Seed** lens-shaped, nearly round, about 5 mm long, ventrally flattened, grayish.

FLOWERING June–October.

WHERE FOUND Waste places and overgrazed pastures; dry sandy and open soils.

ORIGIN Native.

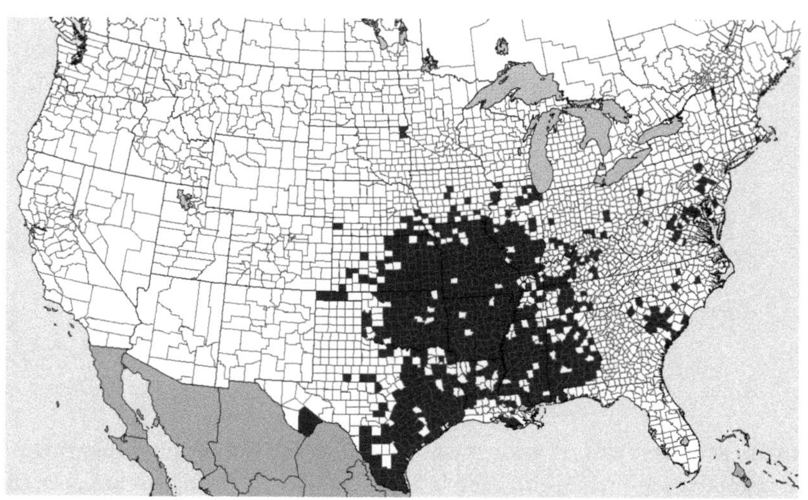

Distribution of **Croton capitatus**, WOOLLY CROTON

Croton capitatus, WOOLLY CROTON. A, habit. B, staminate flower. C, pistillate flower. D, fruit. E, seeds.

Euphorbia corollata L. FLOWERING SPURGE

Perennial herb, reproducing by seed and by short rootstocks.

DESCRIPTION **Stems** single to several from a sturdy, deep root, 3–10 dm tall, erect, light-green, glabrous, branched near the top, with a milky juice. **Leaves** oval, oblong, or linear, light-green, the lower leaves alternate, the upper leaves in whorls on the branches, firm, sessile or subsessile. **Inflorescence** umbellate, many-flowered, - 3to 7-forked and again 2- to 5-forked, the longer pedicels 7–25 cm long. **Flowers** (cyathia) small, 1–1.5 mm high, surrounded by 5 white petal-like bracts (7–10 mm broad) in the form of a cup, in terminal clusters and in the axils of the upper leaves. **Capsule** on short stalks from the cup-like base, smooth, 3-lobed with 3 seeds, 3.5–4.5 mm broad. **Seed** egg-shaped, gray or light-brown, mottled, shallowly pitted, with a dark line on one side.

FLOWERING June–October.

WHERE FOUND Roadsides, waste places, and pastures, especially in dry and sandy areas.

ORIGIN Native.

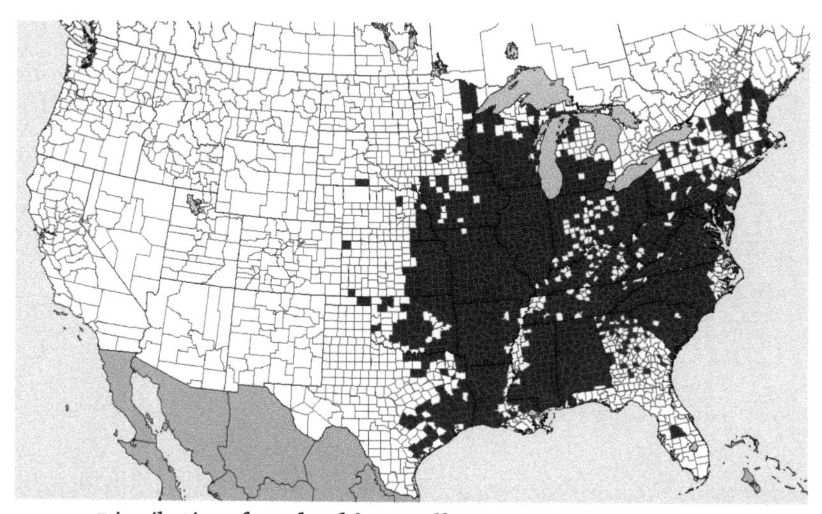

Distribution of **Euphorbia corollata**, FLOWERING SPURGE

Euphorbia corollata, FLOWERING SPURGE. A, habit. B, flowering branchlet. C, capsules. D, seeds.

Euphorbia maculata L. SPOTTED SPURGE

Annual herb, with a shallow taproot, reproducing by seeds, germinating late in the spring or early summer.

DESCRIPTION Stem simple or much-branched, erect or spreading, 0.8–1 m tall, with a milky juice, crisp-pubescent at the young tips, soon becoming glabrous and firm. **Leaves** oblong, oblong-lanceolate or lance-falcate, 0.8–3.5 cm long, the edges slightly toothed, borne on short petioles, with a conspicuous reddish spot or blotch. **Flowers** (cyathia) solitary or clustered, with minute petals in the form of a cup, peduncle 0.5–5 mm long. Involucres 0.7–1 mm in diameter. **Seed pods** on short stalks from the cup-like base, smooth, 3-lobed, ribbed, with 3 seeds. **Seed** 3-sided, obtusely angled, 1.1–1.6 mm long, 0.9–1.1 mm wide, oblong, dark-brown or black, pitted with ridged surfaces.

FLOWERING June–October.

SYNONYMS *Euphorbia supina* Raf.

WHERE FOUND Gardens, cultivated fields, sterile waste places, roadsides; dry, open, sandy and gravelly soils.

ORIGIN Native.

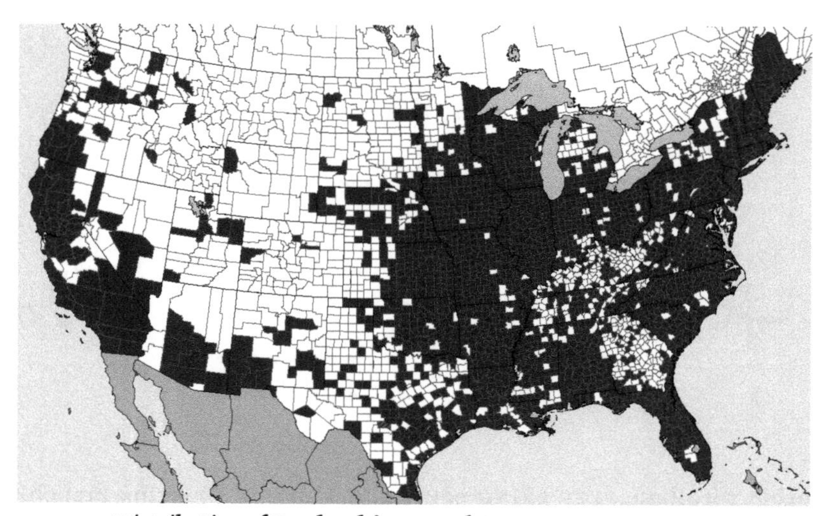

Distribution of **Euphorbia maculata**, SPOTTED SPURGE

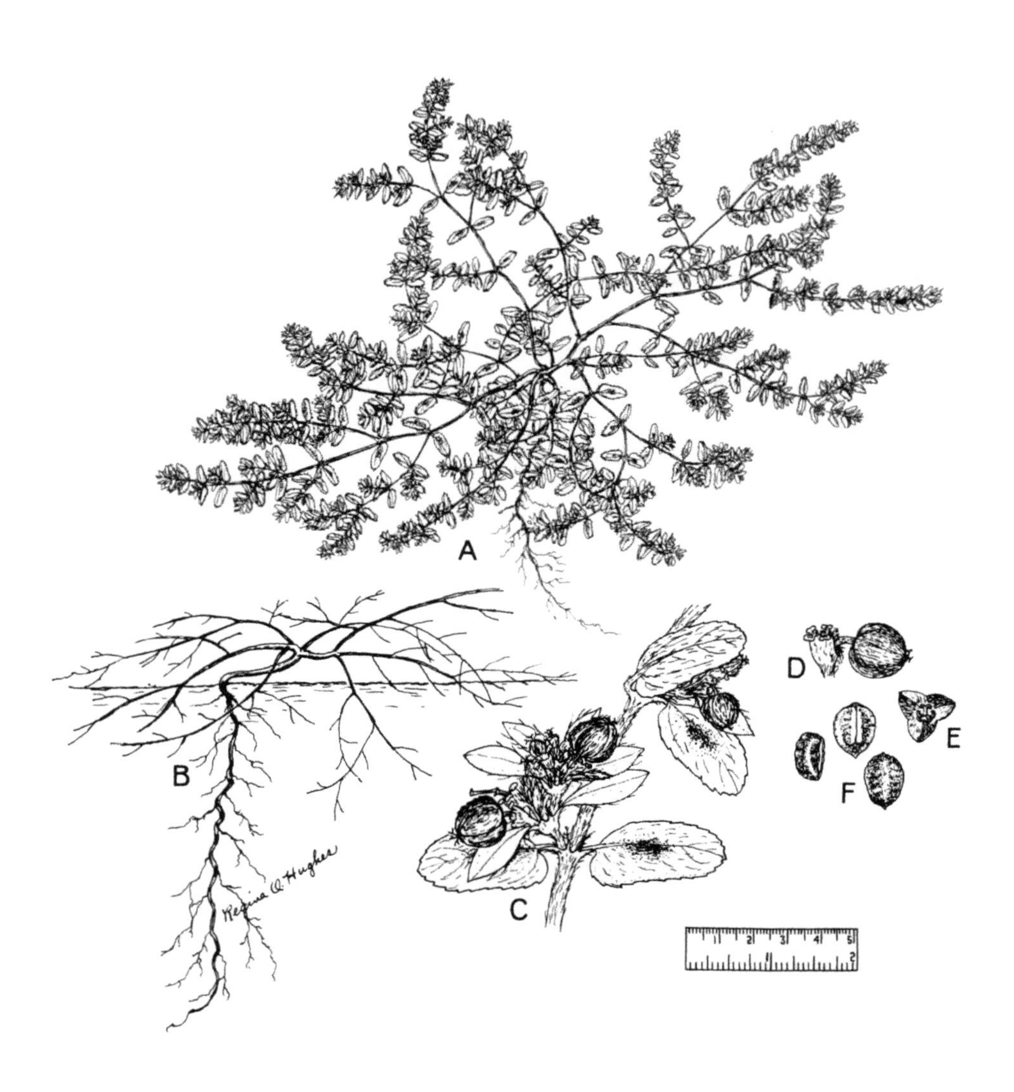

Euphorbia maculata, SPOTTED SPURGE. A, habit. B, staminate flower. C and D, immature capsules of the pistillate flower. E, mature capsule, bottom view to show deep lobes. F, seed, 3 views.

Euphorbia virgata Waldst. & Kit. LEAFY SPURGE

Perennial herb, reproducing by seeds and from extensive slender rootstocks.

DESCRIPTION **Roots** deep and spreading, woody, very persistent. **Stems** erect, glabrous, branched at the top, 3-9 dm tall, with a milky juice. **Leaves** alternate, broadly linear to narrowly oblong-lanceolate or inverted lanceolate, 2-10 mm broad, usually drooping. **Inflorescence** a terminal open umbel, the larger floral leaves kidney-shaped, 1-1.3 cm long, yellow-green. **Flowers** (cyathia) small, 2.5-3 mm high, greenish, the petals fused into a cup-like structure, borne just above the greenish-yellow heart-shaped floral bracts on the top of the stem. **Capsule** on short stalks from the cup-like base, 3-lobed, with 3 seeds. **Seed** elliptic-oval, about 2 mm long, smooth, light-gray to yellow-brown, with a yellow (or white) emarginate caruncle.

FLOWERING May-September.

SYNONYMS *Euphorbia esula* auct non. L.

WHERE FOUND Waste areas, pastures, roadsides, cultivated fields, and sandy banks; a serious weed because of its spreading nature and persistence.

ORIGIN Naturalized from Europe.

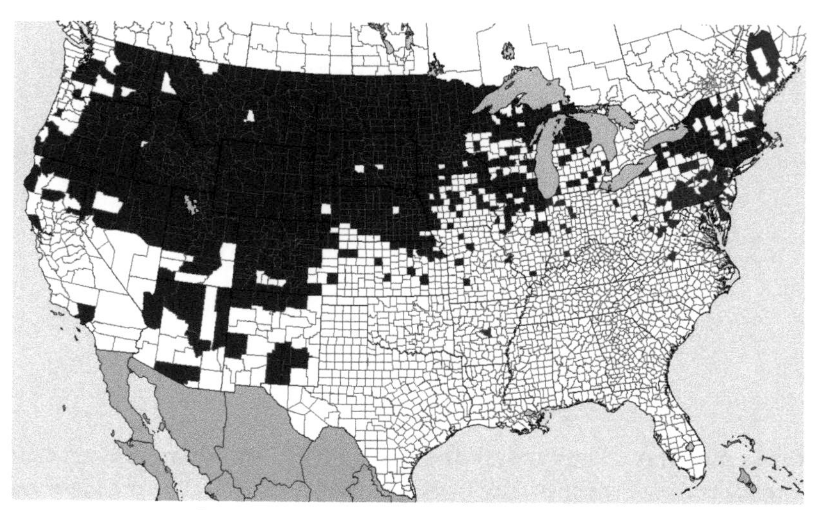

Distribution of **Euphorbia virgata**, LEAFY SPURGE

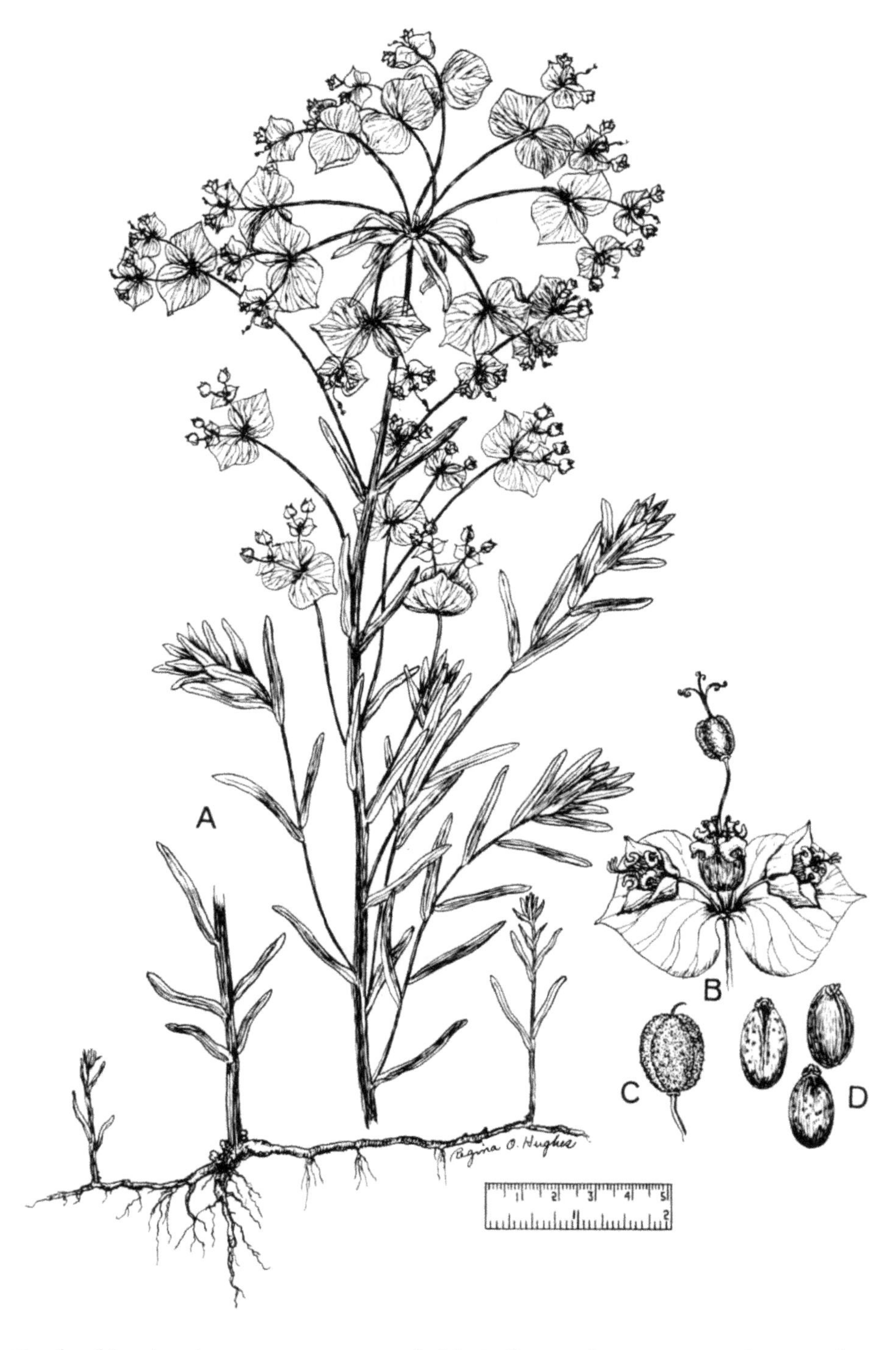

Euphorbia virgata, LEAFY SPURGE. A, habit. B, flower cluster. C, capsule. D, seeds.

Astragalus mollissimus Torr. WOOLLY LOCO

Perennial herb, reproducing by seeds. Root system deep-penetrating, woody, on which are root tubercles.

DESCRIPTION Plants nearly stemless, the internodes few and short. **Stems** less than 10 cm long, somewhat bushy. **Leaves** alternate, 10–20 cm long, pinnately compound, 21–31 leaflets, 10–25 mm long, oval, obovate to broadly elliptic, silky long soft hairs, not reduced upwards. **Flowers** bluish-purple to rose-purple, in dense racemes, almost spike-like in flower, but elongating in fruit. Calyx 10–15 mm long. Corolla 17–21 mm long. **Legume** (pod) 13–20 mm long, narrowly oblong, glabrous.

FLOWERING May–June; fruiting June–August.

WHERE FOUND Dry plains and foothills at lower elevations, east of the Continental Divide.

ORIGIN Native.

Note Poisonous to livestock.

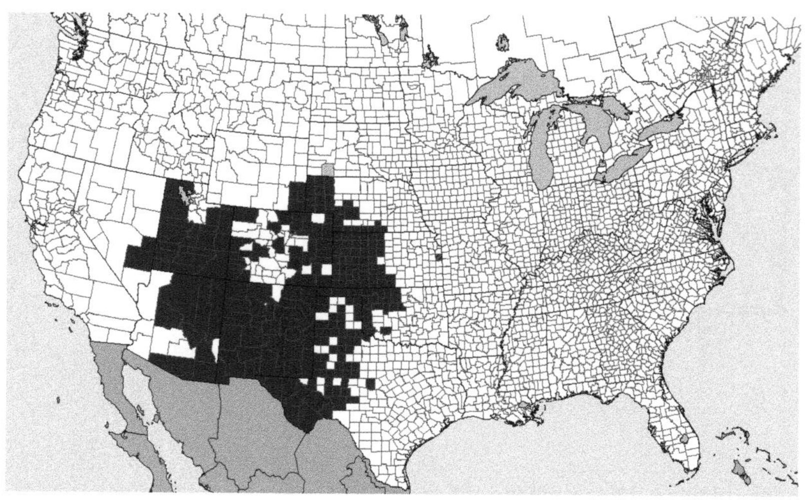

Distribution of **Astragalus mollissimus**, WOOLLY LOCO

Astragalus mollissimus, WOOLLY LOCO. A, Habit. B, enlarged leaves. C, flower. D, pods. E, seeds.

Medicago lupulina L. BLACK MEDIC

Annual, winter annual, biennial, or sometimes acting as a perennial herb, reproducing by seeds. Taproot shallow.

DESCRIPTION Stems slender, procumbent or prostrate, branched at the base, 30–60 dm tall, sparsely pubescent, especially below. **Leaves** alternate, petioled, 3-parted, the center leaflet on a short stalk, the leaflets 5–15 mm long, nearly orbicular to broadly obovate, sparingly hairy, the stipules ovate-lanceolate. Peduncles slender, hairy or glabrate. **Flowers** 3–4 mm long, crowded in spike-like racemes (not over 12 mm long). Corollas yellow, 1.5–2 mm long, longer than the hairy calyx. Petals 5 (standard, 2 lateral wings and the 2 lower fused into a keel). Calyx 5-cleft, persistent. Stamens 9 fused and 1 separate. Pistil solitary. **Legume** (pod) nearly kidney-shaped, with a closed groove, pubescent, hairy or glabrate, net-veined, 1.5–3 mm long, becoming black at maturity, unarmed, 1-seeded. **Seed** 1.5–2 mm long, oval to short kidney-shaped, with a protuberance at the hilum, the sides convex, smooth, dull, yellowish-green, greenish-brown to orange-brown. Pericarp often persisting.

FLOWERING March–September, sometimes to December.

WHERE FOUND Roadsides, waste places, lawns, pastures, and meadows. Grown for forage, but small-yielding; will innoculate soil for alfalfa; may be objectionable as an impurity in alsike clover fields.

ORIGIN Naturalized from Eurasia.

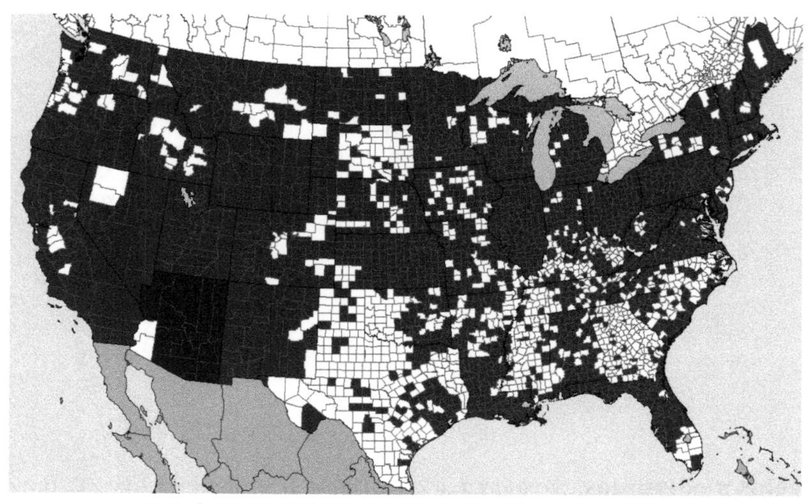

Distribution of **Medicago lupulina**, BLACK MEDIC

Medicago lupulina, BLACK MEDIC. A, habit. B, flower raceme. C, fruiting raceme. D, flower. E, legume. F, seeds.

Oxtytropis lambertii Pursh — STEMLESS LOCOWEED

Perennial herb, reproducing by seeds.

DESCRIPTION Stem forming a crown at the surface of the ground, from a taproot system with root tubercles. Plant low, 1-1.5 dm tall, erect, stout, bushy from the top of the crown (base), densely white-pubescent (var; sericea), intergrading completely with the greener, less pubescent form. **Leaves** alternate, pinnately compound, 8-20 cm long, white hairy, the leaflets 7-13, elliptic or oblong, 1-3 cm long. **Flowers** cream to white, often tinged pink or blue, the keel frequently purple-tipped, in a peduncled, spike-like raceme, 5-10 cm long, the raceme elongating in fruit. Calyx teeth 1-3 mm long, about one-third the length of the tube. Wings of the corolla broad and deeply emarginate. **Legume** (pod) cylindric, erect, about 2 cm long, long-tapering, nearly 2-celled. **Seed** dark-brown to blackish, about 1.5 mm long, kidney-shaped, thinner in the scar region. A variable species, varying in flower color, height of plant, general pubescence, and length of calyx teeth.

FLOWERING May-August; fruiting June-October.

WHERE FOUND Mountain slopes, pasture lands, and grazing lands.

ORIGIN Native.

NOTE Poisonous to livestock.

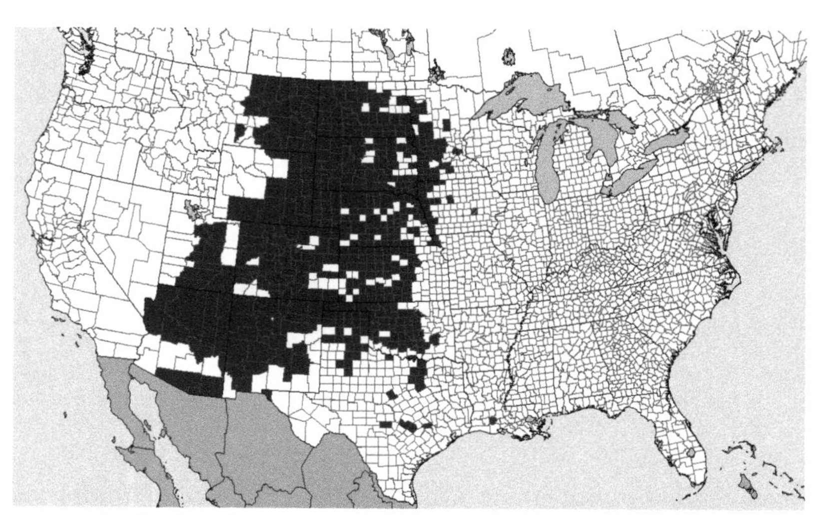

Distribution of **Oxytropis lambertii**, LAMBERT CRAZYWEED

Oxytropis lambertii, LAMBERT CRAZYWEED. A, habit. B, enlarged leaves. C, flower spike. D, flower. E, legumes. F, seeds.

Prosopis velutina Woot. MESQUITE

Perennial, deciduous thorny shrub or small tree, to 12 m tall, trunk to 1.2 m in diameter, bark thick, brown or blackish, shallowly fissured. Roots sometimes penetrate the soil to a depth of 20 m.

DESCRIPTION Leaves compound, commonly many more than 9 pairs, the leaflets mostly 5-10 mm long, linear-oblong, glabrous, often with small hairs, commonly rounded at the apex. Stipular spines, if any, yellowish, often stout. **Flowers** perfect, greenish-yellow, sweet-scented, spike-like. Corolla deeply lobate. **Pods** several-seeded, strongly compressed when young, but thick at maturity, more or less constricted between the seeds, 10-20 cm long, brown or yellowish. **Seed** compressed and oval or elliptic, 2.5-7 mm long, brown, rather glossy, with a central ring on each face.

FLOWERING May–July.

SYNONYMS *Mimosa juliflora* Sw., *Neltuma juliflora* (Sw.) Raf., *Prosopis glandulosa* Torr., *Prosopis juliflora* (Sw.) DC..

WHERE FOUND Trees or shrubs along water courses; smaller shrub on grasslands and lower mountain slopes (often with much of the trunk underground); widely spread on overgrazed grasslands.

ORIGIN Native.

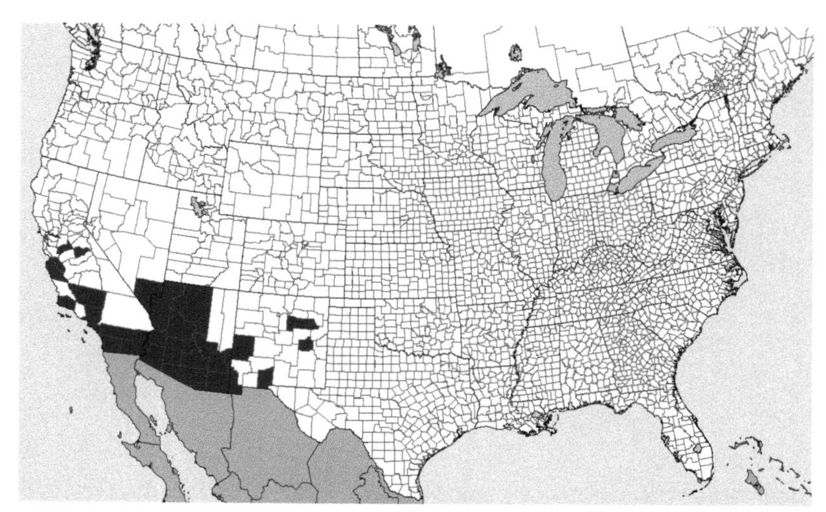

Distribution of **Prosopis velutina**, MESQUITE

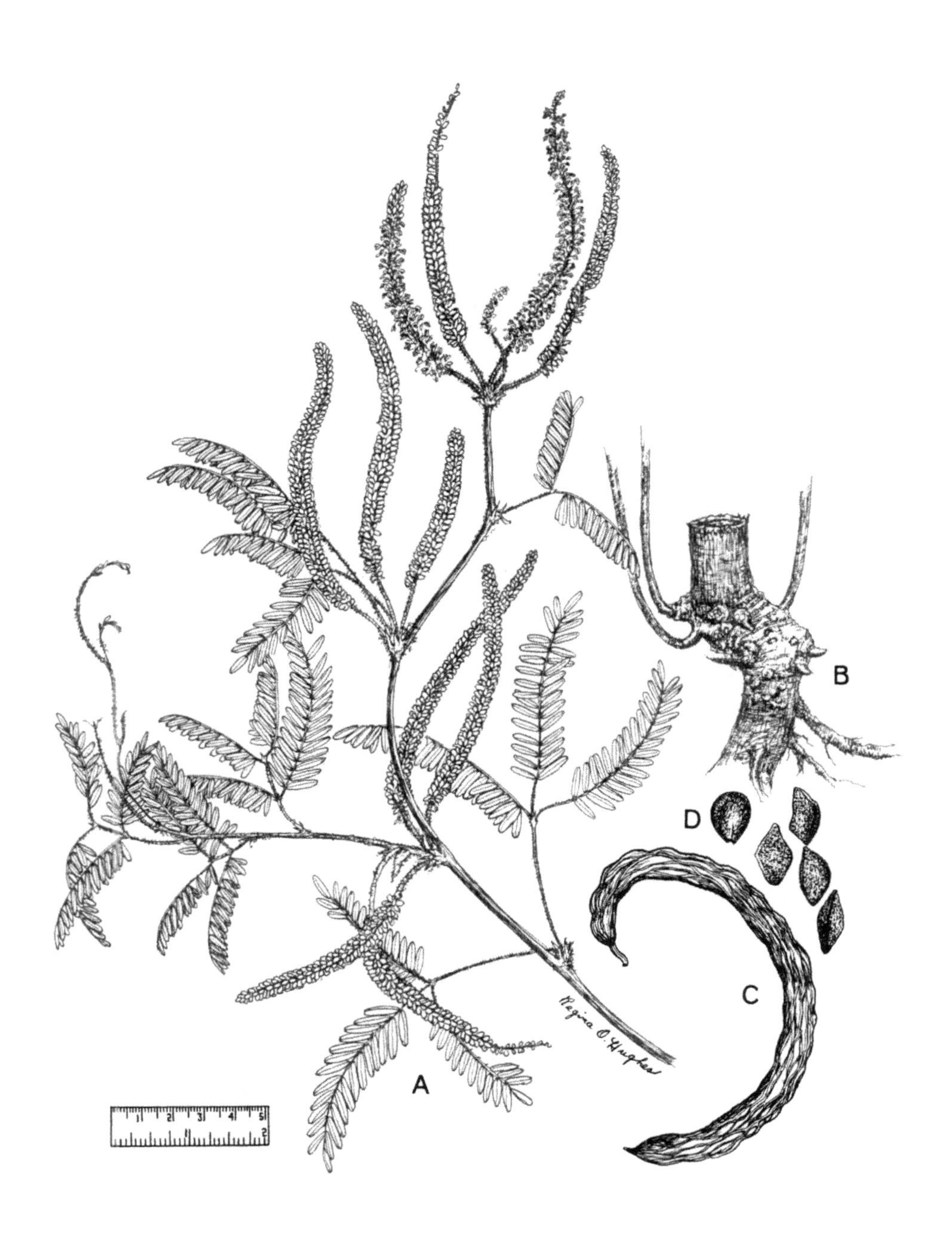

Prosopis velutina, MESQUITE. A, habit. B, rootstock portion. C, pod. D, seeds.

Pueraria montana (Lour.) Merr. KUDZU

Perennial high-climbing vines, reproducing by seeds and from mealy tuberous roots.

DESCRIPTION Stems herbaceous to woody or ligneous, 10–30 m long, up to 2.5 cm thick, high-climbing and twining, the young parts pubescent. **Leaves** pinnately 3-foliolate, the leaflets entire or coarsely and palmately (2–3) lobed, broadly ovate to sub-rotund, up to 1.8 dm long, pubescent beneath, the stipules herbaceous. **Racemes** simple or compound, axillary, peduncled, elongate, 1–2 dm long, the axis and pedicels densely silky, with non-persistent long-tipped bracts. **Flowers** reddish-purple, 2–2.5 cm long, tufted at the nodes at the rachis. Corolla reddish-purple, with the fragrance of grapes. Calyx with the two upper lobes united, the lowest calyx lobe 8–12 mm long. Keel ascending or arched at the tip, about equaling the wings, the standard yellow at the base, nearly round or obovate. Stamens with the filaments united, the axillary stamen free at the base. **Legume** (pod) 4–5 cm long, flattish, continuous or with internal partitions, several-seeded, hirsute. Seed nearly round or transversely ovoid, compressed.

FLOWERING Late August–September.

SYNONYMS *Pueraria lobata* (Willd.) Ohwi

WHERE FOUND Borders of fields and woods and along rivers, roadsides, an embankments, often covering old dwellings and trees.

ORIGIN Native of Japan; introduced from eastern Asia. Throughout approximately all the southeastern area of the United States. Rarely flowering north of Virginia; previously grown as a quick-growing ornamental climber, but easily escaping and rapidly spreading.

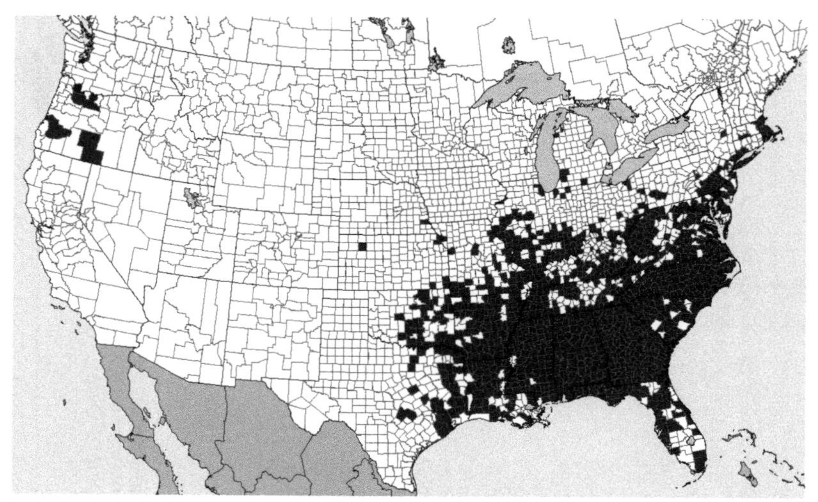

Distribution of **Pueraria montana**, KUDZU

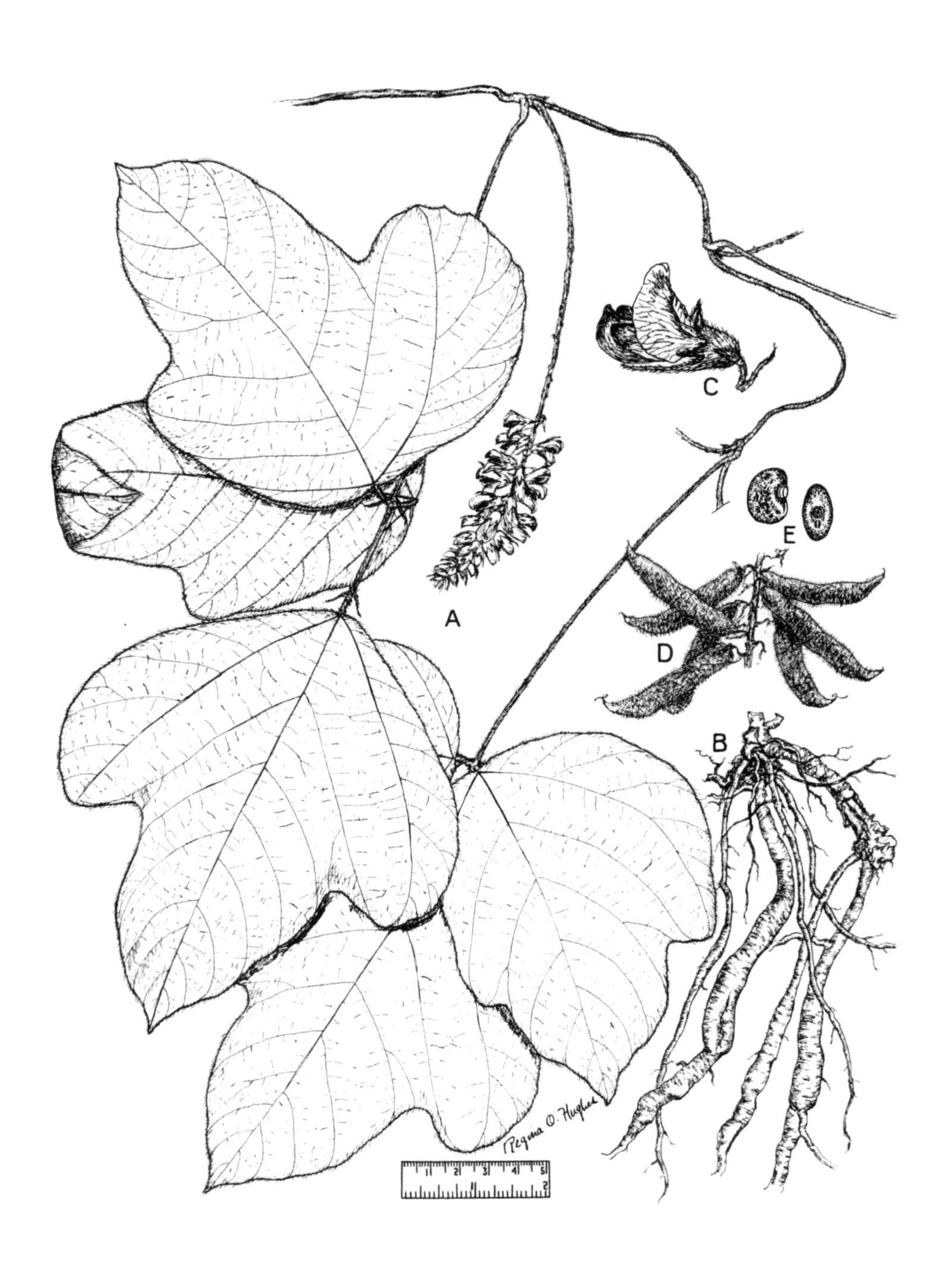

Pueraria montana, KUDZU. A, habit. B, root. C, flower. D, legumes. E, seeds.

Sesbania punicea (Cav.) Benth. COFFEEWEED

Perennial, herbaceous to shrub or small tree, reproducing by seeds.

DESCRIPTION **Stems** branching, becoming bushy, up to 2.5 m tall, usually from a clump. **Leaves** alternate, with oblong stipules, pinnate with 12–60 leaflets, the blades linear-elliptic, rounded or blunt at the apex, 1–2.5 cm long. **Flowers** perfect, in simple axillary racemes shorter than the leaves. Pedicels 5–10 mm long, curved at the apex. Calyx bell-shaped, the lobes acute, ciliate, much shorter than the tube, the lower lobe obtuse. Corolla scarlet to yellow, the standard with a blade 13–18 mm broad, the keel petals with blades tapering into the claw. **Pods** linear-oblong, leathery, 5–8 cm long, the stipe 1–1.5 cm long, compressed with 4 wings arising from the margins of the valves and produced beyond the sutures, indehiscent. **Seed** several, separated from one another by transverse partitions.

FLOWERING July–September.

SYNONYMS *Daubentonia punicea* (Cav.) DC.

WHERE FOUND Waste places, roadsides, and fence rows; in sandy soils of the Coastal Plain.

ORIGIN Native to tropical America.

NOTE Very poisonous to livestock, especially sheep.

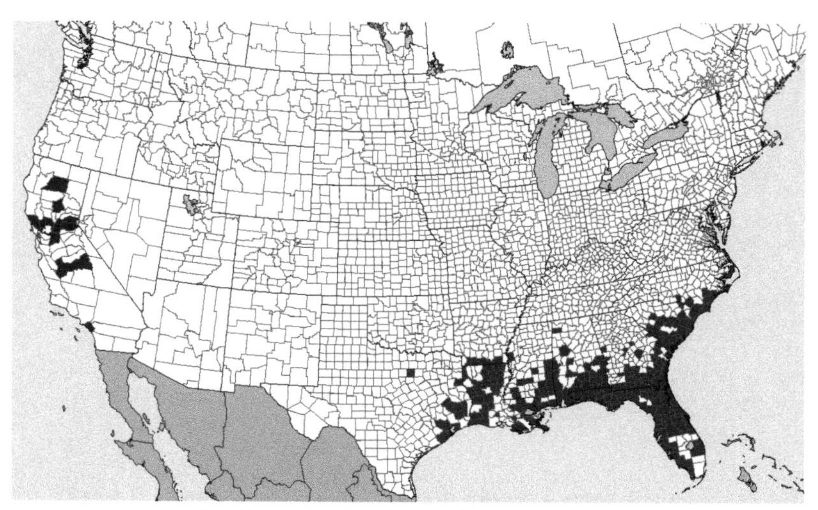

Distribution of **Sesbania punicea**, COFFEEWEED

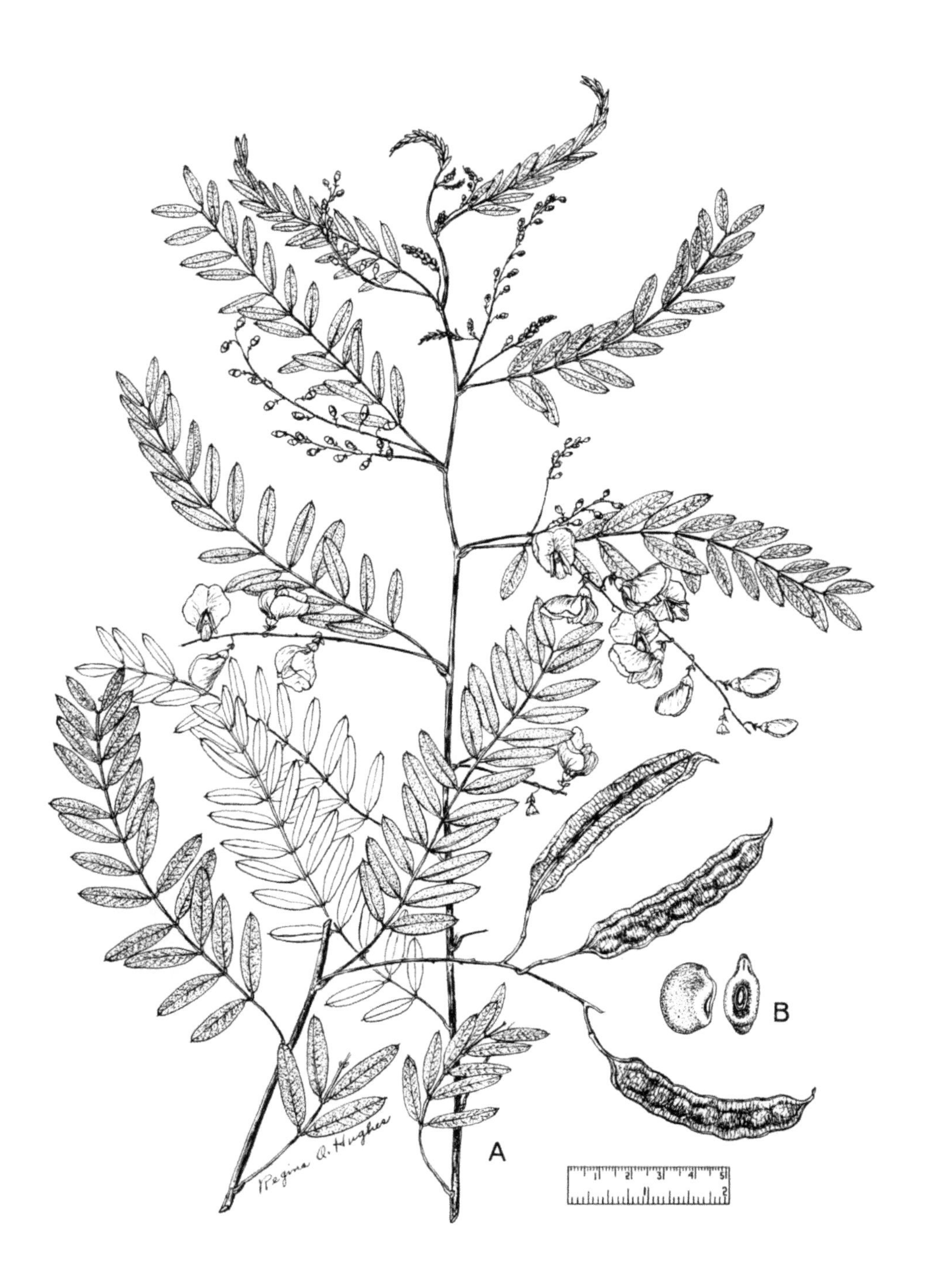

Sesbania punicea, COFFEEWEED. A, habit, upper branches showing flowers and mature pods. B, seed, 2 views.

Vicia sativa L. GARDEN VETCH

Annual or winter annual herb, reproducing by seed.

DESCRIPTION **Stem** slender, glabrous or becoming glabrous, 3-9 dm long, upright but weak, clinging to support by tendrils at the tip of the leaves. **Leaves** pinnately compound, 8-12 leaflets, those of the lower leaves oblong and blunt, those of the upper leaves linear to lance-attenuate, sharp-pointed, 1.5-3 cm long, 1-4 mm wide, all smooth-margined, the leaflet terminating in a branched tendril. Inflorescence in racemes, with 2-9 flowers. **Flowers** perfect, irregular, on very short peduncles or sessile, a few in a leaf axil, 1-1.8 cm long. Calyx with nearly equal teeth, 7-11 mm long. Corolla of 5 unequal purplish petals. Stamens 10, 9 fused by their filaments, 1 separate. **Legume** (pod) smooth, plane, 4-6 cm long, 5-7 mm wide. **Seed** 4-7 per pod, 3 mm broad, round, surface dull, velvety, brown-black or olive-brown, mottled with fine black spots.

FLOWERING June-October.

SYNONYMS *Vicia angustifolia* L.

WHERE FOUND Waste grounds, fields, roadsides, waysides, slopes, and meadows; on rich gravelly soils. Cultivated for forage, but usually a weed, especially in grainfields.

ORIGIN Naturalized from Europe, throughout the United States.

NOTE The most frequent subspecies in the United States is ssp. *nigra* (L.) Ehrh.—Leaflets of upper leaves oblong to oblong-obovate, blunt or notched at the tip, pointed at the apex, 2-9 mm broad. Naturalized from Europe, throughout the range of the species, in wheatfields, roadsides, and waste places.

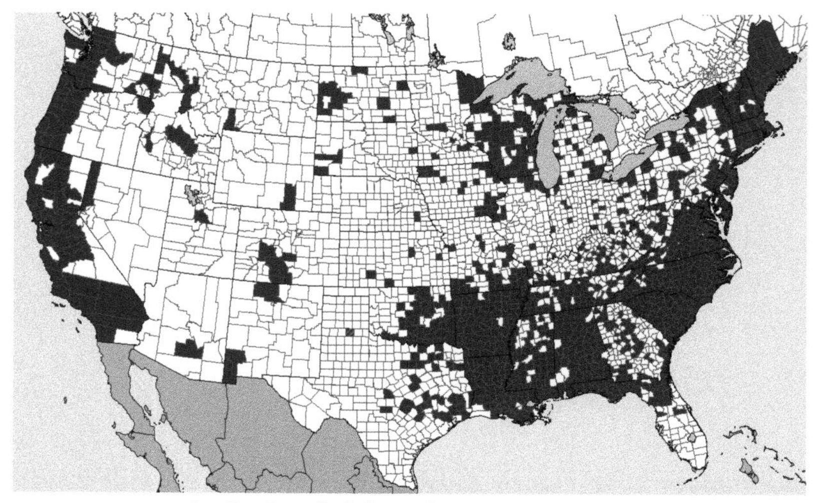

Distribution of **Vicia sativa**, GARDEN VETCH

Vicia sativa, GARDEN VETCH. A, habit. B, leaves. C, flowers. D, bract, showing extra nectary. E, legume. F, seed, 4 views.

Myriophyllum heterophyllum Michx.

BROADLEAF WATER-MILFOIL

Perennial aquatic herb.

DESCRIPTION Stems simple or branching, stoutish, to 8 mm in diameter, 6–12 dm long (or longer). **Leaves** whorled in 4's and 6's, the submersed leaves pinnate, 2–5 cm long with 7–10 pairs of divisions, the **amphibious leaves** slightly divided, the **emersed leaves** and bracts firm, lanceolate or lance-spatulate to elliptic, entire or serrate, 0.4–3 m long, 1.5–5 mm broad. **Spikes** usually above water in summer, 0.3–3.7 dm long. **Flowers** sessile, in whorls of 4–6, chiefly in the axils of the upper leaves, both sexes together, or the lower flowers pistillate and the upper flowers staminate. Bracts whorled, long-persistent and eventually reflexed, lanceolate to oblong, or obovate, sharply denticulate, 4–18 mm long. Bracteoles ovate, acuminate, serrate, 1–1.3 mm long, 0.5–0.7 mm broad. Petals of staminate flowers acutish, 1.5–3 mm long. Stamens 4, 1–2.5 mm long. **Schizocarp** nearly globose, 1–1.5 mm long and wide, minutely papillose. **Mericarps** 2-ridged on the back and rounded on the sides, with conspicuous ascending beaks.

FLOWERING June–September.

WHERE FOUND Ponds, streams, and lakes.

ORIGIN Native

NOTE *Myriophyllum hippurioides* Nutt. is not too dissimilar from *M. heterophyllum*, but restricted to Califronia, Oregon, and Washington. **Stems** simple or branching, 3–6 dm long. **Leaves** whorled in 4's or 5's, the emersed ones linear, about 1 mm wide, conspicuously or obscurely serrate or the uppermost ones nearly entire, 1.5–3 mm long, the submersed leaves pinnately dissected into capillary divisions, 1.5–3 cm long. **Flowers** chiefly in the axils of the emersed leaves. Petals white, obovate. **Mericarps** 0.2 mm long, slightly rounded on the back.

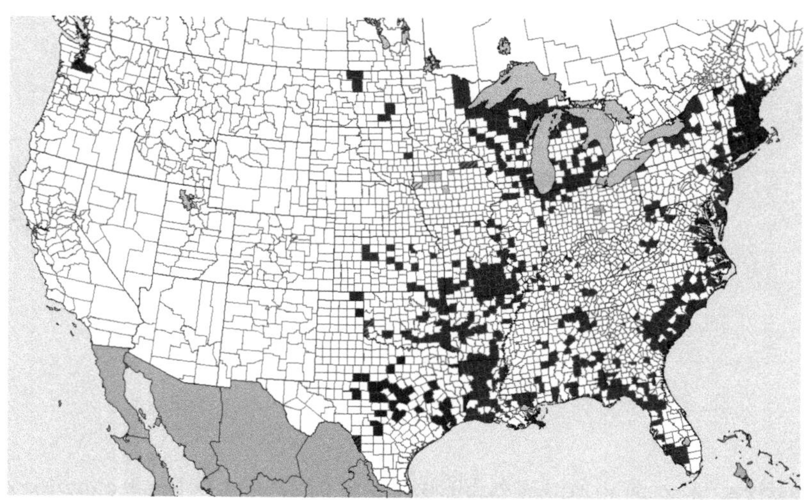

Distribution of **Myriophyllum heterophyllum**, BROADLEAF WATER-MILFOIL

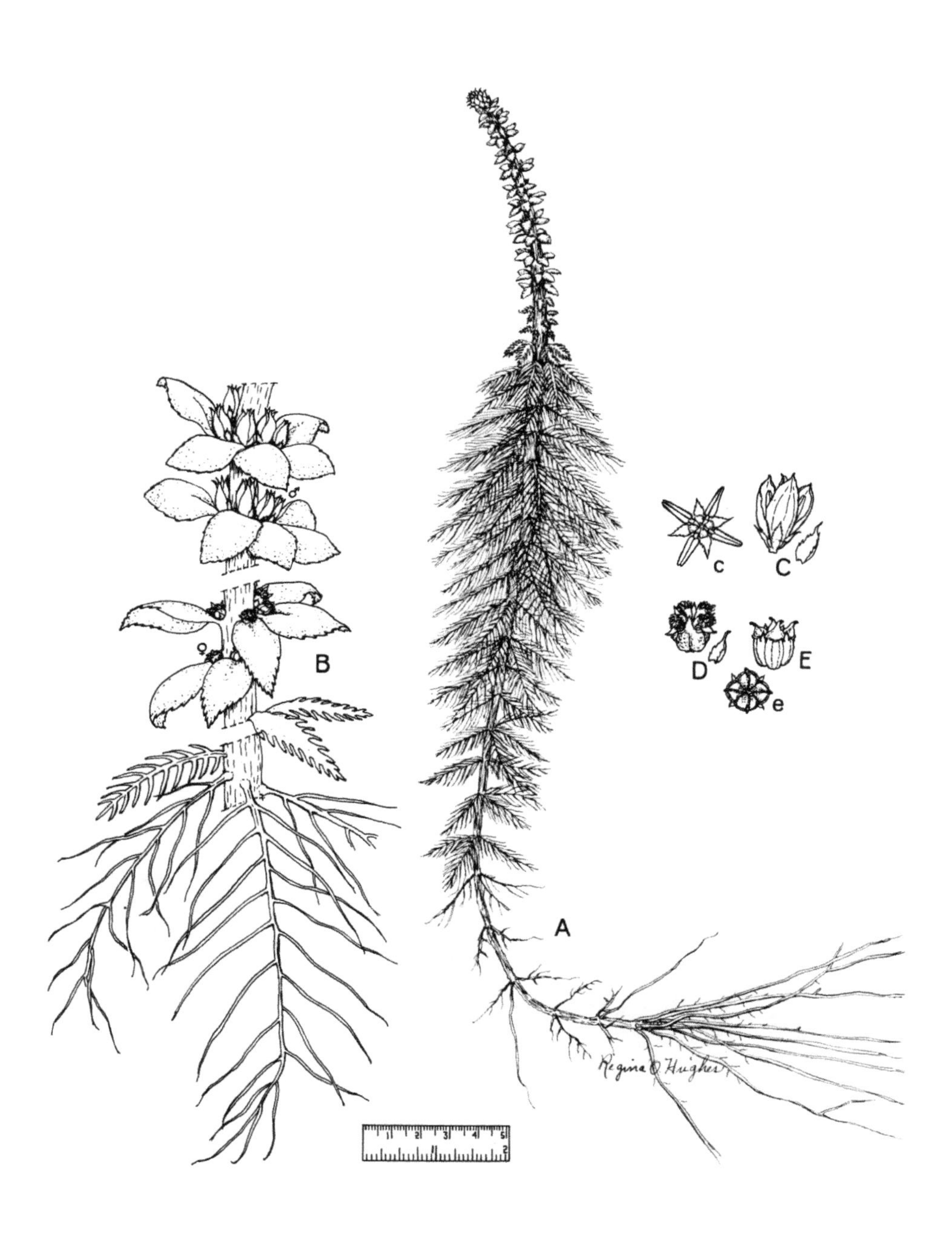

Myriophyllum heterophyllum, BROADLEAF WATER-MILFOIL. A, habit. B, leaf and flower detail. C, staminate flower and bract; c, diagram of staminate flower. D, pistillate flower and bract. E, mature fruits; e, top view of mature fruits.

Myriophyllum sibiricum Komarov NORTHERN WATER-MILFOIL

Perennial aquatic herb, reproducing by seeds, running rhizomes, and fragments of the stems.

DESCRIPTION Stems simple or forking, purplish, when dry becoming white, up to 1 m in length. **Leaves** whorled, in 3's or 4's, 1.2–3 cm long, with 6–11 pairs of capillary flaccid or slightly stiffish divisions, the primary leaves submersed, 1–5 cm long, 1.4 cm broad. **Spikes** almost naked, terminal, with the flowers in whorls, the lower flowers pistillate, the upper staminate. Bracts persistent, rarely equaling the fruit, spatulate-obovate or oblong-shell-shaped, 0.8–1.8 mm long, the lower serrate, the upper entire. Bracteoles ovate, entire, 0.7–1 mm long. Petals oblong-obovate, concave, 2.5 mm long. Anthers 1.2–1.8 mm long. **Schizocarp** nearly globose, very slenderly 4-sulcate, 2.3–3 mm long. **Mericarps** rounded on the back, smooth or roughened.

FLOWERING July–September.

SYNONYMS *Myriophyllum exalbescens* Fern.

WHERE FOUND Lakes, ponds, pools, and quiet waters, often brackish or calcareous: especially troublesome around the edges of lakes.

ORIGIN Native.

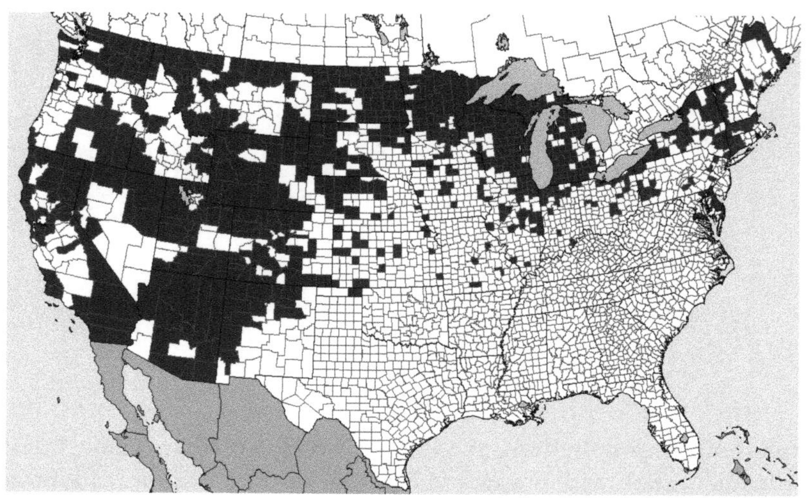

Distribution of **Myriophyllum sibiricum**, NORTHERN WATER-MIFOIL

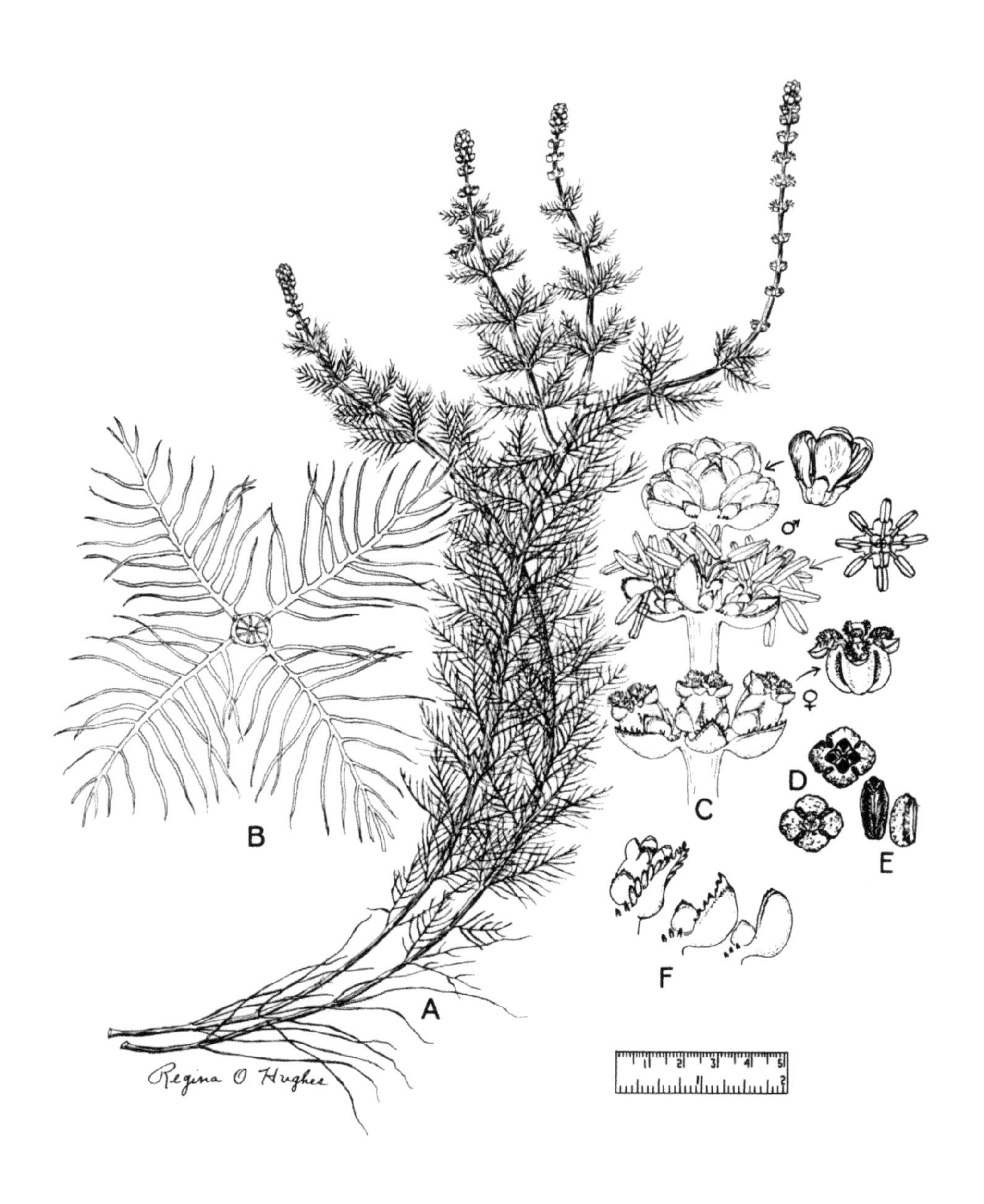

Myriophyllum sibiricum, NORTHERN WATER-MIFOIL. A, habit. B, whorl of leaves. C, flower spike, with male and female flowers. D, schizocarps. E, mericarps. F, bracts.

Myriophyllum spicatum L. EURASIAN WATER-MILFOIL

Perennial aquatic-rooted herb, reproducing by seeds, but very commonly and most efficiently spreading by rhizomes, fragmented stems, and axillary buds that occur throughout the year.

DESCRIPTION Stems long and branching, often from a depth of 5 m (most frequently to 2 m deep), often forming extensive mats at the surface of the water, brick-red or olive-green in dried specimens. **Leaves** whorled in 3's or 4's, to 35 mm long, the principal leaves of the primary stems with 14–21 pairs of rigid slenderly linear divisions. Bracts rhombic-obovate to elongate, the bractlets nearly round or kidney-shaped, broader than long, 0.5–0.8 long. **Spikes** terminal, 2.5–10 cm long, often standing above the water level, after pollination then resubmerging. **Flowers** (after emergence) with the stigmas ripening well in advance of the stamens (favoring cross-pollination). Petals deciduous before ripening of the stamens. Anthers linear, 1.8–2.2 mm long. Floral bracts longer than the fruits. **Schizocarp** 4-locular, with 4 seeds. **Mericarps** spherical, 4-angled, 2.5–3 mm in diameter.

FLOWERING Late July–September.

WHERE FOUND In fresh and saline waters, on muck to hard-packed sand; most common and a nuisance, especially to sportsmen.

ORIGIN Native of Eurasia and parts of Africa.

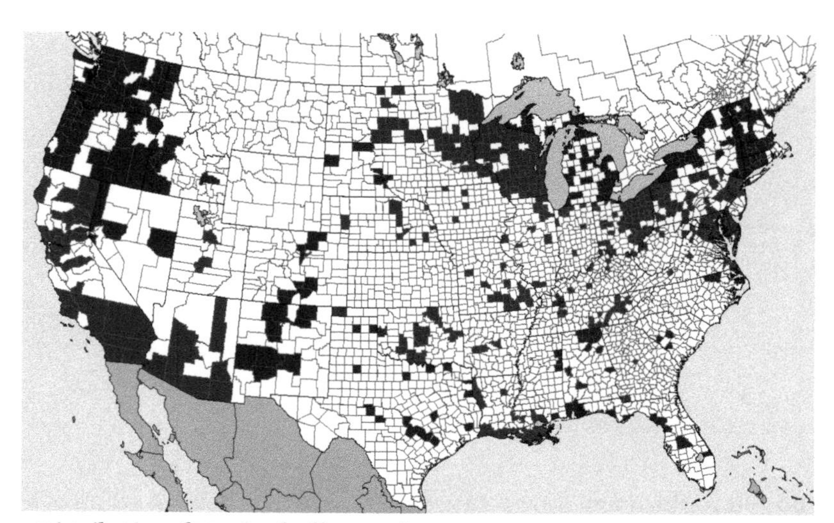

Distribution of **Myriophyllum spicatum**, EURASIAN WATER-MILFOIL

Myriophyllum spicatum, EURASIAN WATER-MILFOIL. A, habit. B, whorl of leaves. C, part of flower spike, with pistillate flowers below and staminate flowers above. D, immature fruits. E, mature fruit.

Egeria densa Planch. BRAZILIAN-WATERWEED

Aquatic herb.

DESCRIPTION Leaves in whorls of 4–6, the principal leaves 2–3.3 cm long, linear-lanceolate, tapering gradually to a point, the lower leaves remote, the upper ones crowded. **Flowers** at anthesis 15–20 mm wide; the staminate spathes with 2 or more exserted flowers with relatively showy petals 9–11 mm long and 6–9 mm broad; the pistillate flowers not known in United States.

FLOWERING June–October.

SYNONYMS *Anacharis densa* (Planch.) Victorin, *Elodea densa* (Planch.) Caspary

WHERE FOUND Commonly cultivated in aquaria; occasionally established in ponds, pools, and quiet streams.

ORIGIN Native of Argentina.

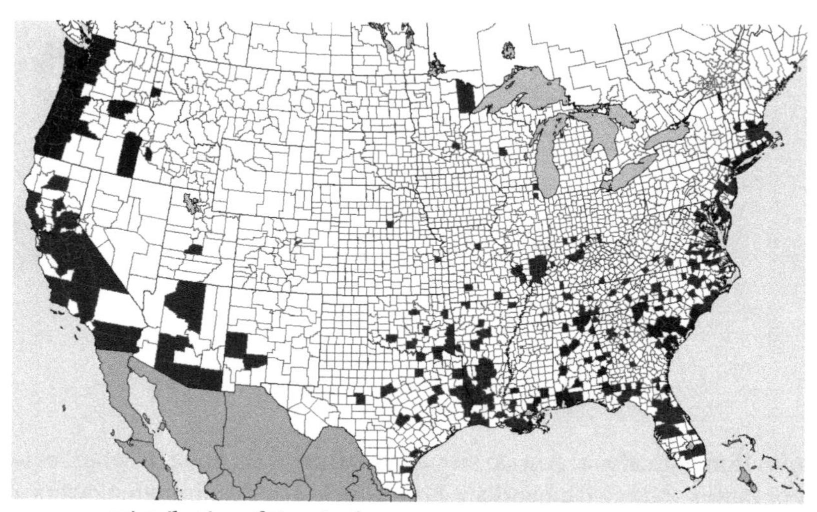

Distribution of **Egeria densa**, BRAZILIAN-WATERWEED

Egeria densa, BRAZILIAN-WATERWEED. A, habit. B, spathe detail. C, flower detail.

Elodea canadensis Michx. CANADIAN WATERWEED

Aquatic herb, perennial by means of abundant vegetation. Plants dioecious.

DESCRIPTION **Pistillate plants** with slender branched stems, forking regularly by pairs, from creeping threadlike stolons, often forming large masses; the **lower leaves** opposite, ovate, small, the **median and upper leaves** whorled in 3's, oblong-ovate or ovate-lanceolate, with minute teeth pointing forward, firm, dark-green, crowded and strongly overlapping toward the summit of the branches, 6-13 mm long, 1-5 mm broad; pistillate spathes cylindric, in upper axils, with 2 broad apical teeth; **pistillate flowers** appearing before the staminate flowers, exserted by the threadlike base of the flower receptacle prolonged to 2-15 cm, the dark-striate and oblong-elliptic sepals falling at anthesis without flattening out, 2-2.2 mm long, 1.1 mm broad, the delicate broadly elliptic-spathulate white petals 2.6 mm long, 1.3 mm broad, the 3 slender needle-shaped staminoides 0.7 mm long; stigmas 3, broad, 2-cleft at the apex for one-third of their total length; ovary lance-ovoid, 3 mm long, with 3-4 erect ovules; **capsule** sessile, 6-9 mm long, ovoid, long acuminate by reason of the persistence of the base of the style, about 2-3 mm thick, with 1-2 (rarely 3) seeds. **Seed** 4.5 mm long, slenderly cylindric, acuminate at the summit, subglabrous. **Staminate plants** rare, with thin leaves, linear to lance-oblong; staminate spathes in upper axils, peduncular-based, inflated and ellipsoidal or ovoid above, 7 mm long, 4 mm thick, gaping at the summit, with 2 acute teeth; **flowers** carried to surface of water by elongated threadlike base of flower receptacle (10-20 cm long), not detaching at anthesis, with dark-striate elliptic sepals 3.5-5 mm long, 2-2.5 mm broad, concave-convex, and delicate, slender-clawed lanceolate petals about 5 mm long, 0.3-0.7 mm broad. Stamens 9, with almost sessile anthers, the 6 outer ones (3 mm long) falling backward onto the perianth at anthesis, the 3 inner ones (4 mm long) elevated on a common stalk, becoming petaloid after emission of the pollen.

FLOWERING July-September.

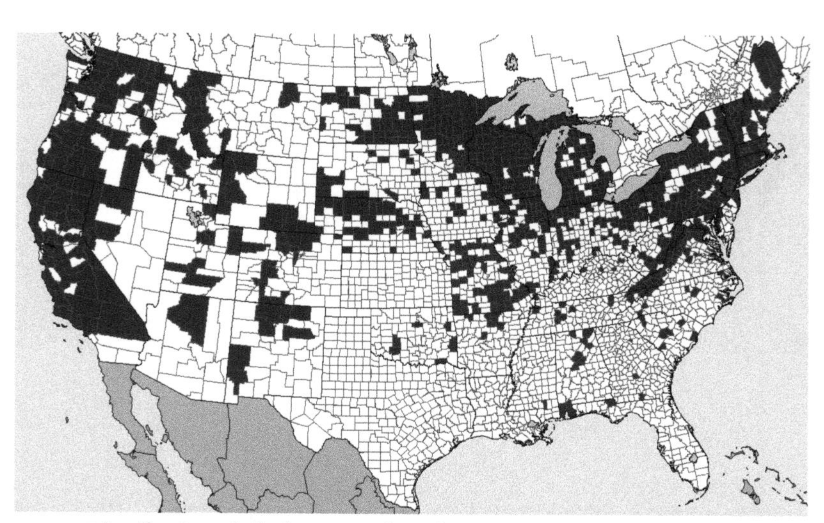

Distribution of **Elodea canadensis,** CANADIAN WATERWEED

SYNONYMS *Anacharis canadensis* Planch.
WHERE FOUND In quiet waters, often calcareous, and in quarries, lakes, and ponds.
ORIGIN Native.

Elodea canadensis, CANADIAN WATERWEED. A, habit, pistillate plant. B, flower and leaf detail, pistillate plant. C, flower and leaf detail, staminate plant, flower fully expanded, sepals and outside anthers fallen. D, flowers, female: male, before full expansion, with all anthers upright. E, capsules. F, seeds.

Najas flexilis (Willd.) Rostk. & Schmidt — SLENDER NAIAD

Annual herb, aquatic, monoecious.

DESCRIPTION Stems fragile, bushy-branched, with crowded nodes or elongate and slender light-green to reddish or olive-green. **Leaves** linear or narrowly lanceolate, triangularly dilated at the base, unlobed, often with slightly inrolled margins and tapering to long, fine, recurving tips, 1–4 cm long, 0.2–2 mm wide, each margin with 20–40 minute 1-celled spinules; each edge of the gradually tapering to rounded herbaceous sheath with 6–13 teeth. **Flowers: staminate** 2.5–3.2 mm long, borne near the tips of the fertile branches, the anthers 1-locular; **pistillate** 1.6–2.5 mm long, borne in the middle and lower axils, the style and 2 commonly spinulose-based stigmas 0.8–2 mm long. **Fruit** slenderly to broadly ellipsoid, 2–3.5 mm long. **Seed** lustrous, obscurely reticulate with 30–40 rows of hexagonal areolae, closely covered by the yellowish to purplish pericarp.

FLOWERING July–October.

WHERE FOUND Shallow fresh to brackish water; lakes and bays.

ORIGIN Native.

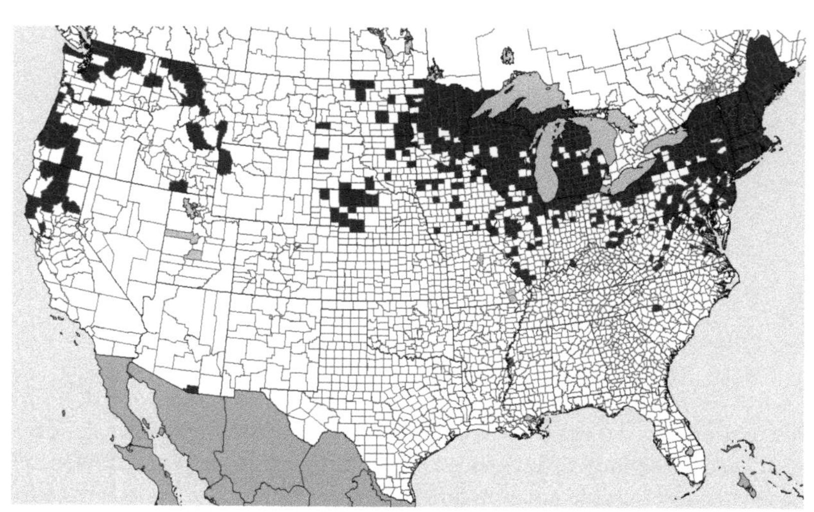

Distribution of **Najas flexilis**, SLENDER NAIAD

Najas flexilis, SLENDER NAIAD. A, habit, as it grows in running water B, leaf detail. C, flower detail, pistillate flower. D, seed, with enlargement showing rows of areolae.

Najas guadelupensis (Spreng.) Magnus — SOUTHERN NAIAD

Annual aquatic herb.

DESCRIPTION Stems very leafy, often much elongate, rather firm, deep-green to purple. **Leaves** linear, flat or slightly crisped, dark-green, olive-green or tawny, 1-2 cm long, 0.4-0.8 mm wide, with smaller ones clustered in the axils, obtuse or acute at the apex, the base ovate, the margins beset at intervals of 0.3-1 mm with 3-10 minute spines. Anther 4-locular. Style and 2 or 3 stigmas 0.1-0.6 mm long. **Fruit** purple-brown at maturity, 1.5-2.5 mm long. **Seed** dull, straw-colored, 2.5-3 mm long, marked with 15-20 longitudinal rows of squarish areolae.

FLOWERING August-October.

WHERE FOUND Fresh and brackish waters.

ORIGIN Native. Throughout most of the United States; south into Central and South America; West Indies.

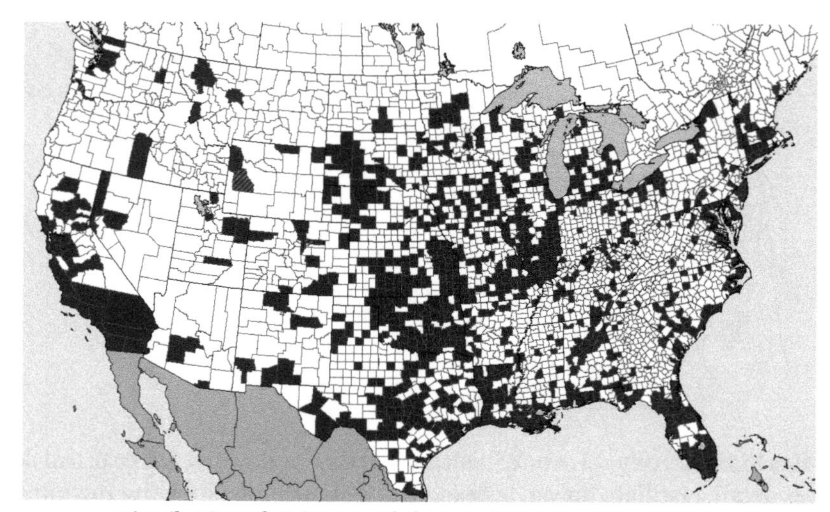

Distribution of **Najas guadelupensis**, SOUTHERN NAIAD

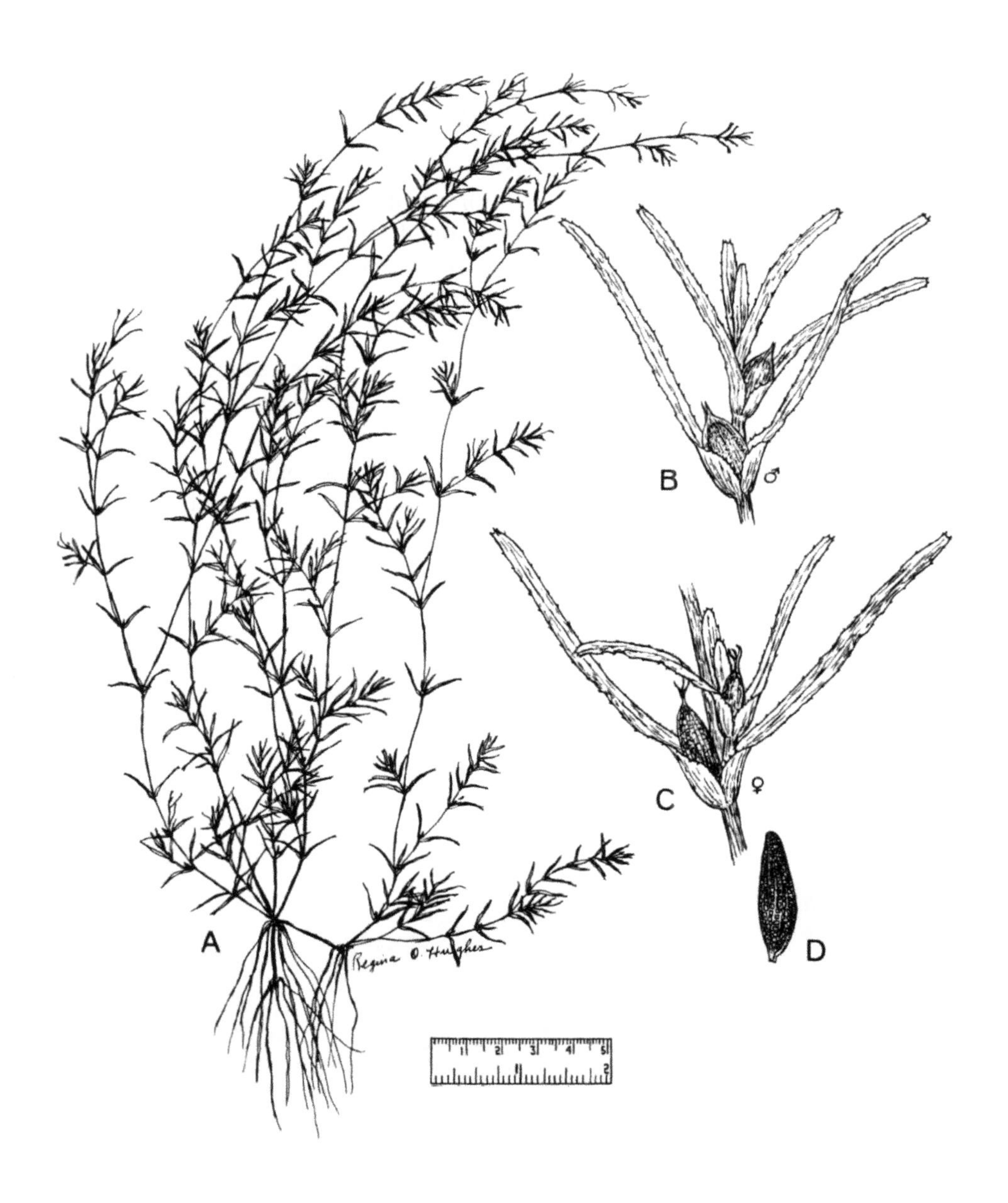

Najas guadelupensis, SOUTHERN NAIAD. A, habit. B, staminate flowers, in spathe; C, pistillate flowers, lower one mature, upper one immature. D, seed, showing square areolae.

Hypericum perforatum L. COMMON ST. JOHN'S-WORT

Perennial herb, reproducing by seeds and rootstocks. Root system branched and extending to considerable depth, shallow short rootstocks extending out a few decimeters from the crown.

DESCRIPTION Stems smooth, branched, erect, somewhat 2-edged, 3–9 dm tall, woody at the base, producing leafy basal outshoots. **Leaves** opposite, elliptic to oblong, covered with small translucent dots. **Cymes** leafy, with numerous flowers. **Flowers** about 2 cm in diameter, 5-petaled, orange-yellow with occasional black dots along the edge of the petals. **Seed pods** rounded, pointed, with 3 parts and many seeds. **Seed** about 1.5 mm long, cylindrical, blackish, shiny with a rough, pitted, resinous surface.

FLOWERING June–September.

WHERE FOUND Meadows, dry pastures, rangelands, and neglected fields and along roadsides.

ORIGIN Naturalized from Europe.

NOTE Difficult to eradicate; not relished by grazing animals; may cause skin irritation and loss of condition in livestock, especially in white-skinned animals.

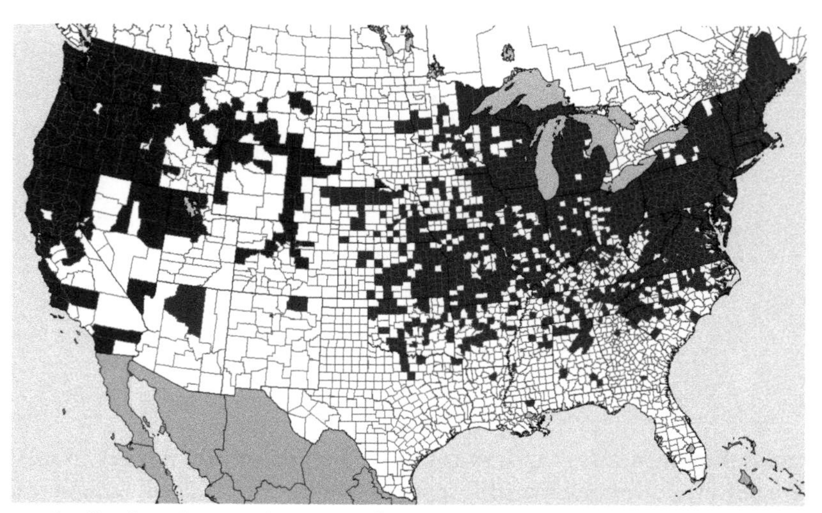

Distribution of **Hypericum perforatum**, COMMON ST. JOHN'S-WORT

Hypericum perforatum, COMMON ST. JOHN'S-WORT. A, habit. B, enlarged leaves. C, flower and bud. D, capsule. E, seeds.

Galeopsis tetrahit L. HEMP-NETTLE

Annual herb.

DESCRIPTION Stems simple or branching, bristly-hairy with long, straight, somewhat reflexed hairs, 3-8 dm tall, swollen at the nodes. **Leaves** ovate to lanceolate, mostly rounded at the base, pointed, 5-10 cm long, on petioles 1-3 cm long, with rounded or pointed teeth, pubescent on both sides. **Flowers** borne in 2-6 dense whorls, in the axils of the upper foliage leaves. Calyx teeth 7.5-11 mm long, enlarged in fruit. Corolla strongly 2-lipped, the tube exceeding the calyx, the upper lip entire, erect, concave, the lower lip 3-lobed, bearing 2 protuberances (nipples) at its base, the petals white or pink suffused with purple, rarely purple completely, commonly with 2 yellow spots, about 2 cm long, the middle lobe rarely longer than broad, usually rounded, rarely notched, the margins flat. **Nutlet** 3-4 mm long, broadly obovate, smooth.

FLOWERING June-September.

WHERE FOUND Waste places, roadsides, and forests; introduced as a weed of gardens and cultivated fields; abundant only northward.

ORIGIN Naturalized from Eurasia.

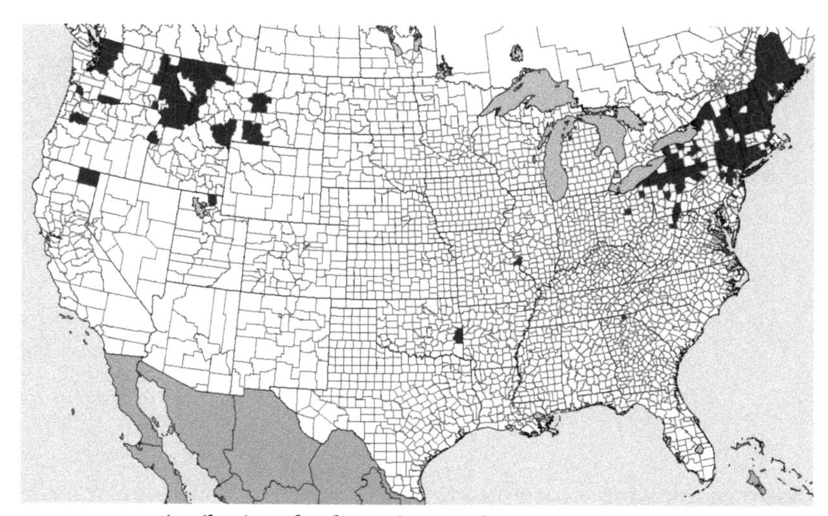

Distribution of **Galeopsis tetrahit**, HEMP-NETTLE

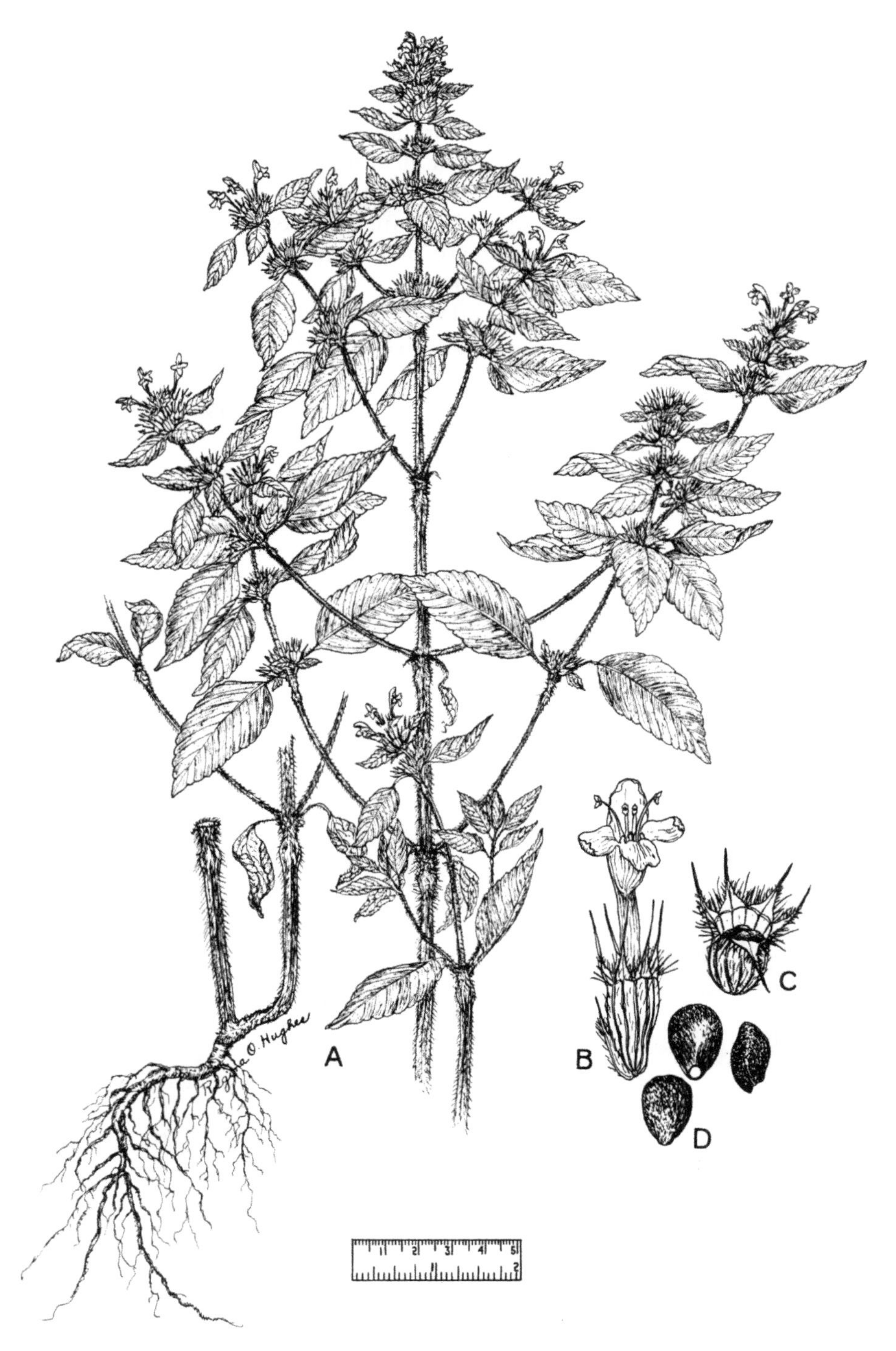

Galeopsis tetrahit, HEMP-NETTLE. A, habit. B, flower. C, fruit, showing nutlets. D, nutlets.

Glechoma hederacea L. GROUND IVY

Perennial, reproducing by seeds and creeping stems. Roots shallow.

DESCRIPTION Stems 3–7 dm long, creeping or trailing, rooting at the nodes, with numerous erect flowering branches, 4-angled, glabrous or nearly so. **Leaves** opposite, palmately veined, petioled, rounded kidney-shaped, round-toothed edges, bright-green, glabrous, 1–3 cm in diameter, with a minty odor. **Flowers** small, in axillary clusters. Calyx tubular, with 5 equal teeth, pubescent, persistent. Corolla bluish-purple to purplish, 2-lipped, the upper lip erect, rather concave, 2-cleft, the lower lip 3-lobed. **Nutlet** ovoid, 1.5–2 mm long, in 4's, flat on 2 sides and round on the 3rd side, granular, dark-brown with a small whitish hilum at the base.

FLOWERING April–June.

WHERE FOUND In lawns, gardens, orchards, damp rich and shaded areas, and waste places.

ORIGIN Introduced from Eurasia.

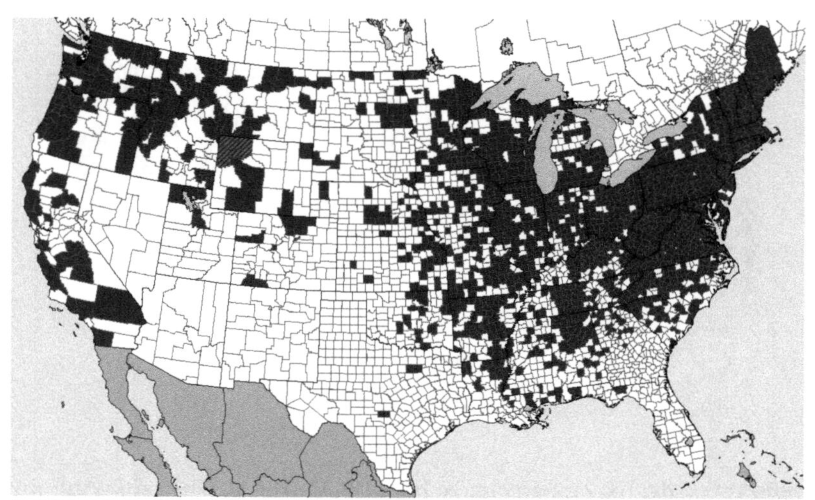

Distribution of **Glechoma hederacea**, GROUND IVY

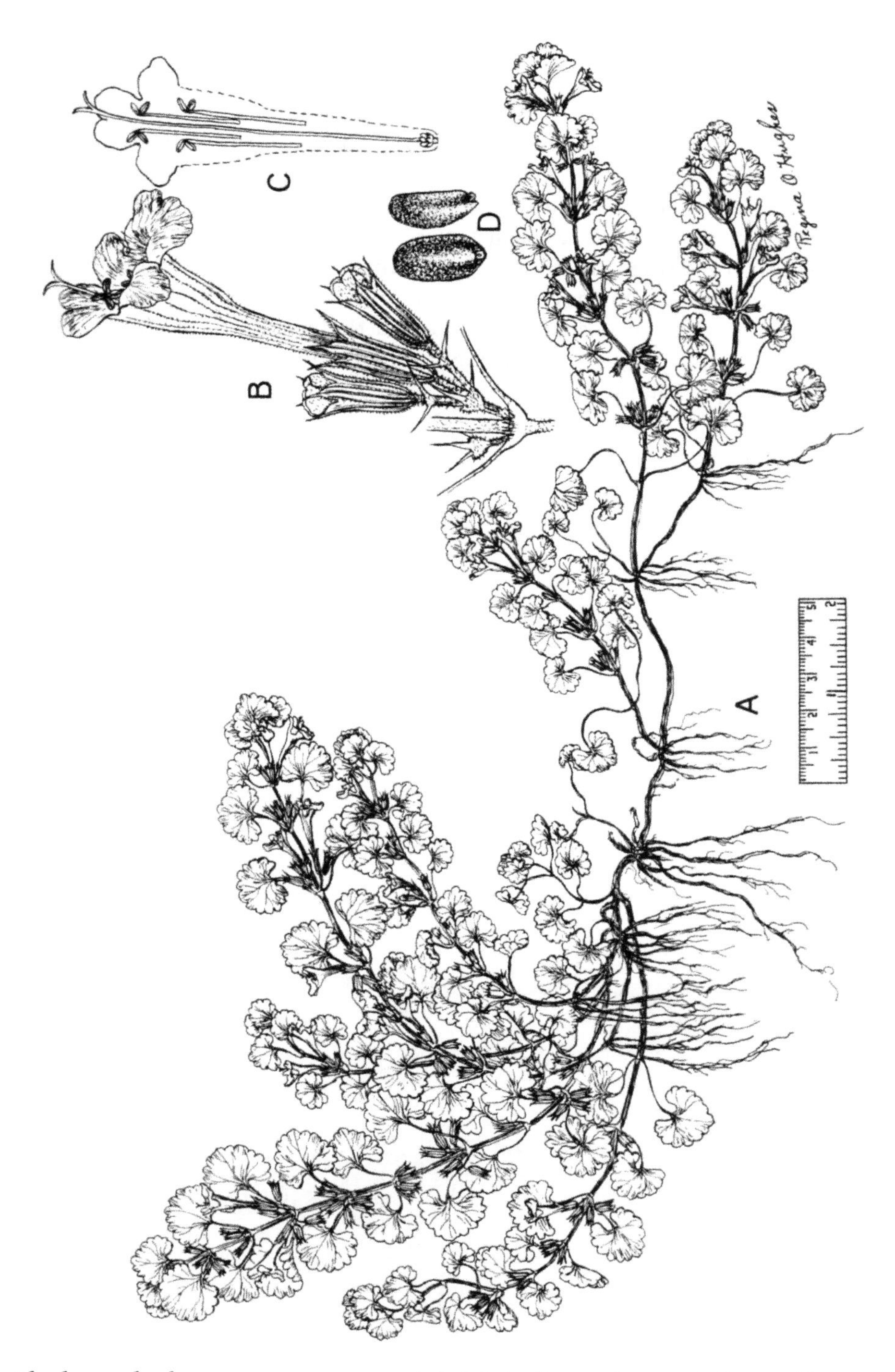

Glechoma hederacea, GROUND IVY. A, habit. B, flower cluster. C, flower diagram, showing the four ascending stamens. D, nutlets.

Lamium amplexicaule L. HENBIT

Biennial or winter-annual herb, reproducing by seeds and rooting systems. Roots fibrous.

DESCRIPTION Stems decumbent with numerous ascending branches, frequently rooting at the lower nodes, 10–40 cm tall, slender, nearly smooth, 4-angled. **Leaves** opposite, circular, with palmate venation, hairy, with rounded teeth, the lower leaves petioled, doubly crenate-lobed, 1–2 cm long, the upper leaves sessile and clasping the stem. **Flowers** in whorls in the axils of the upper leaves. Corolla tubular but 2-lipped, about 1–1.5 cm long, pinkish to purple, surrounded at the base by the calyx (5–6.5 mm long) with 5 sharp teeth, spotted. **Nutlets** borne 4 in a pod, sharply 3-angled, 1.5–2.4 mm long, obovate-oblong, the apex blunt, grayish-brown speckled with silvery-gray granules.

FLOWERING April–June; September.

WHERE FOUND Waste places, cultivated fields, and gardens, especially in rich soils.

ORIGIN Native of Eurasia and Africa; introduced from Eurasia.

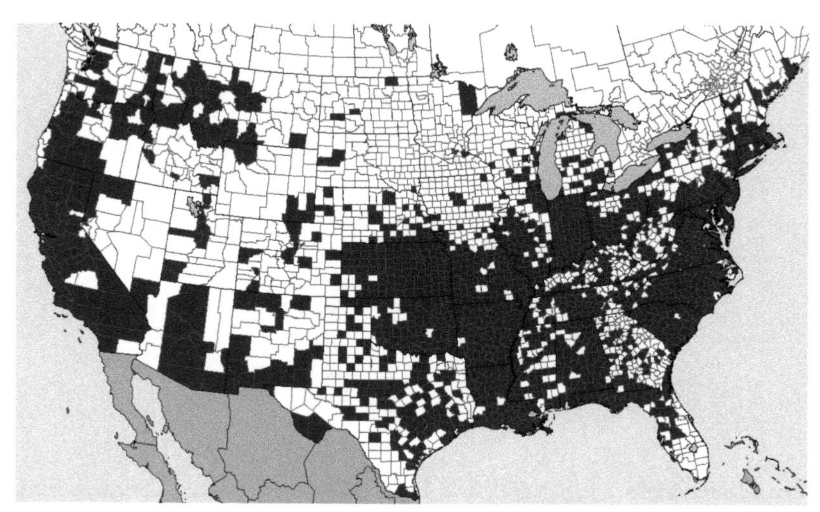

Distribution of **Lamium amplexicaule**, HENBIT

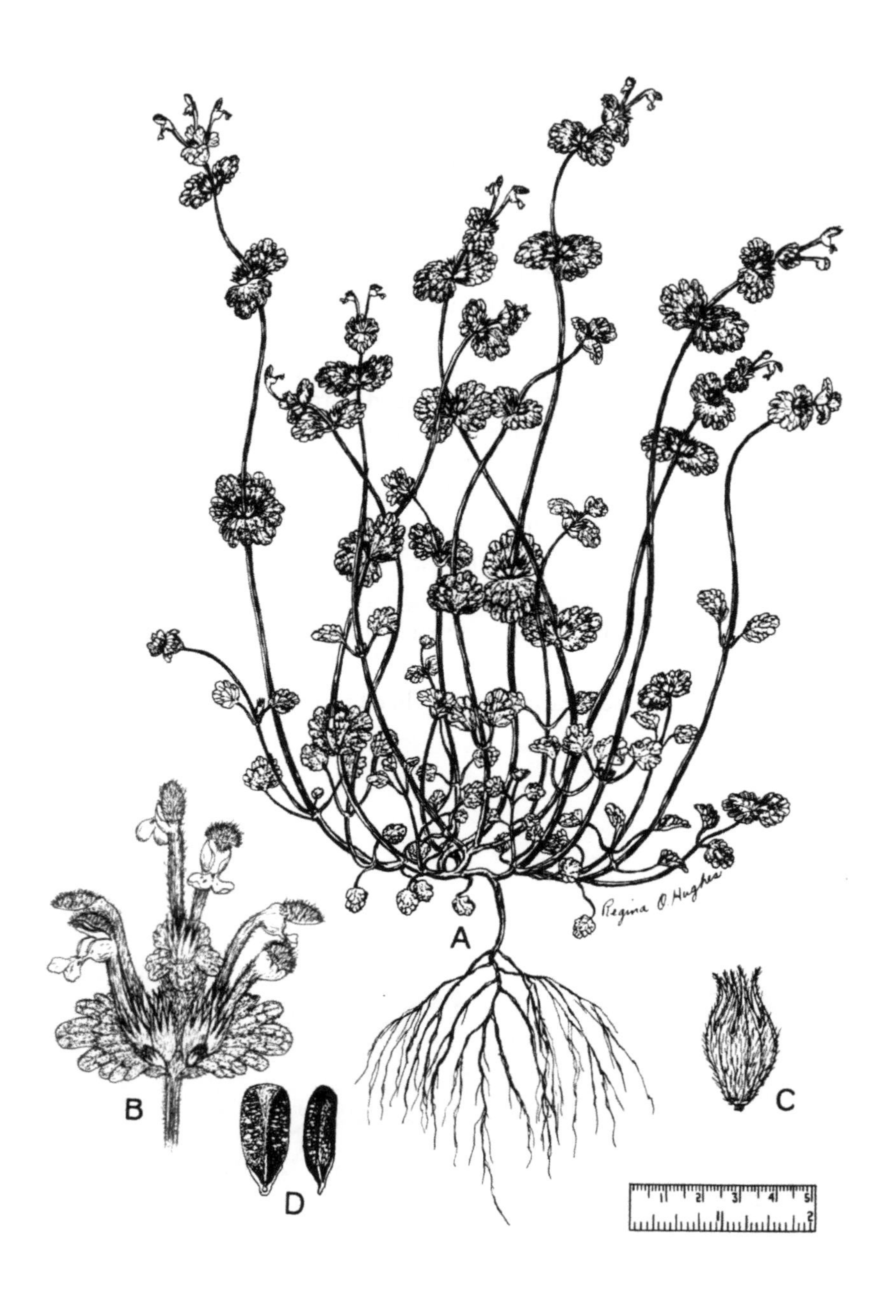

Lamium amplexicaule, HENBIT. A, habit. B, flower clusters, showing very short upper internodes. C, calyx surrounding nutlets. D, nutlets.

Prunella vulgaris L. SELFHEAL

Perennial herb, reproducing by seeds and short runners that root freely at the nodes.

DESCRIPTION Stems erect, 0.5–6 dm tall, ascending or prostrate, mostly tufted, simple or branched, 4-angled, pubescent becoming nearly glabrous with age. **Leaves** opposite, pinnately veined, with moderately long petioles, margins entire or irregularly dentate, ovate-oblong, pubescent or glabrous, 2.5–10 cm long. **Flowers** sessile, in a close thick spike, 3 in the axils of each rounded membranaceous bract, the bracts mostly bristly-ciliate. Calyx irregularly 10-nerved, 7–10 mm long, green or purple, 2-lipped, the lips longer than the tube, the upper lip broad, shallowly 3-toothed, the lower lip deeply cleft, with two narrow segments, the teeth spinulose-tipped. Corolla blue or purple to white or pink, 1–2 cm long, 2-lipped, the tube equaling or surpassing the calyx, the lips short, the upper lip hood-shaped and entire or nearly so, the lower lip shorter and 3-lobed. Stamens 4, the upper pair shorter than the lower pair, exserted. **Nutlets** 4, in the persistent calyx, obovate, about 1.5 mm long, slightly flattened on 2 sides, brown with dark vertical lines, the base tapering to a pointed white outgrowth, slightly roughened and glossy. A variable species with several named varieties.

FLOWERING May–September.

WHERE FOUND In lawns, fields, waste places, pastures, grasslands, and roadsides. When repeatedly mowed, trampled, or grazed, the plants become densely matted, depressed, and small-leaved.

ORIGIN Native, but some populations naturalized from Eurasia (ssp. *vulgaris*).

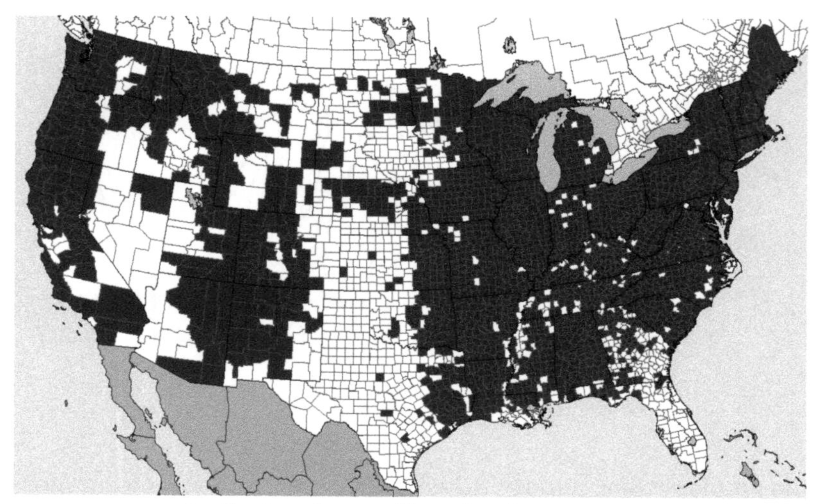

Distribution of **Prunella vulgaris**, SELFHEAL

Prunella vulgaris, SELFHEAL. A, habit. B, flower. C, persistent calyx. D, nutlets.

Abutilon theophrasti Medic. VELVETLEAF

Annual herb, 6–12 dm high.

DESCRIPTION Stem smooth, with short velvety hairs. **Leaves** alternate, round to cordate at the base, taper-pointed at the apex, velvety, hairy-surfaced. **Flowers** with peduncles shorter than the petioles, corolla of 5 yellow petals, 2 cm wide. **Seed pod** cup-shaped, 2.5 cm in diameter, with a ring of prickles about the upper edge, the carpels 12–15, hairy, beaked, 5–15 seeded. **Seed** grayish-brown, flattened, notched, 3 mm long.

FLOWERING August–September.

WHERE FOUND Waste places, vacant lots, gardens, and cultivated fields, especially corn and soybean fields, and along fence rows.

ORIGIN Naturalized from India.

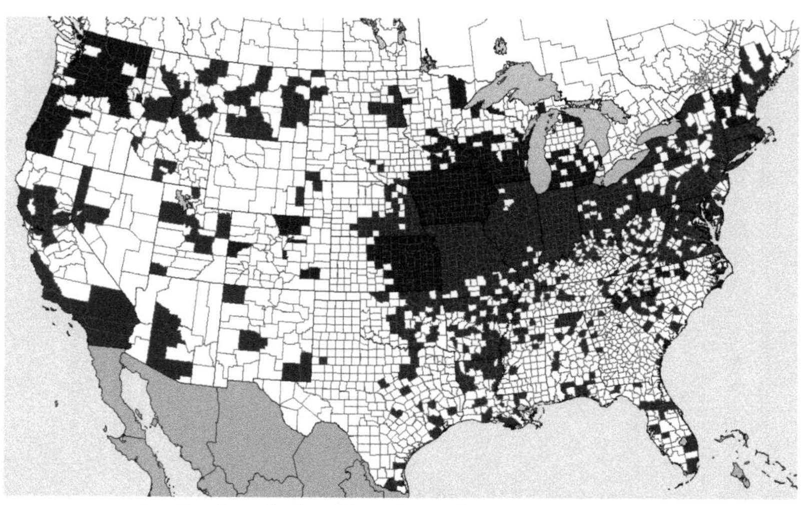

Distribution of **Abutilon theophrasti**, VELVETLEAF

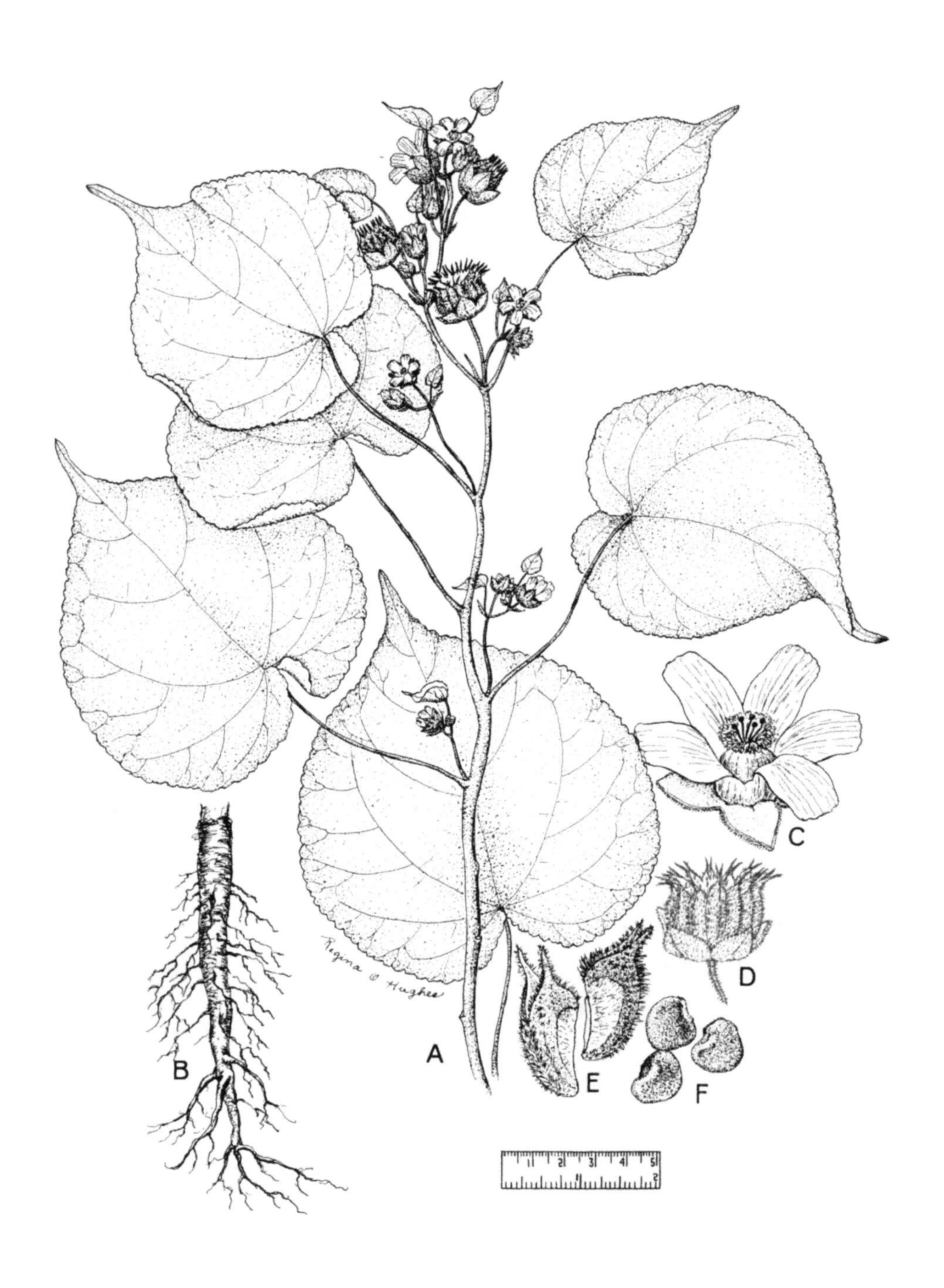

Abutilon theophrasti, VELVETLEAF. A, habit. B, rootstock. C, flower. D, capsule. E, carpels. F, seeds.

Malva neglecta Wallr. COMMON MALLOW

Annual or biennial herb, reproducing by seeds, with a short straight taproot.

DESCRIPTION Stems 10–30 cm long, procumbent or branching at the base, nearly erect, or spreading on the ground with the tips generally turned up, more or less pubescent, especially when young, often with star-shaped pubescence. **Leaves** alternate, round-heart-shaped to round kidney-shaped, 2–6 cm wide, simple, on very long slender petioles, toothed with 5–9 shallow rounded lobes or lobeless, more or less pubescent on both sides. **Flowers** small, with 5 whitish or pale-lilac petals, borne singly or in clusters in the axils of many leaves. Petals about 1 cm long, twice as long as the calyx. Bractlets linear to linear-lanceolate. Carpels round-margined, smooth, not reticulated on the back. **Seed pod** a flattened disk, when ripe breaking up into 10–20 small, hairy, 1-seeded sections. **Seed** nearly round, flattened, reddish-brown, about 1 mm in diameter, notched.

FLOWERING April–October.

WHERE FOUND Cultivated fields, waste places, gardens, lawns, roadsides, and barnyards.

ORIGIN Naturalized from Europe. Thoughout the United States.

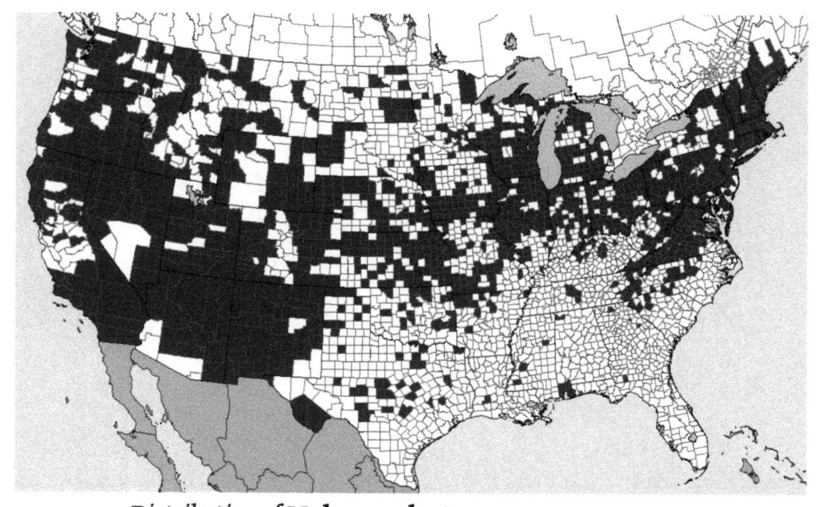

Distribution of **Malva neglecta**, COMMON MALLOW

Malva neglecta, COMMON MALLOW. A, Habit. B, enlarged branchlet. C, flower diagram. D, carpel. E, seeds.

Sida spinosa L. PRICKLY SIDA

Annual herb, reproducing by seeds. Taproot slender, branching, rather long.

DESCRIPTION Stems erect, 0.2–1 m tall, softly hairy, bearing 2–3 short, blunt, spiny projections below each node, much-branched. **Leaves** alternate, simple, ovate-lanceolate or oblong, 2–4 cm long, toothed, rather long-petioled. Peduncles axillary, 1-flowered, shorter than the petiole. **Flowers** solitary or clustered in the axils of the leaves, on pedicels 2–12 mm long. Calyx thin, star-shaped. Petals 5, pale-yellow, 4–6 mm long. Carpels 5, pubescent at summit, combined into an ovoid fruit. **Seed pod** splitting at the top into 5 1-seeded sections, each with 2 sharp, spreading spines at the top. **Seed** 1–2 mm long, 3-angled, egg-shaped, dull, dark reddish-brown.

FLOWERING June–October.

WHERE FOUND Waste places, cultivated fields, open ground, gardens, and pastures.

ORIGIN Naturalized from the Tropics.

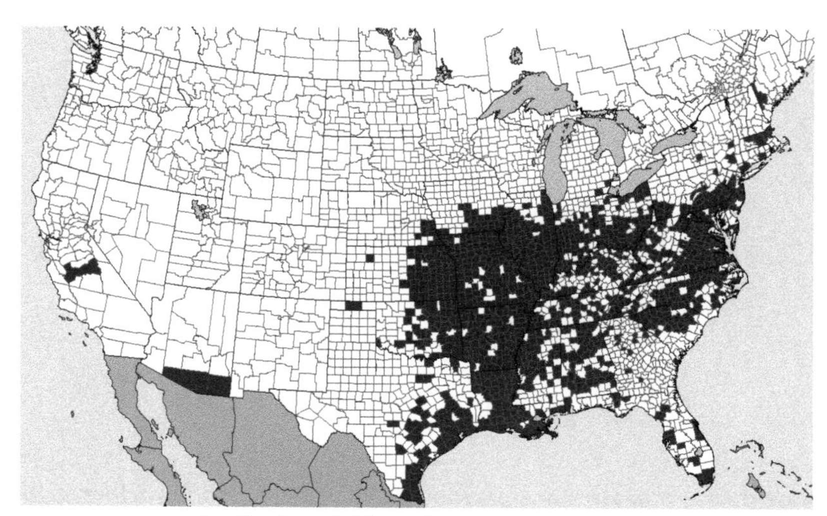

Distribution of **Sida spinosa**, PRICKLY SIDA

Sida spinosa, PRICKLY SIDA. A, habit. B, upper part of branch, showing tubercles at base of leaves C, flower, with oblique petals and united stamens. D, carpels with 5 styles. E, capsule, enclosed in the persistent calyx. F, carpels. G, seeds.

Mollugo verticillata L. CARPETWEED

Annual herb, reproducing by seeds. Taproot little-branched.

DESCRIPTION Stems 5–30 cm long, green, smooth, branched at the base, prostrate along the ground in all directions from the root, making large flat mats on the soil surface. **Leaves** whorled, 5 or 6 (occasionally 3–8) at each joint of the stem, smooth, 1–2.5 cm long, spathulate to linear-oblanceolate, obtuse to acute at apex, narrowly triangular below to a short petiole. Stipules obsolete. **Flowers** small, 2–5 from each node, on slender pedicels, 3–14 mm long. Sepals 5, 1.5–2.5 mm long, white inside, oblong. **Capsule** 3-locular, 3-valved, the partitions breaking away from the many-seeded axis. **Seed** small, orange-red, somewhat kidney-shaped, ridged on the back and sides.

FLOWERING June–November.

WHERE FOUND A late-starting but quick-growing summer annual weed, quickly covering any fertile bare soil; gardens, tilled crops, lawns, waste places, and sandy riverbanks.

ORIGIN Native of tropical America.

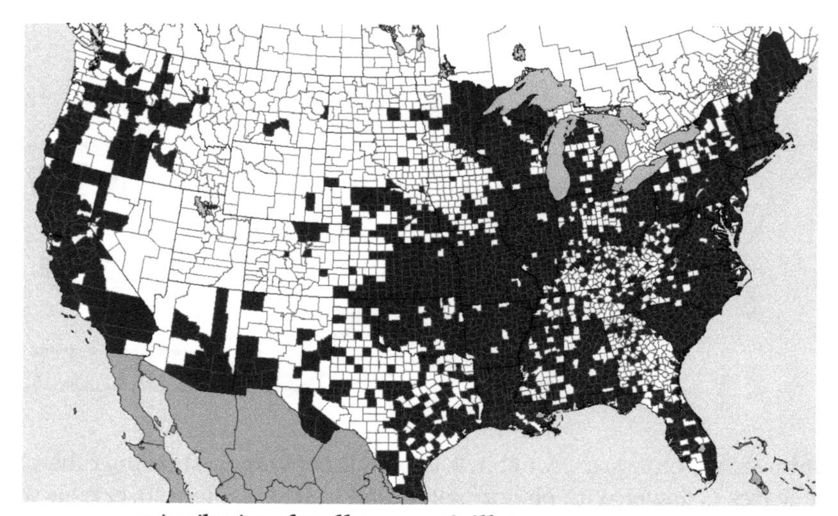

Distribution of **Mollugo verticillata**, CARPETWEED

Mollugo verticillata, CARPETWEED. A, habit, as seen from above. B. flowers and fruits. C, seeds. D, taproot.

Nelumbo lutea (Willd.) Pers. AMERICAN LOTUS

Perennial, aquatic herb, with thick rhizomes, or tubers that are starch-like and edible. Rhizome underwater in the mud.

DESCRIPTION Leaves circular in outline, centrally shield-shaped, 3–7 dm broad, with the center depressed or cupped, normally raised above the water surface. **Flowers** pale-yellow, solitary, 15–15 cm wide. Sepals and petals numerous, usually 20 or more, grading into each other, the outermost in the bud green and sepal-like. Stamens numerous, closely surrounding the pistils, the anthers tipped with a slender hooked appendage. **Fruiting receptacle** about 1 dm broad, prolonged, inversely conical, the numerous 1-celled ovaries sunk in small pits on its abrupt-ending summit. **Fruit** large, indehiscent, nut-like, each separately embedded in the receptacle, about 1 cm in diameter.

FLOWERING July–September.

WHERE FOUND Quiet water along ponds, lakes, estuaries, and sluggish rivers; usually local in distribution; along Illinois River most abundant, forming colonies many acres in extent. Tubers and seed edible.

ORIGIN Native.

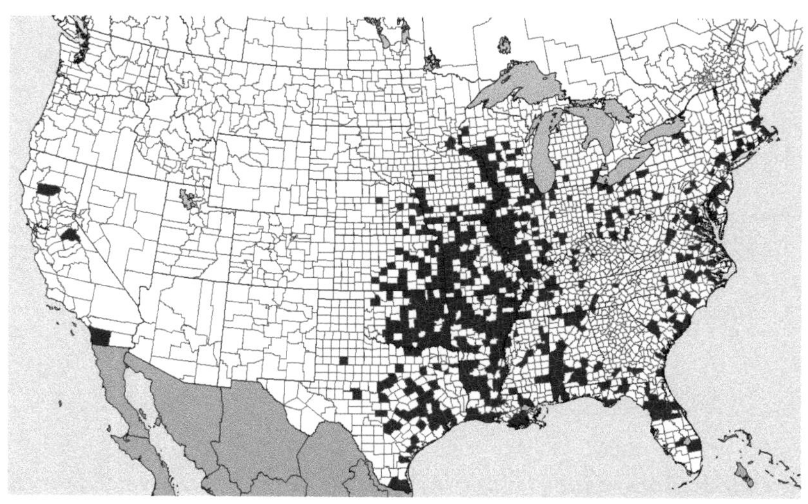

Distribution of **Nelumbo lutea**, AMERICAN LOTUS

Nelumbo lutea, AMERICAN LOTUS. A, habit. B, pod, vertical section. C, seed. D, stamens.

Nuphar advena (Ait.) Ait. f. SPATTERDOCK

Perennial from rhizomes, aquatic herb. Rhizome creeping in the mud, cylindrical.

DESCRIPTION Leaves erect (floating or submersed leaves uncommon and only in very deep water), the stout petiole nearly circular in cross section, rising above the water level, blades erect, ovate to rounded-oblong, 1-4 dm long, with broad V- or U-shaped cleft between the two lobes 4-15 cm wide between the nearly triangular basal lobes. **Flowers** usually on erect peduncles, raised above the water, 3-4 cm high, when spread open 6-10 cm broad, outer sepals green outside, inner sepals yellow with yellowish tips, only rarely tinged with red. Stamens in 5-8 circles. Disk pale-green to yellowish, with 9-23 stigmatic rays. **Fruit** erect or drooping into the mud, furrowed, green, 2-5 cm high, with very stout and thick neck, the mature disk 1.3-2.5 cm across.

FLOWERING May-October.

SYNONYMS *Nuphar lutea* (auct. non L.) Small.

WHERE FOUND Tidal waters, pond margins, swamps.

ORIGIN Native.

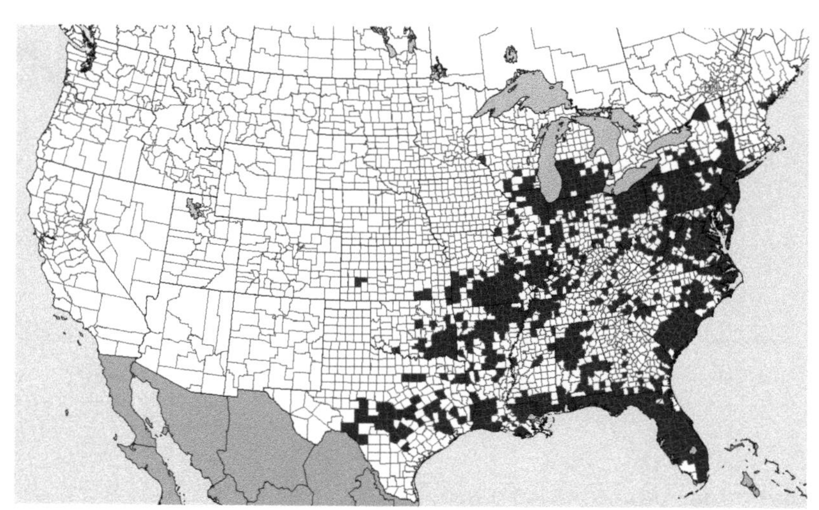

Distribution of **Nuphar advena**, SPATTERDOCK

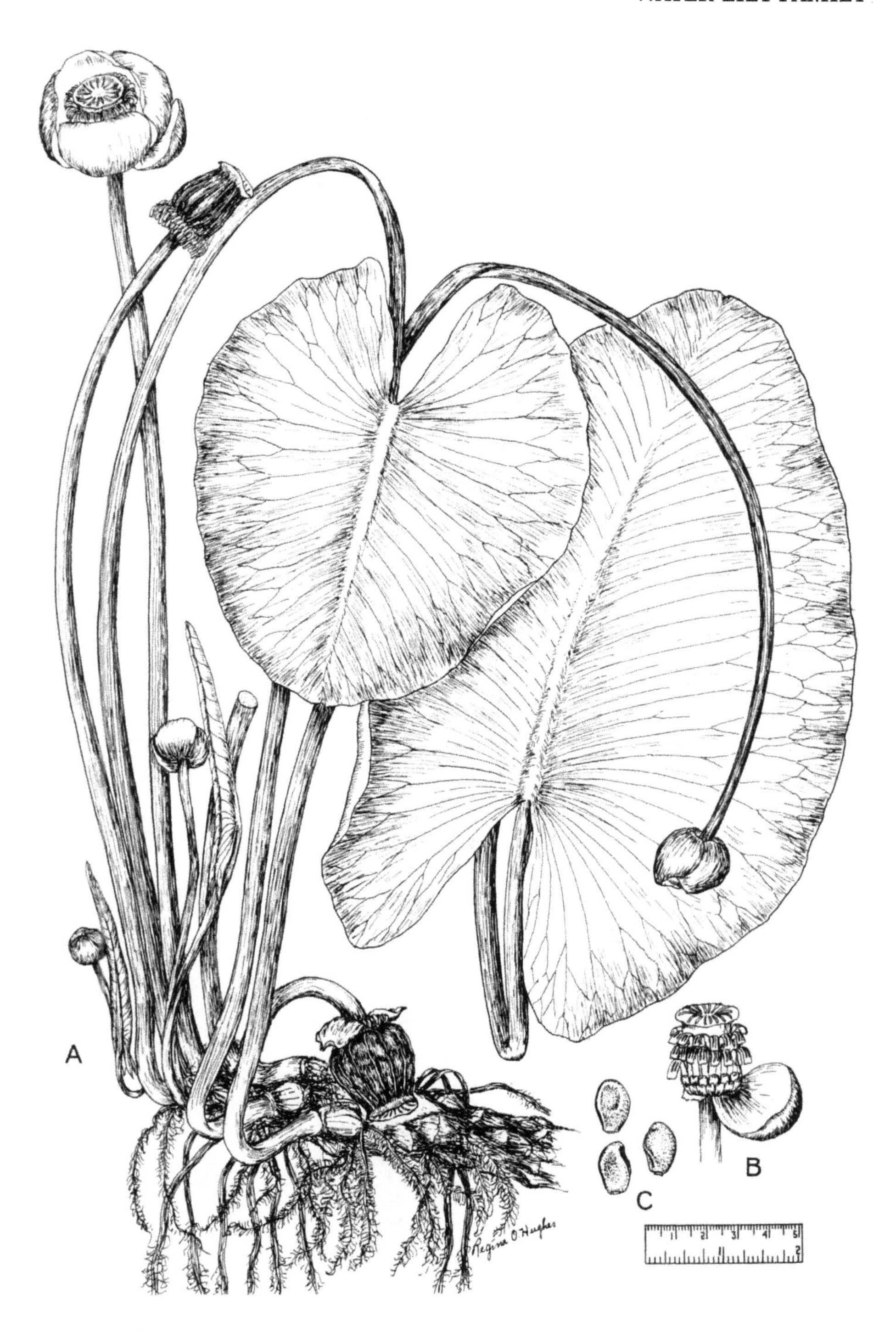

Nuphar advena, SPATTERDOCK. A, habit. B, petal and stamen detail. C, seeds.

Nymphaea odorata Ait. FRAGRANT WATERLILY

Perennial, aquatic, with horizontal, elongate, rhizome without tubers. Rhizome forking, the branches not strongly constricted at the base nor readily detached.

DESCRIPTION Leaves arising along the rhizome, the non-striped petioles purplish-green to red, the blades nearly circular, floating, depressed in the mud or ascending, flat, green above, usually purple to purplish-red beneath, 0.5-2.5 dm in diameter, with a narrow cleft between the two lobes. **Flowers** very fragrant, expanded on 3 or 4 days from early morning until about noon, 5-15 cm broad. Sepals often purplish on the back, ovate to ovate-lanceolate, 2.8-8 cm long, 1-2.5 cm broad, with rounded tips. Petals 17-32, 1-2.2 cm broad, white to roseate, gradually tapering above to ovate rounded tips. Stamens 36-100, the inner filaments narrower than their anthers. **Fruit** depressed-globular, usually covered by the bases of the decayed petals, maturing under water. **Seed** enveloped in a sac-like aril, ellipsoid, 1.5-2.3 mm long, exceeded by the aril. A variable species with some named varieties and forms.

FLOWERING June-September.

WHERE FOUND Quiet waters, ponds, bogs, edges of lakes.

ORIGIN Native.

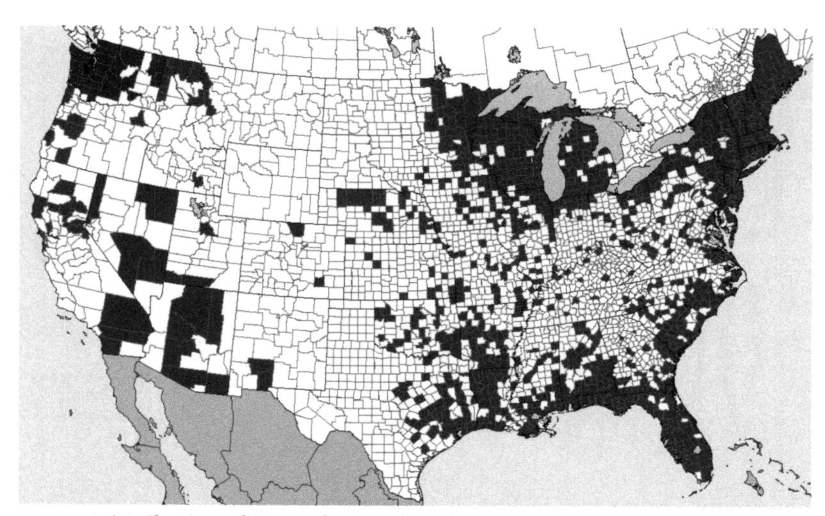

Distribution of **Nymphaea odorata**, FRAGRANT WATERLILY

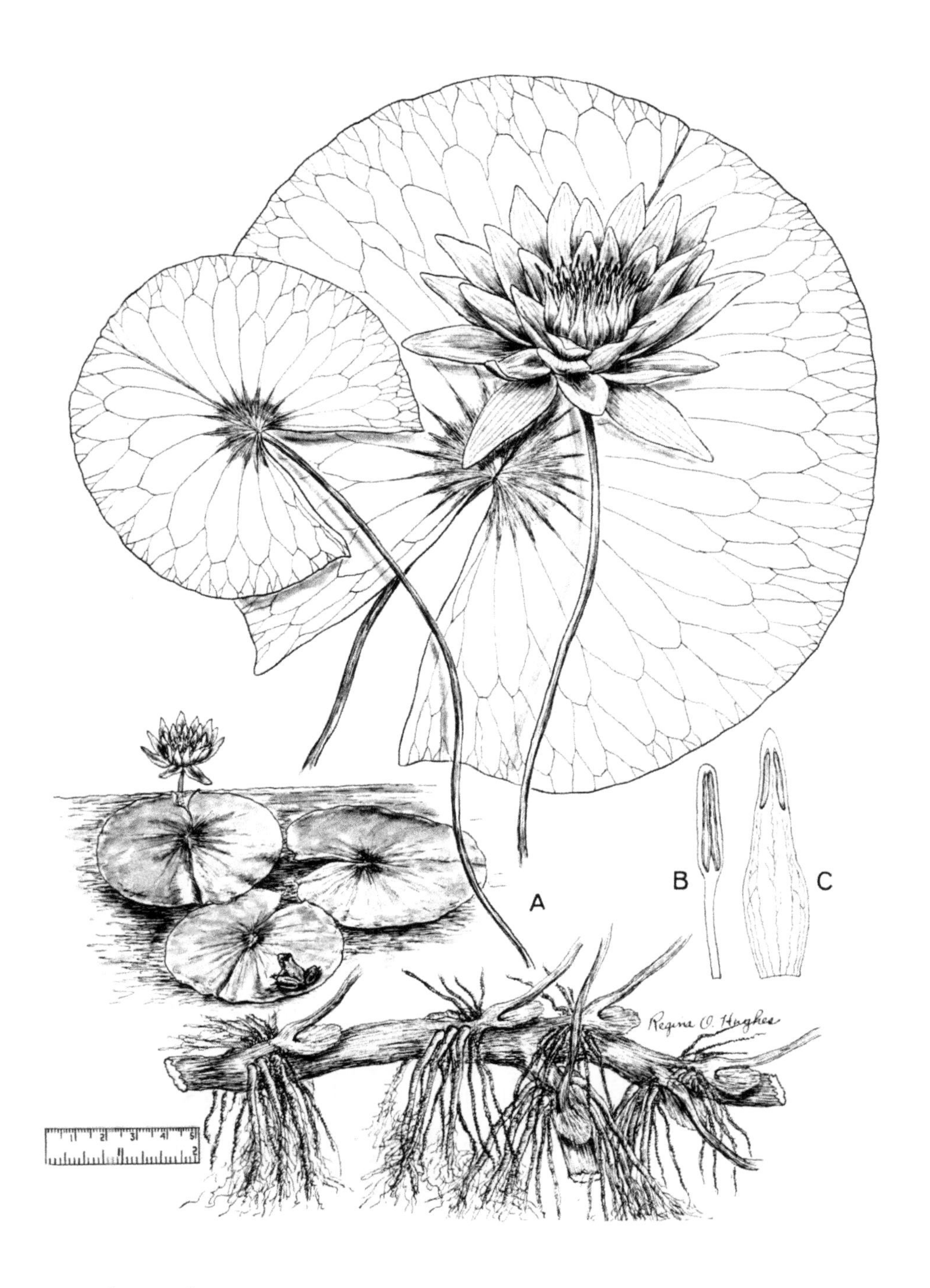

Nymphaea odorata, FRAGRANT WATERLILY. A, habit. B, inner stamen. C, outer stamen.

Ludwigia decurrens Walt. WINGED WATER-PRIMROSE

Annual herb, with clustered roots, reproducing by seeds.

DESCRIPTION Stems erect, nearly simple to freely ascending branched, up to 2 m tall, glabrous, 4-angled, and sometimes 4-winged by the decurrent leaf bases. **Leaves** alternate, lanceolate or linear-lanceolate, nearly sessile, 5-18 cm long, membranaceous, the leaves on the threadlike branches greatly reduced in size. **Flowers** from many axils, on pedicels 1-10 mm long. Hypanthium at anthesis cone-shaped, 4-angled, 1-2 cm long. Calyx segments and petals 6-12 mm long. **Capsule** slenderly pyramidal, 4-angled or narrowly 4-winged, 1-2 cm long. **Seed** ellipsoid, free from the endocarp, 0.3-0.4 mm long, in several rows in each locule.

FLOWERING May-October.

SYNONYMS *Jussiaea decurrens* (Walt.) DC.

WHERE FOUND Swamps, wet wastelands, ditches, and wet fields.

ORIGIN Native.

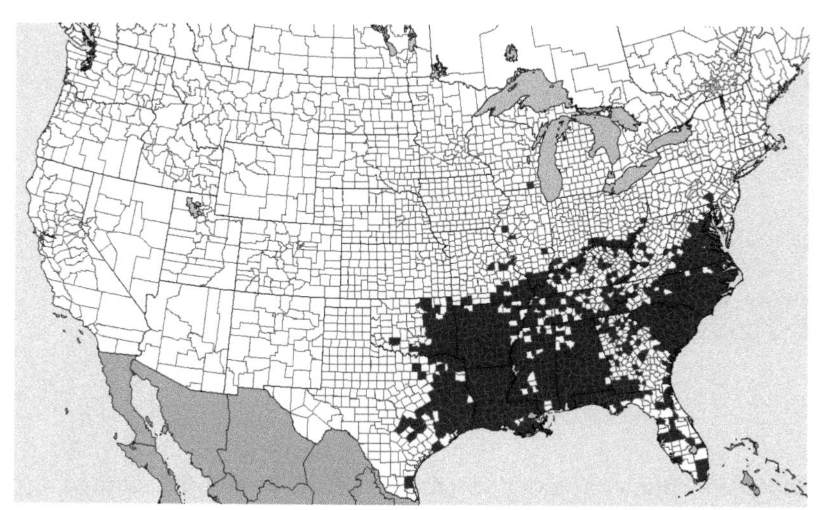

Distribution of **Ludwigia decurrens**, WINGED WATER-PRIMROSE

Ludwigia decurrens, WINGED WATER-PRIMROSE. A, habit. B, flower. C, capsule. D, seeds.

Ludwigia peploides (Kunth) Raven CREEPING WATER-PRIMROSE

Perennial herb, reproducing by seeds and spreading by stolons.

DESCRIPTION Stems prostrate or floating or ascending at the tips, rooting at the nodes, glabrous or very sparsely finely hairy. **Leaves** glabrous, or almost so, 3–9 cm long, lanceolate to oblanceolate or obovate, obtuse or acute, narrowed at the base into a petiole 2–5 cm long. Hypanthium at time of flowering cylindric, 7–12 mm long, its pedicel 3–8 cm long, smooth. Petals 10–15 mm long. Calyx segments glabrous or slightly pubescent, about 1 cm long. **Capsule** subcylindric, 3–5 cm long, its pedicel 3–8 cm long. **Seed** in one series in each locule, longer than thick with a blunt end, the true seed completely covered.

FLOWERING June–October.

SYNONYMS *Jussiaea repens* L. var. *glabrescens* Kuntze

WHERE FOUND Swamps, ponds, wet areas, muddy places, and ditches.

ORIGIN Native.

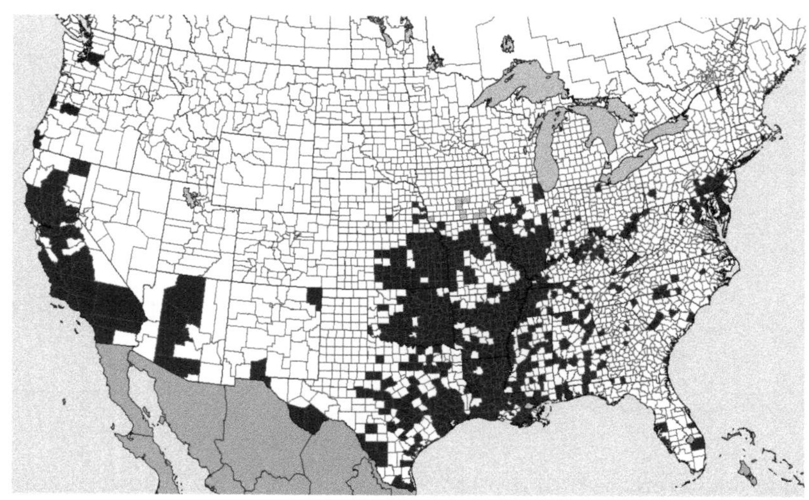

Distribution of **Ludwigia peploides**, CREEPING WATER-PRIMROSE

Ludwigia peploides, CREEPING WATER-PRIMROSE. A, habit, showing bladder on root. B, enlarged flower. C, capsule. D, seeds with endocarp. E, seeds.

Orobanche minor Sm. CLOVER BROOMRAPE

Parasitic herb, reproducing by seeds.

DESCRIPTION Stems pubescent, pale yellowish-brown, 1-4.5 dm tall. **Leaves** scale-like, the scales ovate to lanceolate, 6-20 mm long, acute, sessile, without chlorophyll. **Spike** loosely flowered, interrupted below, continuous above, 1-2 dm long, each purple-tinged flower in the axil of a basal bract. Bracts lanceolate, as long as the flowers or longer. Calyx pubescent, cleft before and behind almost or quite to the base, the lateral lobes often 2-cleft, lanceolate awl-shaped. Corolla glandular-pubescent especially along the back, irregular, the tube slightly curved, yellowish, the lips bluish with 2 rounded lobes, the upper lip erect or incurved, the lower lip spreading. **Capsule** oblong, less than 1 cm long, 2-valved.

FLOWERING April-July.

WHERE FOUND Parasitic on roots of clover and tobacco.

ORIGIN Adventive and naturalized from Europe.

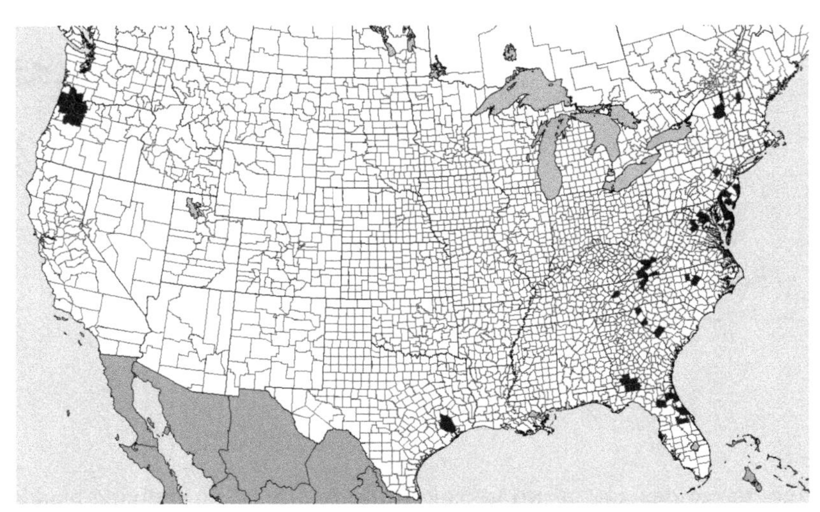

Distribution of **Orobanche minor**, CLOVER BROOMRAPE

Orobanche minor, CLOVER BROOMRAPE. A, habit. B, enlarged flower spike. C, flower diagram. D, capsules. E, seeds.

Striga asiatica (L.) Kuntze — WITCHWEED

Annual herb, reproducing by seeds, parasitizing the roots of many crops.

DESCRIPTION **Root** watery, white, round in cross section, no root hairs, the underground portion of the stem with a series of scales placed similarly to the leaves aboveground. **Stems** above the third node on the stem square, 20–30 cm high, rarely up to 45 cm, plants aboveground bright-green; multiple branches both near the ground and higher on the plant. **Leaves** nearly opposite, linear-lanceolate, alternating at about 90 degrees in pairs, slightly hairy, the upper and lower surfaces alike, the aerial leaf flattened with stomata more numerous on the lower side. **Flowers** small, 6–9 mm wide, usually brick-red or scarlet, but may vary to red, yellow, or almost white, in the axils of the leaves, more flowers coming into bud, blooming and setting capsules throughout the season. **Capsules** 5-sided, each side terminating in a characteristic spur; each capsule containing about 1,350 tiny brown seeds, about 0.2 mm in length, about three-fifths as wide as long (up to 500,000 seeds per plant), deeply reticulated, striate. **Seeds** may lie dormant 15–20 years; seed germination after 15–18 months' rest period, flowering and seed production continuous from July until frost (70- to 90-day cycle).

FLOWERING June–until frost.

SYNONYMS *Striga lutea* Lour.

WHERE FOUND Parasitizes the roots of approx. 60 species of grasses in many genera (including corn, sugarcane, sorghum, wheat, oats, barley, rice, and crabgrass), some sedges, and many broadleaf plants.

ORIGIN Pest in South Africa; native of India. Introduced and spreading in North Carolina and South Carolina.

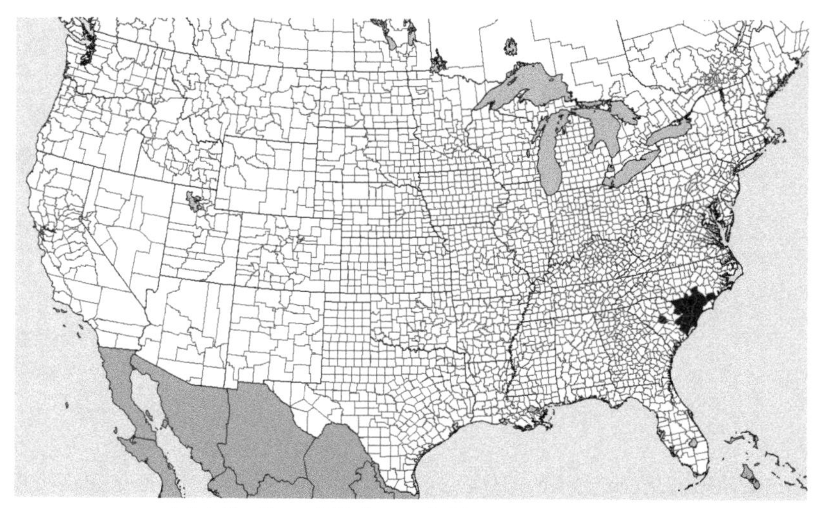

Distribution of **Striga asiatica**, WITCHWEED

Striga asiatica, WITCHWEED. A, habit. B, flowers. B, flower diagram, showing stamens. C, persistent calyx. D, capsule; d, capsule, opened, showing great number of seeds. E, seeds; e, seed, greatly enlarged, showing coarse reticulations over striae. F, roots; f, haustoria.

Oxalis stricta L. COMMON YELLOW WOOD-SORREL

Perennial herb, reproducing by seeds. Taproots without subterranean stolons.

DESCRIPTION Stems erect but weak, becoming decumbent, gray-green, 10–50 cm tall, branching at the base, bushy, pubescent with whitish hairs, sometimes rooting at the joints. **Leaves** alternate, on long petioles, divided into 3 heart-shaped leaflets, 1–2 cm broad, sour-tasting. Stipules oblong, firm, pale. **Flowers** 7–11 mm long, with 5 yellow, rarely green petals, sometimes red at the base. Peduncles pubescent, mostly equaling or over-topping the leaves, umbellately 1- to 4-flowered. Pedicels 1–2.5 cm long, in fruit. Sepals 3.5–7 mm long. **Capsules** 1.2–2.5 cm long, slender, cylindrical, 5-ridged, pointed, hoary pubescent. **Seed** small, 1–1.3 mm long, flat, brown, thrown from the bursting pods.

FLOWERING May–October.

WHERE FOUND Dry open soil, in pastures, lawns, and waste places; a troublesome weed in lawns and gardens.

ORIGIN Native, but considered adventive in the western states.

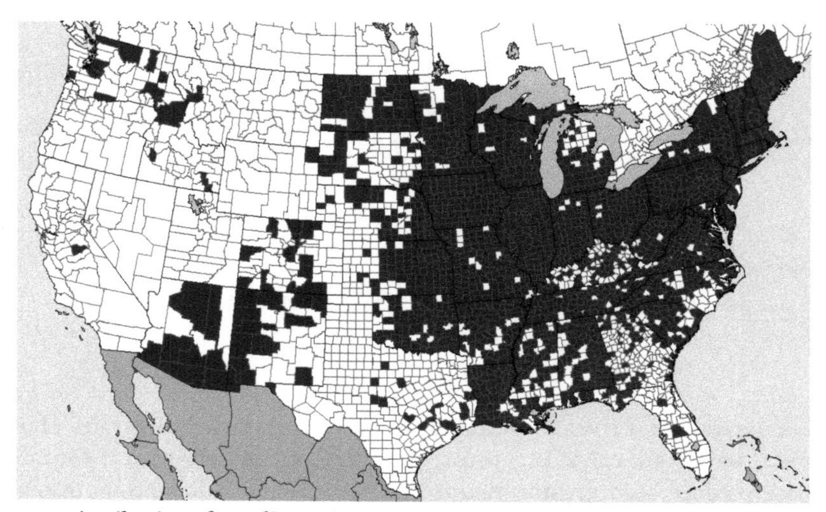

Distribution of **Oxalis stricta**, COMMON YELLOW WOOD-SORREL

Oxalis stricta, COMMON YELLOW WOOD-SORREL. A, habit. B, leaves enlarged. C, flower diagrams. D, capsule. E, seeds.

Phytolacca americana L. POKEWEED

Perennial herb, reproducing by seeds or from a very large (up to 15 cm across) poisonous taproot.

DESCRIPTION Stems erect, stout, glabrous, 2-3 m tall, branches above, often reddish, single or several from the large fleshy, white root. **Leaves** alternate, simple, entire, oblong-lanceolate to egg-shaped, 1-3 dm long, with long petioles 1-5 cm long, glabrous. **Inflorescences** long, narrow, in terminal peduncled racemes, 1-2 dm long, becoming lateral and opposite to the leaves. **Flowers** small, about 6 mm wide. Sepals 5, greenish-white to white, or suffused with pink, petal-like and rounded. Stamens 10. Pistil 1, of about 10 united carpels, each with 1 vertical seed. **Fruit** a nearly globose berry, dark-purple, with a crimson juice, depressed, 10-seeded, about 1 cm in diameter. **Seed** lens-shaped, about 3 mm in diameter, glossy, black.

FLOWERING July-September.

SYNONYMS *Phytolacca decandra* L.

WHERE FOUND Rich pastures, fields, waste places, gardens, and open places in woodlands, and along fence rows; mostly on deep, rich, gravelly soils. Young shoots used for greens or potherbs when thoroughly cooked; roots, leaves, and berries poisonous, formerly used in preparation of herbal medicines.

ORIGIN Native.

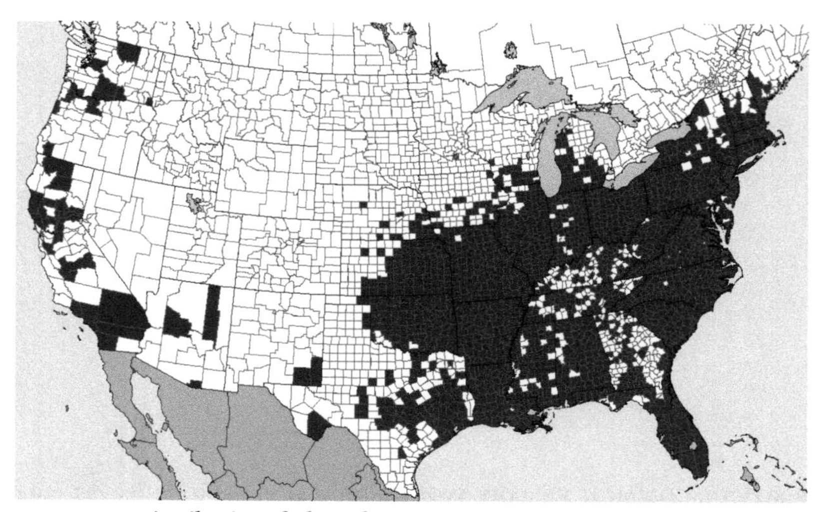

Distribution of **Phytolacca americana**, POKEWEED

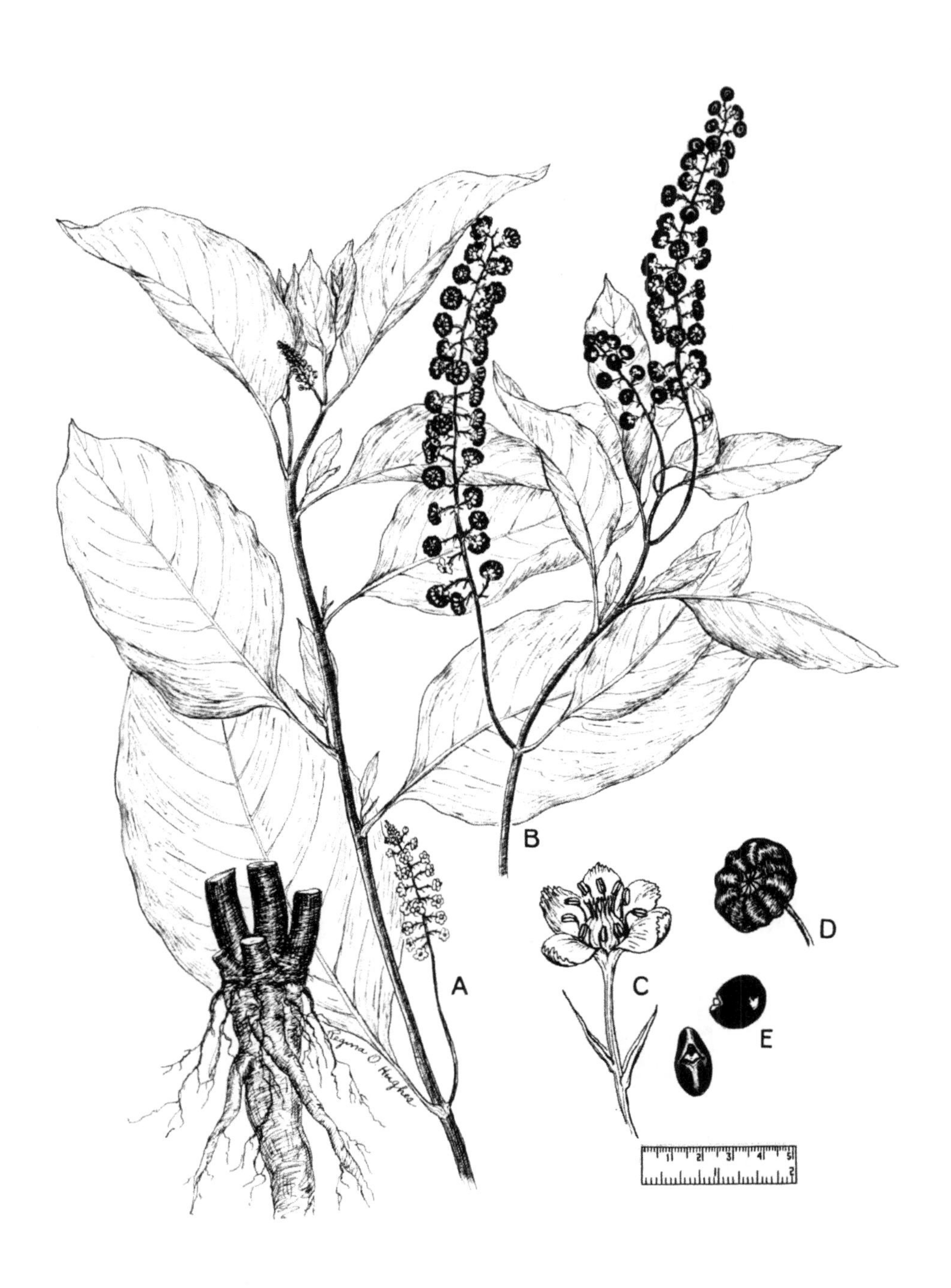

Phytolacca americana, POKEWEED. A, Habit, in flower. B, fruiting raceme. C, flower. D, berry. E, seeds.

Linaria dalmatica (L.) P. Mill. DALMATIAN TOADFLAX

Perennial herb, reproducing by seeds, sometimes developing horizontal or creeping rootstocks.

DESCRIPTION Stems erect, 8–12 dm tall, robust, branched above, glabrous, more or less glaucous, from a woody branching base. **Leaves** alternate, but crowded and sometimes appearing opposite, entire, 3–8 cm long, 1–2 cm broad, ovate, ovate-lanceolate or even lanceolate below, sessile and cordate-clasping at the base. **Flowers** in terminal elongate racemes, short-pedicellate or nearly sessile, bright-yellow but often purplish-red at the apex in the bud. Sepals 5, partly united, 6–8 mm long. Corolla 1.7–3.5 cm long, exclusive of the spur (1.3–2 cm long), strongly 2-lipped, upper lip 2-lobed, the lower one 3-lobed. Stamens 4, in pairs, included. **Capsule** broadly ovoid-cylindric, 7–8 mm high, opening near the summit by pores. **Seed** numerous, irregularly wing-angled.

FLOWERING July–September.

WHERE FOUND Roadsides and near dwellings, spreading to valleys and sagebrush flats.

ORIGIN Introduced; native of the Mediterranean region.

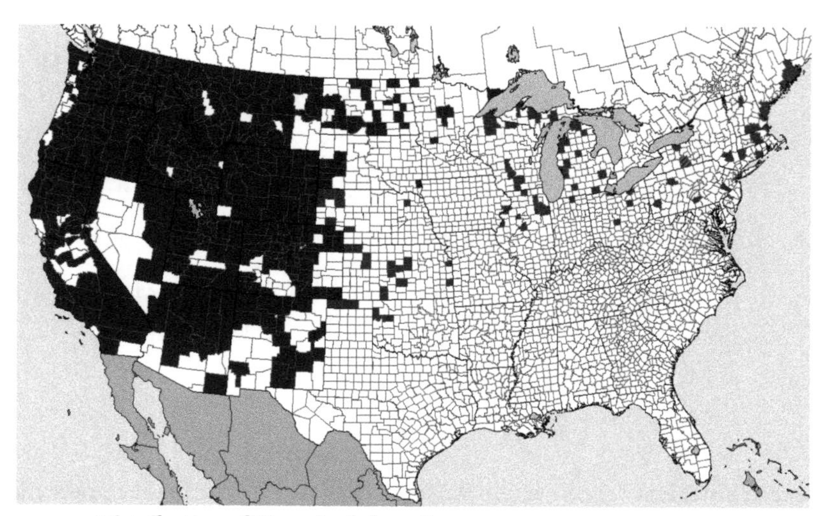

Distribution of **Linaria dalmatica,** DALMATION TOADFLAX

Linaria dalmatica, DALMATION TOADFLAX. A, habit. B, flower. C, capsule. D, seeds.

Linaria vulgaris P. Mill YELLOW TOADFLAX, BUTTER-AND-EGGS

Perennial herb, reproducing by seeds and spreading by roots and creeping rhizomes, forming colonies. Root system well-branching.

DESCRIPTION Stems ascending, glabrous or somewhat glandular above, up to 1.3 m tall, sparingly branched. **Leaves** nearly opposite, but mostly alternate, simple, sessile, pale-green, 2.5 cm long, 2-4 mm wide, linear or linear-lanceolate, narrowed below to a petiole-like base. **Flowers** in a dense, terminal raceme, finally becoming more or less lax. Corolla including the awl-shaped spur 2-3 cm long, bright-yellow, with a rounded orange palate, or the corolla whitish or creamy. **Capsule** round-ovoid, 8-12 mm long, 2-celled, many-seeded opening by 2-3 pores or slits just below the apex. **Seed** with a nearly circular, flattened, notched wing, the body minutely warty, dark-brown or black, 1.5-2 mm in diameter. Aberrant forms with regular flowers or with 2-5 spurs are occasionally found.

FLOWERING May-October; fruiting August-November.

WHERE FOUND Roadsides, dry fields, grainfields, waste places, pastures, and railroad yards; grown as an ornamental and escaping; mildly poisonous to stock.

ORIGIN Naturalized from Eurasia; throughout temperate North America.

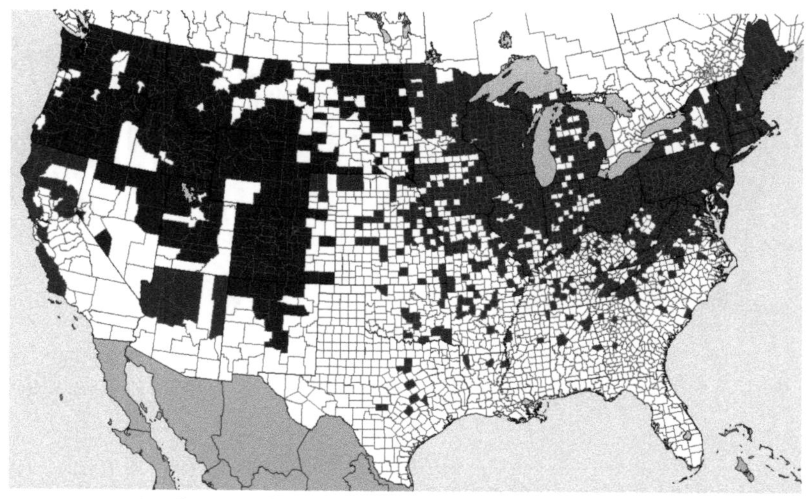

Distribution of **Linaria vulgaris,** YELLOW TOADFLAX

Linaria vulgaris, YELLOW TOADFLAX. A, habit. B, flowers. C, capsules. D, seeds.

Plantago lanceolata L. ENGLISH PLANTAIN

Annual herb, becoming perennial, reproducing by seeds. Caudex strong with tough slender rootlets.

DESCRIPTION Stems erect, leafless, 10–30 cm tall, terminating with a flower spike. Leaves all basal in a rosette, the blades lanceolate to lance-oblong, ascending or spreading, 5–30 cm long, 0.6–4 cm wide, with 3 to 5 prominent veins running lengthwise, tapering into the petiole. **Scape** tough, grooved-angled, elongating, 2–8 dm tall, stiff-haired above. **Spikes** dense, at beginning of flowering slenderly ovoid-conic, tapering at the apex, at maturity becoming cylindric and obtuse, 1.5–10 cm long. Bracts thin and papery, broadly ovate, the margin wavy. **Flowers** numerous, about 5 mm broad, inconspicuous, the forward sepals united, 3–3.5 mm long, the corolla lobes 2–3 mm long. **Capsule** ellipsoid, 3–4 mm long, 2-seeded, splitting across the middle or toward the base. **Seed** 1–2, small, 2–3 mm long, brown or black, shining, smooth, nearly ellipsoid, deeply concave on the inner face, sticky when damp.

FLOWERING May–October.

WHERE FOUND In lawns, meadows, pastures, and waste places; common and troublesome weed of grasslands.

ORIGIN Naturalized from Europe; throughout the United States.

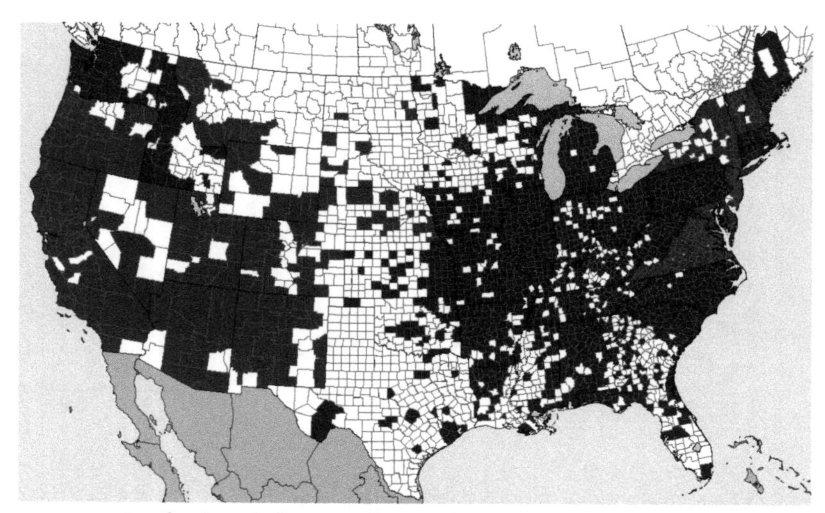

Distribution of **Plantago lanceolata**, ENGLISH PLANTAIN

Habit for all. A, **Plantago lanceolata**, ENGLISH PLANTAIN, a, flower. b, capsule. c, seed. B, **Plantago major**, BROADLEAF PLANTAIN, a, flower. b, capsule. c, seeds. C, **Plantago rugelii**, BLACK-SEED PLANTAIN, a, flower. b, capsule. c, seeds.

Plantago major L. BROADLEAF PLANTAIN

Perennial herb (sometimes annual), reproducing by seeds.

DESCRIPTION Leaves alternate in rosettes, all basal, 0.5–3 dm long, the blades thick, roughish on one or both sides when dry, with minute hairs, elliptic or lanceolate to broadly ovate, strongly ribbed, wavy or angular-toothed, the petioles broad, usually green (no purple tinge) and pubescent at the base. **Spikes** dense, obtuse, at the ends of the stems, 0.1–5 dm long. **Flowers** sessile, the bracts glabrous, broadly ovate with a slender keel. Sepals glabrous, elliptic to elliptic-round, the rounded keel about as wide as the papery margins, 1.5–2 mm long. Corolla lobes definitely less than 1 mm long. **Capsule** stoutly ellipsoid, 2–4 mm long, splitting across the upper half (near the middle), brown or purple. **Seed** 6–15, angled, net-veined, light- to dark-brown, about 1–1.7 mm long.

FLOWERING June–October.

WHERE FOUND In lawns, roadsides, and waste places.

ORIGIN Native of Eurasia; naturalized throughout the United States.

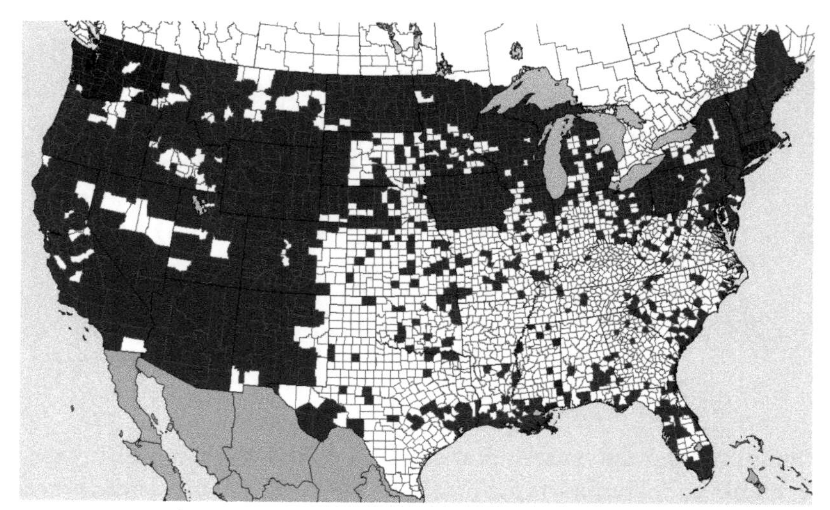

Distribution of **Plantago major**, BROADLEAF PLANTAIN

Habit for all. A, **Plantago lanceolata**, ENGLISH PLANTAIN, a, flower. b, capsule. c, seed. B, **Plantago major**, BROADLEAF PLANTAIN, a, flower. b, capsule. c, seeds. C, **Plantago rugelii**, BLACK-SEED PLANTAIN, a, flower. b, capsule. c, seeds.

Plantago rugelii Decne. BLACK-SEED PLANTAIN

Perennial herb, reproducing by seeds. Roots mostly fibrous.

DESCRIPTION Stems erect, leafless, 15-30 cm tall, terminating in the flower spike. **Leaves** alternate in rosettes, erect or spreading, all basal, blades simple, thin, broadly elliptic to oval, 5-20 cm long, a little over half as wide, pale, glabrous or slightly hairy, usually wavy-edged, veins conspicuous, petiole margined, at base usually glabrous and tinged with purple. **Spikes** slender, dense to alternate-flowered, tapered at the apex, up to 3 dm tall, about 5 mm wide, rather loose, pedicels about 0.5 mm long, bracts narrowly triangular-lanceolate, one-half to three-fourths as long as the calyx, tapering regularly from the base to the apex, prominently elevated into an acute keel. Sepals ovate or oblong, acute, sharp keel much wider than papery margin. Corolla inconspicuous, lobes less than 1 mm long, reflexed after flowering. **Capsule** nearly cylindrical, about 4-6 mm long, splitting across the lower half, 4- to 9-seeded. **Seed** dark-brown or black, oval, angular, with a scar (minute hilum) near the center on one side, 1.5-2.5 mm long, not reticulated.

FLOWERING July-October.

WHERE FOUND Damp, rich soils; roadsides, damp shores, lawns, gardens, and waste places.

ORIGIN Native.

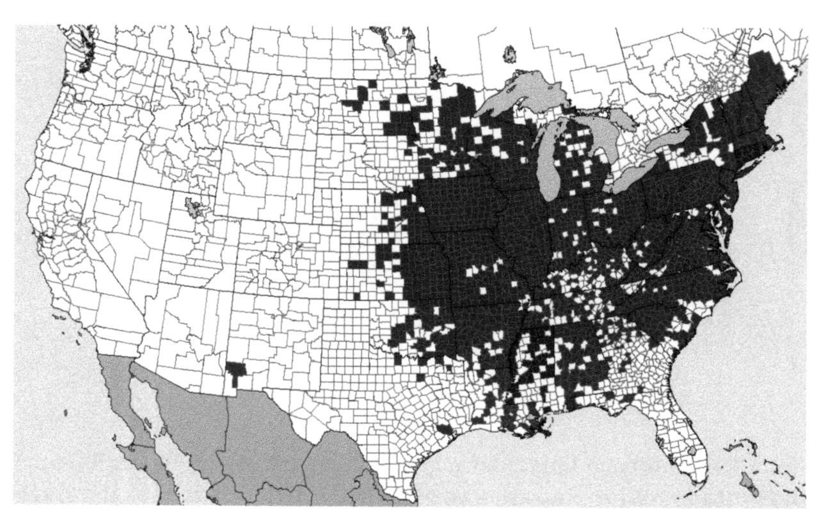

Distribution of **Plantago rugelii**, BLACK-SEED PLANTAIN

Habit for all. A, **Plantago lanceolata**, ENGLISH PLANTAIN, a, flower. b, capsule. c, seed. B, **Plantago major**, BROADLEAF PLANTAIN, a, flower. b, capsule. c, seeds. C, **Plantago rugelii**, BLACK-SEED PLANTAIN, a, flower. b, capsule. c, seeds.

Veronica arvensis L. CORN SPEEDWELL

Annual herb, reproducing by seeds. Root system fibrous.

DESCRIPTION Stems erect or nearly so, simple to diffusely branched, with more or less long soft hairs, 0.5–4 dm tall. **Leaves:** the **lower leaves** rounded or ovate, obtuse, 6–12 mm long, with 2–4 blunt teeth on each side, palmately veined, short-petioled; the **upper leaves** smaller, sessile, lanceolate to linear. **Inflorescence** terminal racemes, constituting about two-thirds of the plant, the bracteal leaves progressively smaller upward, narrower and mostly entire. Pedicels up to 1.5 mm long. Calyx lobes oblong or oblanceolate, very unequal, 3–5 mm long. Corolla violet-blue to blue, about 2 mm wide. Style about 8 mm long, extending about as far as the summit of the capsule. **Capsule** 3–4 mm wide, nearly as long, hairy, deeply notched. **Seed** 1 mm long.

FLOWERING March–August.

WHERE FOUND Weed in gardens, lawns, fields, waste open ground, rocky and sterile pastures, and woodlands.

ORIGIN Naturalized from Eurasia.

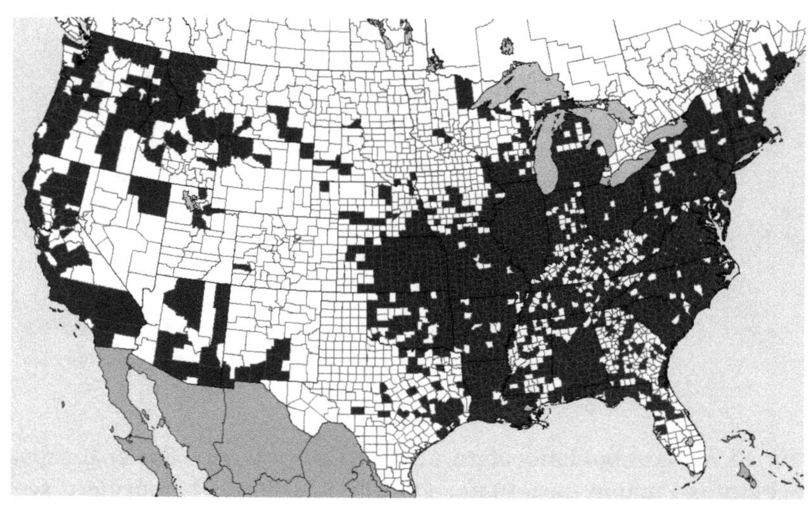

Distribution of **Veronica arvensis**, CORN SPEEDWELL

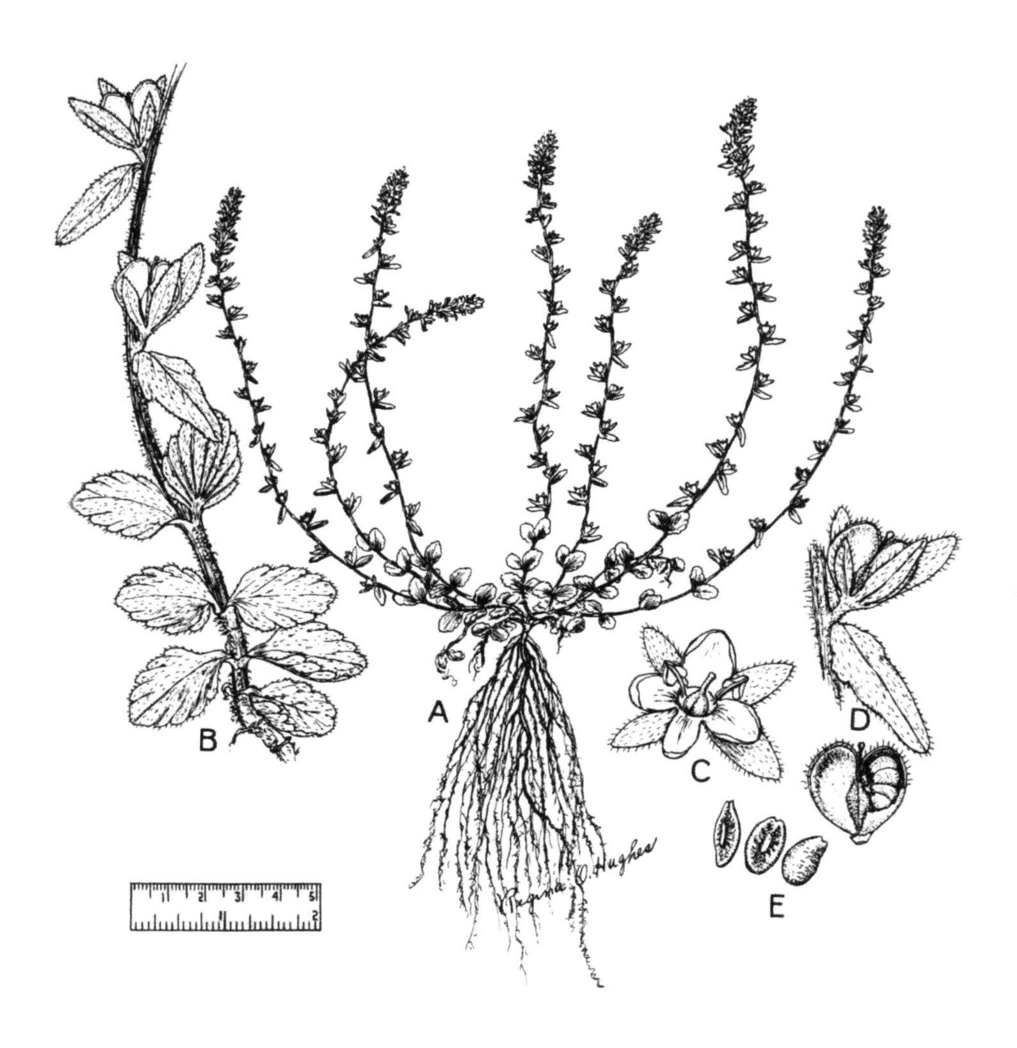

Veronica arvensis, CORN SPEEDWELL. A, habit. B, enlarged branch. C, flower. D, capsules. E, seeds.

Veronica officinalis L. COMMON SPEEDWELL

Perennial herb, reproducing by seeds and creeping stems.

DESCRIPTION Stems prostrate or creeping, rooting at the base, stout, pubescent; the flowering branches erect or ascending. **Leaves** opposite, simple, obovate-elliptic or oblong, obtuse, toothed, hairy or nearly rough, 2.5-6 cm long, 1-3 cm broad, short-petioled. Peduncles stout, ascending, range from shorter than the subtending leaves to longer than these leaves. **Flowers** in dense axillary racemes, the pedicels shorter than the calyx. Calyx lobes obtuse, nearly equal in size. Corolla pale-blue to lilac or lavender, or white, marked with darker lines, 5-6 mm in diameter, with obtuse lobes. **Capsule** obovate-triangular, 3-4 mm in diameter, about as broad as long, flattened, broadly notched, pubescent. **Seed** about 1 mm long, oval to elliptic, flattened, minutely granular, lemon-yellow.

FLOWERING May-July.

WHERE FOUND In pastures, open woodlands, and old fields; mostly on gravelly or stony acid soils.

ORIGIN Native (in part); also native to Eurasia and introduced and naturalized from Europe.

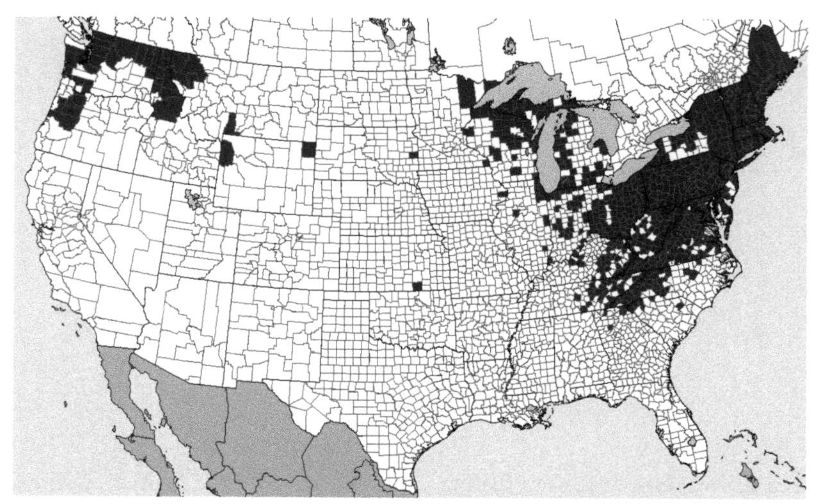

Distribution of **Veronica officinalis**, COMMON SPEEDWELL

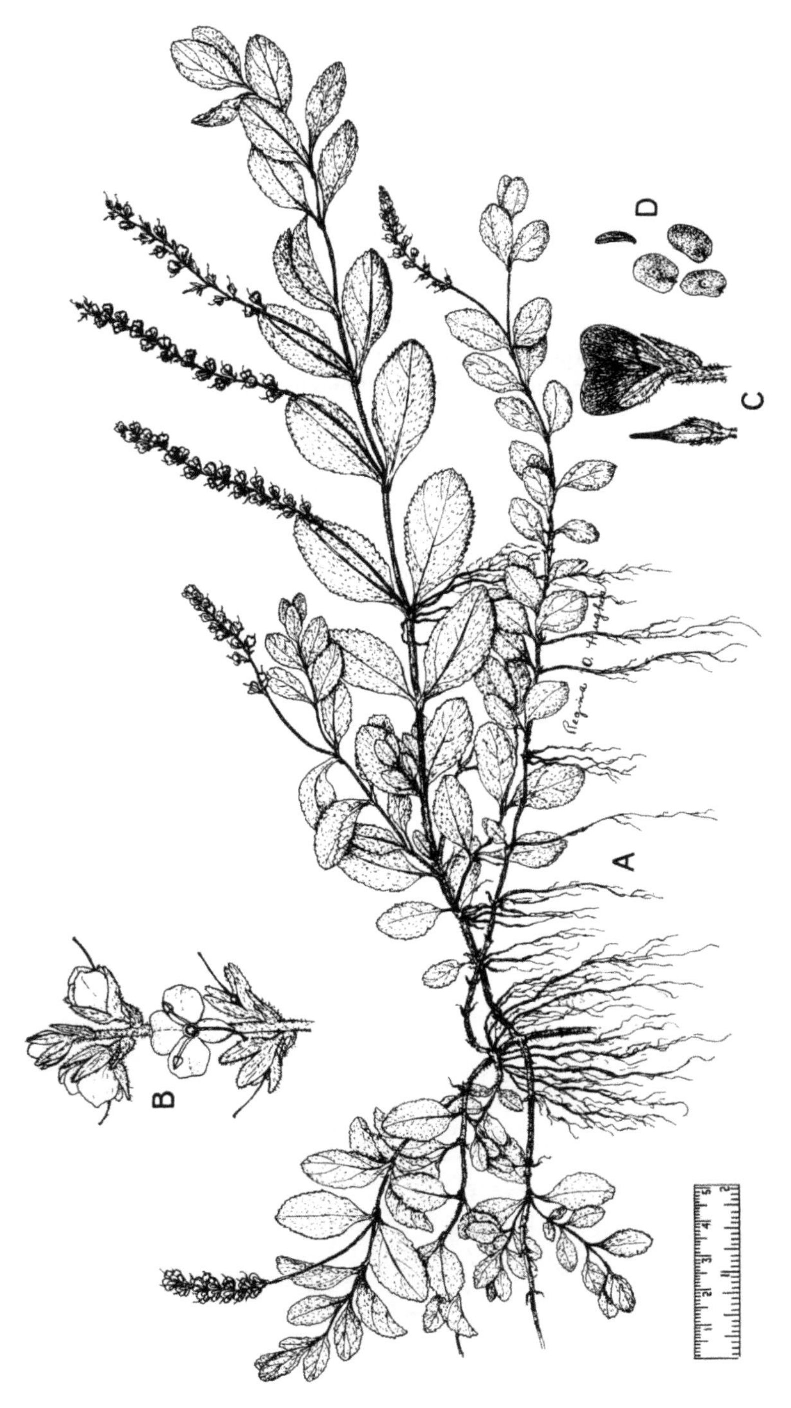

Veronica officinalis, COMMON SPEEDWELL. A, habit. B, flowers. C, capsules. D, seeds.

Veronica peregrina L. PURSLANE SPEEDWELL

Annual or winter annual herb, reproducing by seeds. Root system fibrous.

DESCRIPTION Stems erect, simple, or branching from the base, glabrous or glandular-puberulent, 1–4 dm tall, the lower half with opposite leaves, the upper half bearing flowers. **Leaves** simple, narrowly oblong to oblanceolate, 1.5–3 cm long, often obtuse, entire or with a few low teeth, sessile or narrowed to a petiole-like base, the lower leaves opposite, the upper leaves alternate. Bracteal leaves similar to the stem leaves but progressively smaller. **Flowers** small, in axils of bracts, sessile or short-stalked. Sepals sub-equal. Corolla white. Style very short, the stigma appearing sessile in the notch of the capsule. **Capsule** flattened, heart-shaped, glabrous, 3–4 mm long, 4–5 mm wide, conspicuously notched. **Seed** long oval, flattened, translucent, glossy, orange-yellow, with a scar on one side.

FLOWERING March–August.

WHERE FOUND In lawns, gardens, fertile fields, and waste places; damp open soil; a roadside weed.

ORIGIN Native, but frequently appearing to be introduced.

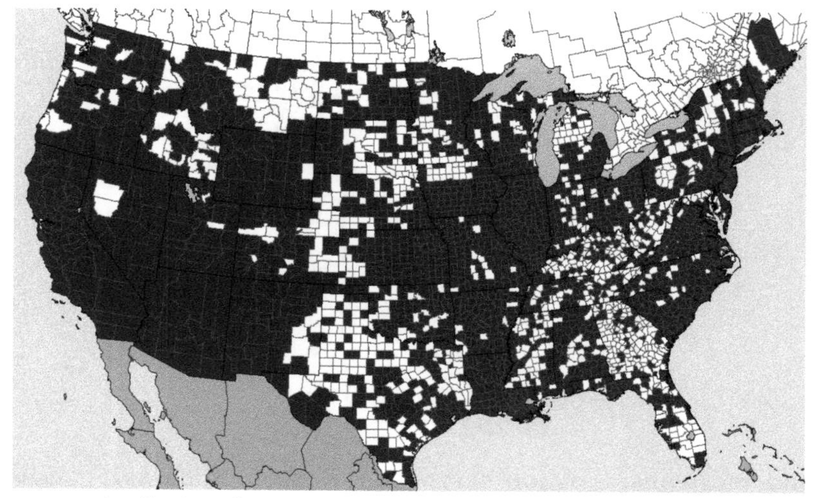

Distribution of **Veronica peregrina**, PURSLANE SPEEDWELL

Veronica peregrina, PURSLANE SPEEDWELL. A, habit. B, basal leaf. C, flower; c, corolla, showing stamens; cc, calyx, showing subequal sepals. D, capsule. E, seeds.

Aegilops cylindrica Host — JOINTED GOATGRASS

Annual grass. Culms erect, branching at the base, 2-6 dm tall. Root system fibrous.

DESCRIPTION **Leaves** alternate, simple, with auricles at the base of the blade, smooth or hairy. **Spike** 5-8 cm long. **Spikelets** few, 2-5 flowers placed with the side against the articulated rachis and closely appressed to it, the rachis joints upwardly enlarged. **Glumes** with lateral keel prolonged into an awn. **Lemmas** of upper spikelets with harsh awns 4-5 cm long, those of the lower spikelets shorter. **Seed** ripening before wheat, shattering easily.

FLOWERING June-August.

WHERE FOUND Waste places, railroads, and fields, mainly in wheatfields, spreading to roadsides and wastes.

ORIGIN Naturalized from Europe.

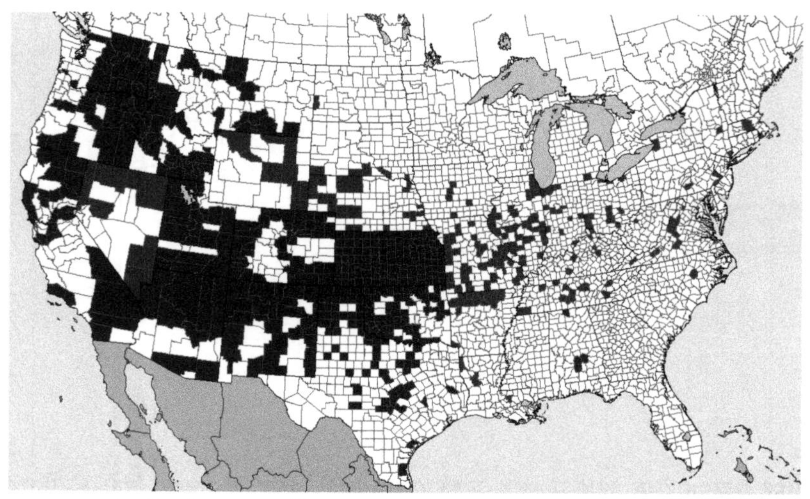

Distribution of **Aegilops cylindrica**, JOINTED GOATGRASS

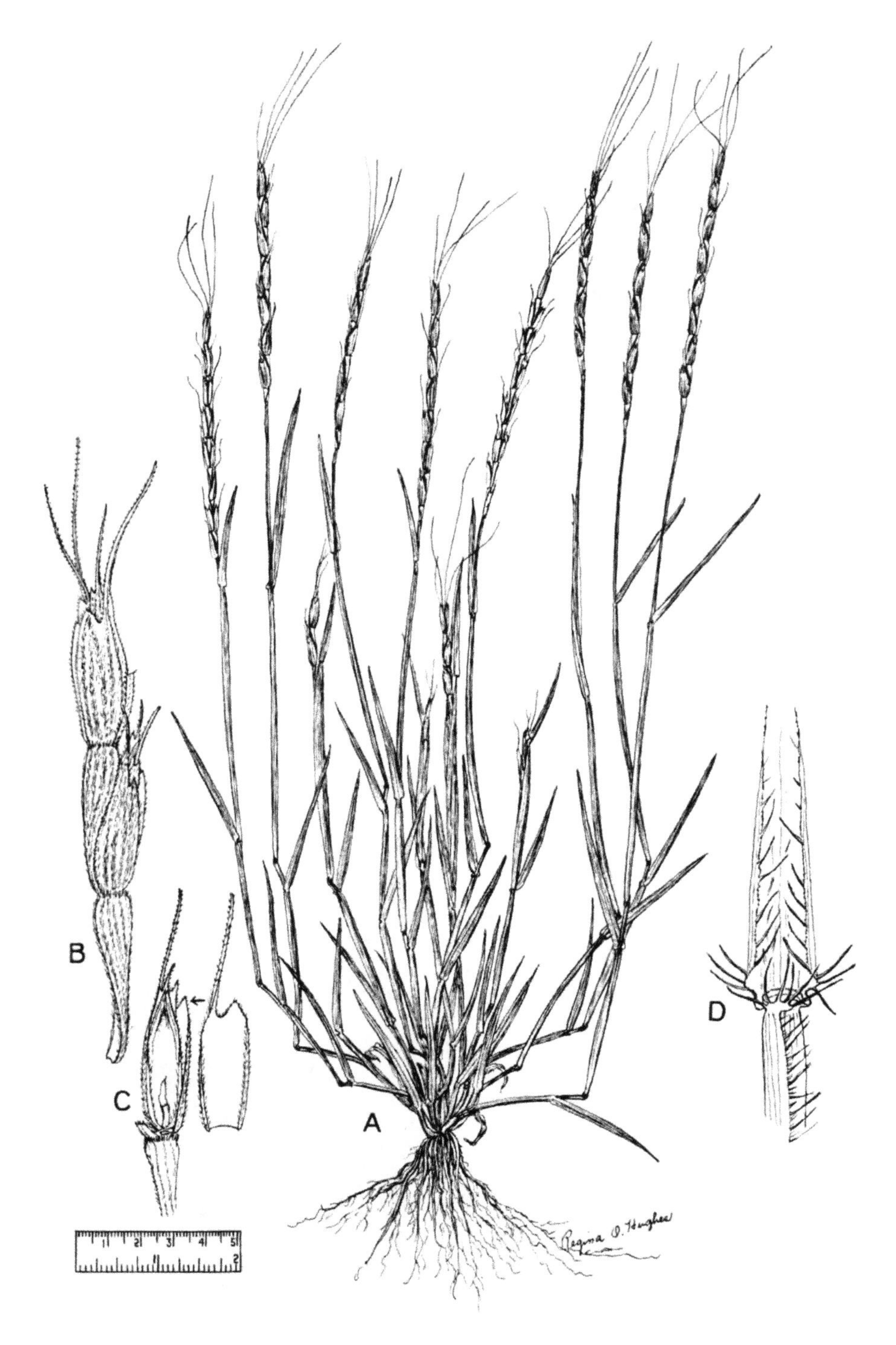

Aegilops cylindrica, JOINTED GOATGRASS. A, habit. B, spike detail. C, floret. D, ligule.

Andropogon virginicus L. BROOM-SEDGE

Perennial grass, growing erect in small clumps, reddish-brown when dry. Roots densely fibrous.

DESCRIPTION Culms slender to stout, several in a tuft, slightly flattened, 30–90 cm tall, upper part branched. **Leaves** green or glaucous, 15–30 cm long, flat or folded, very hairy on upper surface where attached to stem, the sheaths strongly compressed, keeled, and enfold each other. **Racemes** mostly paired of 2–4 finger-like clusters, each 1–4.5 cm long, bearing tufts of conspicuous white hairs, enclosed by reddish-brown leaves. **Sessile spikelets** 3–4 mm long, the rachis joints long and slender, the awn straight or spiraling only at the base, 1–2 cm long. **Pediceled spikelets** represented by an awl-like scale or absent, the divergent beard longer than the axis of the pedicel. **Seed** about 3 mm long, brown, bearded. A highly variable species.

FLOWERING September-January.

WHERE FOUND Wastes, plains, old fields, and sterile areas. A serious pasture weed on marginal lands.

ORIGIN Native

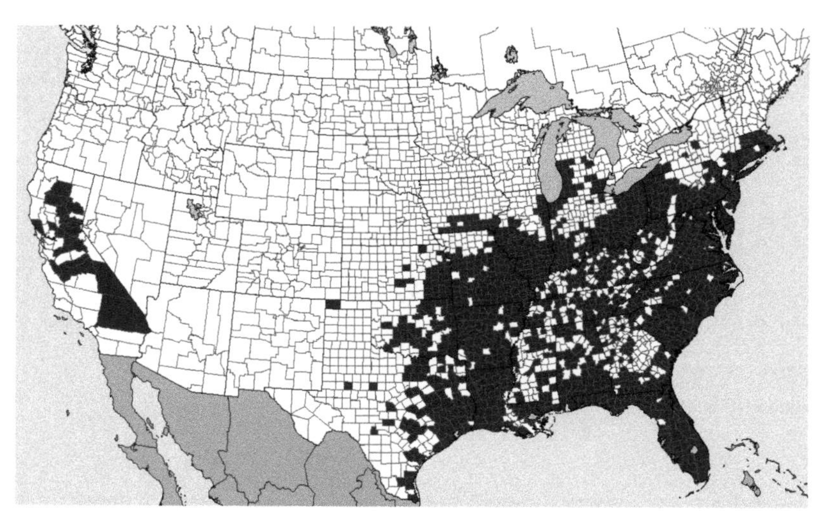

Distribution of **Andropogon virginicus**, BROOM-SEDGE

Andropogon virginicus, BROOM-SEDGE. A, habit. B, ligule (left, opened; right, compressed as in nature). C, inflorescence. D, florets.

Avena fatua L. WILD OAT

Annual grass, reproducing by seeds. Root system extensive and fibrous.

DESCRIPTION Culms smooth, erect, stout, 4-12 dm tall, in small tufts. **Leaf blades** 7-20 cm long, 5-8 mm wide, resembling those of tame oats. **Panicle** loose and open, the slender branches ascending. **Spikelets** pendulous, 2.2-2.5 cm long. **Glumes** smooth, striate, acuminate. **Lemmas** with long, dark awns, 3 cm long or more, the lower parts twisted, the upper parts bent sharply at right angles to the twisted parts, and with a ring of hairs at the base and more or less appressed-pubescent with long stiff brownish hairs, or glabrous. **Caryopsis** varying from white to yellow, brown, gray or black, usually hairy, especially near the base.

FLOWERING June-October.

WHERE FOUND In fields under continuous cultivation to small grains and flax; probably the most serious annual weed in the hard red spring wheat area; seeds usually ripen earlier than most cereals and many drop to ground before time to harvest cultivated cereals.

ORIGIN Naturalized from Europe.

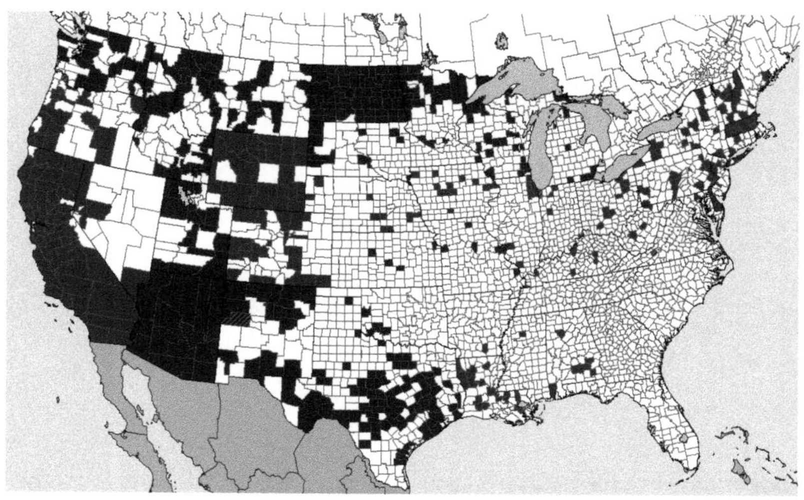

Distribution of **Avena fatua**, WILD OAT

Avena fatua, WILD OAT. A, habit. B, spikelet. C, florets, 3 views.

Bromus arvensis L. JAPANESE BROME

Winter-annual grass, reproducing by seeds.

DESCRIPTION Culms erect, 0.2–1 m tall. **Leaves,** both blades and sheaths, covered with soft hairs. **Panicle** open, the divergent branches with drooping tips, 2–2.5 cm long, somewhat hairy, borne on long, slender pedicels. **First glume** 3-nerved, acute. **Second glume** 5-nerved, obtuse. **Lemmas** glabrous, 7–9 mm long, obtuse, firm, obscurely 9-nerved, the margins somewhat inrolled at maturity, with a twisted or divaricate awn, 8–12 mm long, with stiff beards, 0.7–1.3 cm long, bent conspicuously outward at maturity. **Palea** distinctly shorter than its glume.

FLOWERING Maturing in May–August.

SYNONYMS *Bromus japonicus* Thunb. ex Murr.

WHERE FOUND Grainfields, meadows, wasteland, and roadsides.

ORIGIN Naturalized from Eurasia.

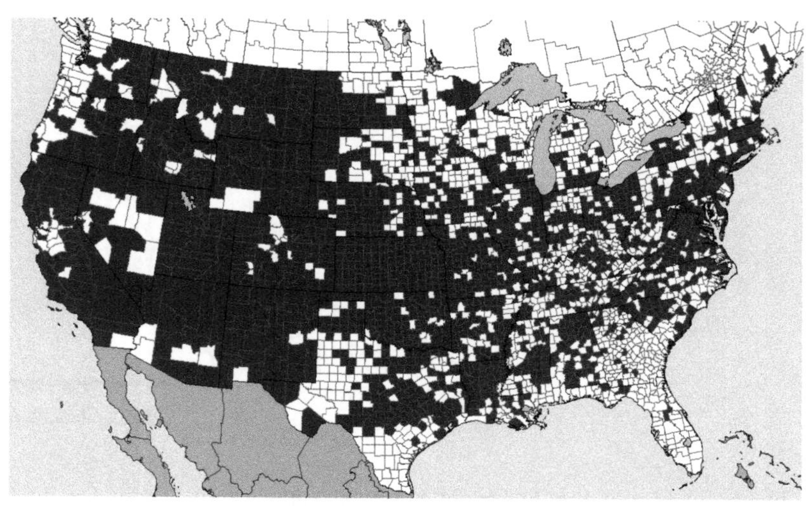

Distribution of **Bromus arvensis,** JAPANESE BROME

Bromus arvensis, JAPANESE BROME. A, habit. B, spikelet. C, ligule. D, floret. E, caryopsis.

Bromus racemosus L. HAIRY CHESS

Winter-annual grass, plants mostly 3–10 dm tall, erect or decumbent at the base.

DESCRIPTION **Leaf blades** sparsely to densely pubescent, 2–7 mm wide. **Sheaths** sparsely to densely pubescent or short-pilose, the hairs usually retrorse. **Ligules** 1–2 mm long. **Panicle** 10–12 cm long, open, with 2–6 rather stiffly ascending or drooping branches. **Spikelets** 1–2.5 cm long, 6- to 10-flowered, somewhat flattened with rather closely imbricated florets. **First glume** 4.5–6 mm long, 3- to 5-nerved. **Second glume** 6–8 mm long, 7- to 9-nerved, both glumes glabrous to scaberulous. **Lemmas** 7–11 mm long, 7- to 9-nerved, minutely scabrous, thin, and acute, awns 7–9 mm long.

FLOWERING June–early August.

SYNONYMS *Bromus commutatus* Schrad.

WHERE FOUND Dry roadsides, waste places, pastures, and fields.

ORIGIN Introduced from Europe.

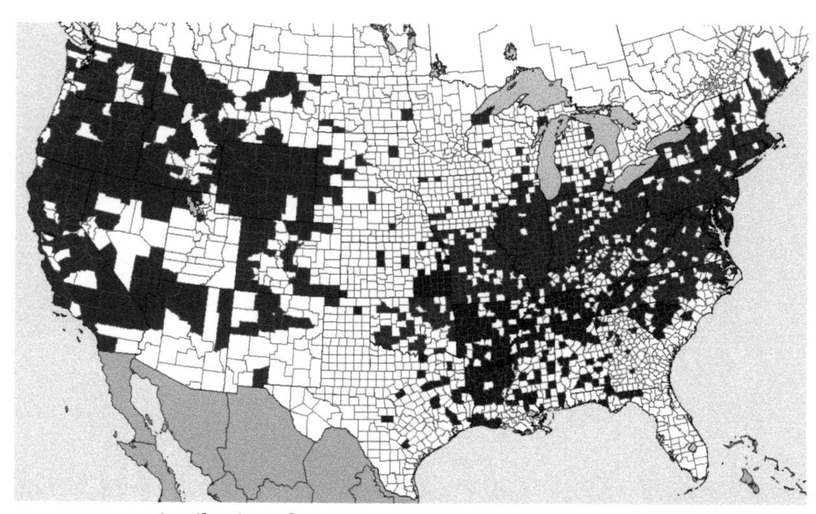

Distribution of **Bromus racemosus**, HAIRY CHESS

Bromus racemosus, HAIRY CHESS. A, habit. B, spikelet. C, floret, 3 views. D, seed. E, ligule.

Bromus secalinus L. CHEAT, CHESS

Winter-annual grass, reproducing by seeds.

DESCRIPTION Culms erect, 0.1–1.3 m tall. **Leaves**, upper sheaths smooth and strongly nerved, blades with moderately stiff hairs above and harsh. **Panicle** 0.3–2 dm long, branches ascending or the lowest becoming divergent. **Spikelets** borne on elongate pedicels, 5- to 15-flowered, glabrous or scabrous, in maturity lax. **Lemmas** spreading-ascending at maturity, strongly inrolling, firm, sub-equal, obscurely 7-nerved, mostly 5–8 mm long; awn straight or flexuous, 1–6 mm long, deciduous, sometimes slightly projecting in maturity. **Caryopsis** plumper and heavier than those of other brome species.

FLOWERING June–September.

WHERE FOUND Grainfields, meadows, and waste places.

ORIGIN Naturalized from Europe.

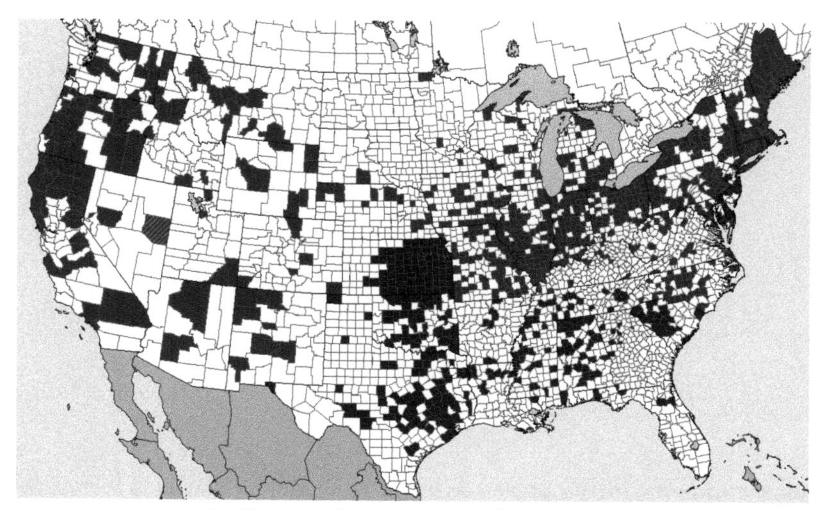

Distribution of **Bromus secalinus**, CHEAT

Bromus secalinus, CHEAT. A, habit. B, spikelet. C, ligule. D, floret, 3 views. E, caryopsis.

Bromus tectorum L. CHEATGRASS

Winter-annual grass, reproducing by seeds.

DESCRIPTION Culms erect or spreading, slender, 0.2–1 m tall. **Leaves**, both blades and sheaths, light-green, covered with long soft hairs. **Panicle** rather dense, soft, very drooping, often purplish, 0.5–2 dm long with spreading or recurving flexuous branches, the branches often with several spikelets. **Spikelets** 2–3.5 cm long, including awns (1–1.7 cm long). **Glumes** sparsely pilose. **Lemmas** slenderly 5- to 7-nerved, hispid, bearing long beards, 1–1.5 cm long. **Caryopsis** long and narrow.

FLOWERING April–May; maturing in May–June.

WHERE FOUND Meadows, pastures, ranges, small grainfields, wastelands, and roadsides; often too abundant.

ORIGIN Introduced.

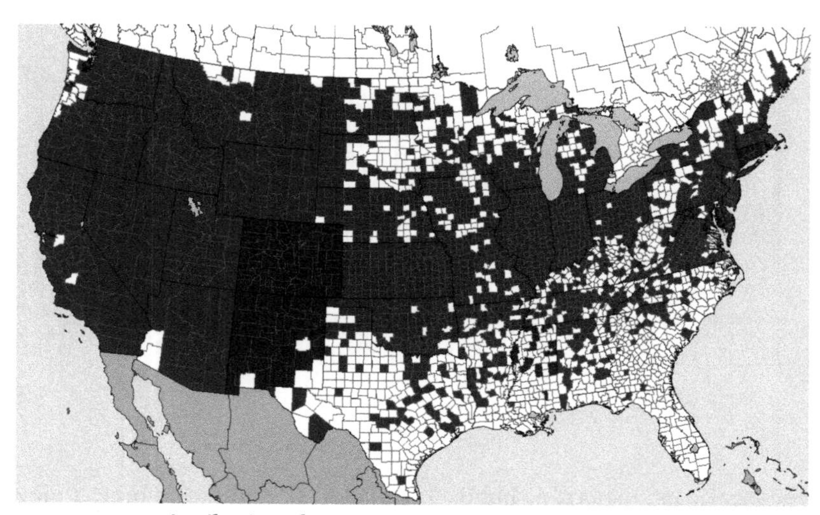

Distribution of **Bromus tectorum**, CHEATGRASS

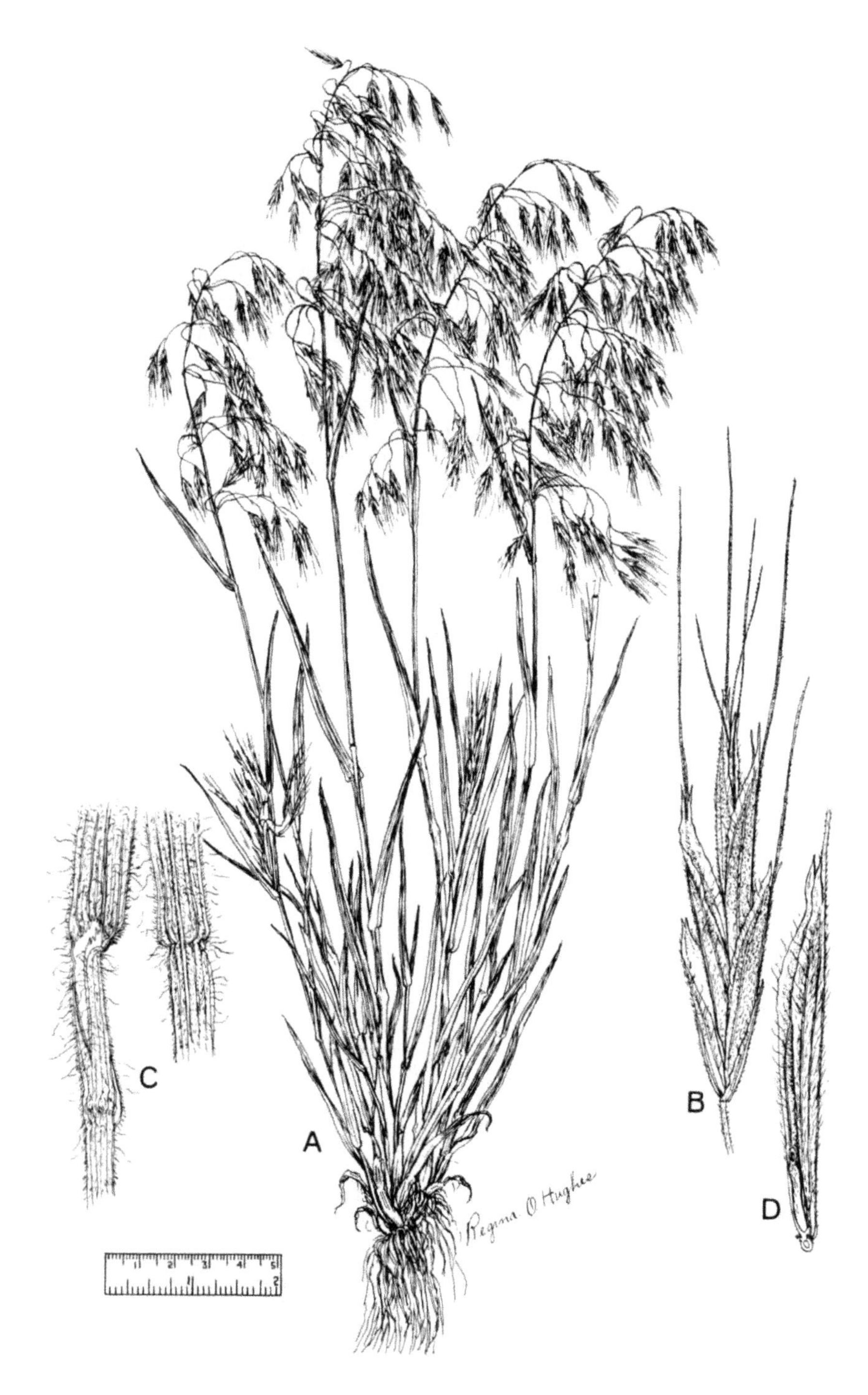

Bromus tectorum, CHEATGRASS. A, habit. B, spikelet. C, ligules. D, floret.

Cenchrus echinatus L. SOUTHERN SANDBUR

Annual grass.

DESCRIPTION Culms round in cross section, ascending from an abruptly bent base, 15-85 cm long; pubescence variable. **Leaves** glabrous to pubescent, 4-26 cm long, 3.5-11 mm wide. **Sheaths** flattened, with moderately stiff hairs on the margins near the summit. **Ligule** with marginal hairs, 0.7-1.7 mm long. **Inflorescence** open, 2-10 cm long, 0.8-1.8 cm wide. Rachis strongly flexuous, scabrous, the internodes 2-3.0 mm long. **Spikelets** 2 or 3 per bur, sessile, 5-7 mm long. Burs truncate at the base, globose, 5-10 mm long, 3.5-6 mm wide, the spine tips usually turning purple with age. **Seed** ovoid, 1.6-3.2 mm long, 1.3-2.2 mm wide.

FLOWERING June-September.

WHERE FOUND Open ground and waste places; a common weed in tropical areas.

ORIGIN Native.

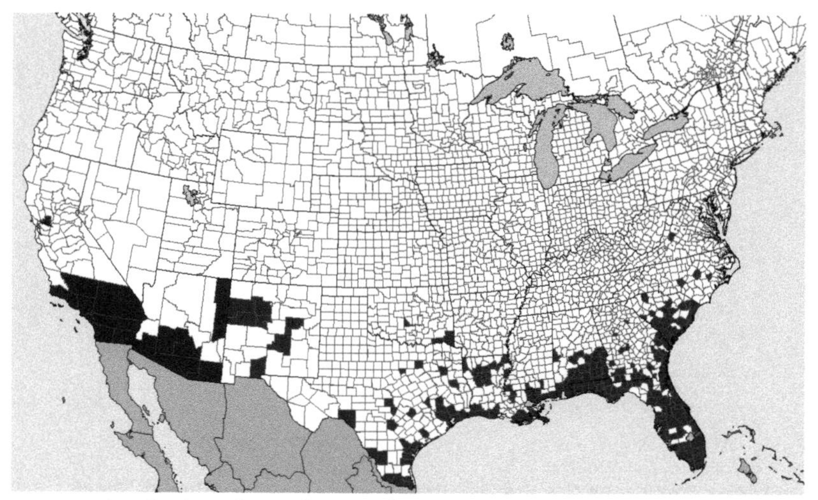

Distribution of **Cenchrus echinatus**, SOUTHERN SANDBUR

Cenchrus echinatus, SOUTHERN SANDBUR. A, habit. B, ligule. C, inflorescence. D, bur. E, spikelet. F, seed, 3 views.

Cenchrus incertus M.A. Curtis — COASTAL SANDBUR

Annual grass, often biennial and overwintering, reproducing by seeds. Roots fibrous, rooting sometimes at the nodes of the stems when they contact the soil.

DESCRIPTION Culms decumbent or erect, sometimes with many spreading branches from the base, 5-80 cm tall, smooth. **Leaves** glabrous, the margins scabrous, 2-18 cm long, 2-6 mm wide. **Ligule** with marginal hairs, 0.5-1.5 mm long. **Spike** short, composed of 2-4 sessile, glabrous spikelets, enclosed in sharp spiny burs (8-40 spines). **Burs** each containing 1-3 seeds, usually 2, straw-colored to mauve or purple.

FLOWERING June-September.

SYNONYMS *Cenchrus pauciflorus* Benth.

WHERE FOUND Mostly on sandy soil, fields, wastes, and roadsides; a troublesome weed in garden crops, lawns, and fields, mainly because of the spiny burs.

ORIGIN Native.

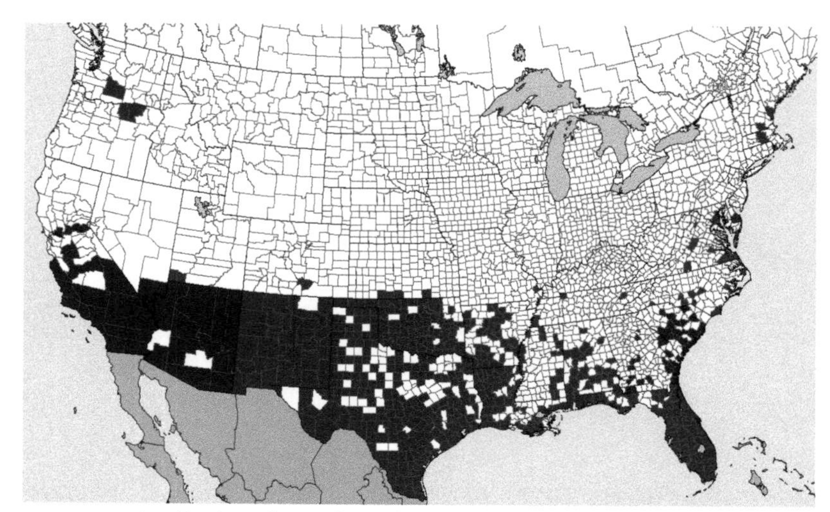

Distribution of **Cenchrus incertus**, COASTAL SANDBUR

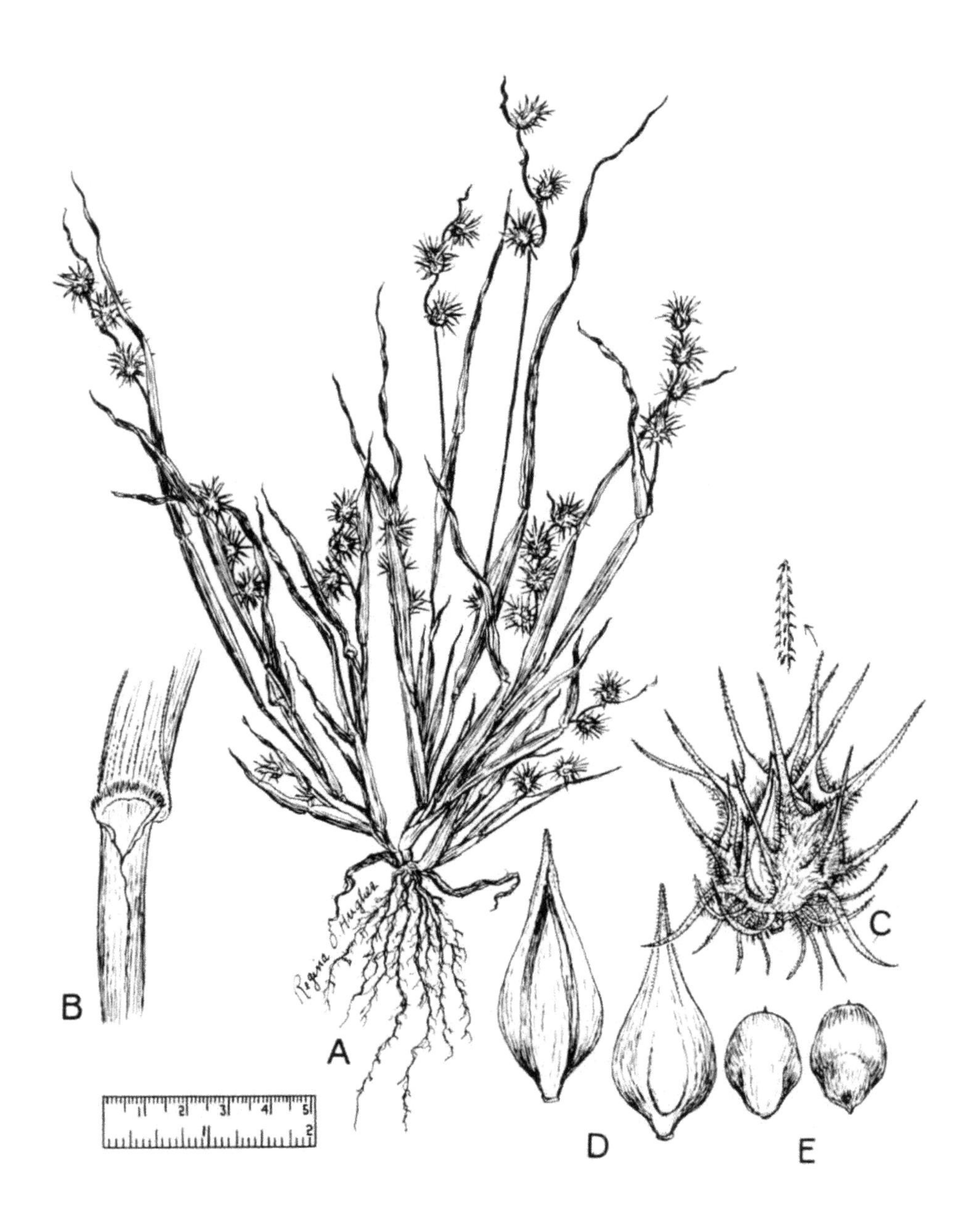

Cenchrus incertus, COASTAL SANDBUR. A, habit. B, ligule. C, bur. D, florets. E, seeds.

Cenchrus longispinus (Hack.) Fern. LONG-SPINE SANDBUR

Annual grass, often forming large clumps.

DESCRIPTION Culms round in cross section, 10–90 cm tall. **Leaf blades** rough to the touch to sparsely pilose, 6–19 cm long. 3–8 mm wide. **Sheaths** strongly flat-keeled, pilose on the margins and at the throat. **Ligule** a rim of fine hairs, 0.7–1.7 mm long. Inflorescence compact, 4–10 cm long, 1.2–2.2 cm wide. **Rachis** angled, curved alternately in opposite directions and glabrous, more often minutely rough to the touch, the internodes 2–5 mm long. **Burs** somewhat globose, medium to short-pubescent, 8.3–12 mm long, 3.5–6 mm wide. Spines slender, retrorsely barbed, and often purple-tinged, 3.5–7 mm long, 0.7–1.4 mm wide, the margins often narrowly grooved, the lower margins sometimes long-pubescent. Spines at the base of the bur numerous and pointing downward, shorter than those on the body of the bur. **Spikelets** sessile, 2–3 per bur, rarely 4, 6–7.8 mm long. **Seed** ovoid, 2.2–3.8 mm long, 1.5–2.6 mm wide.

FLOWERING June–August.

WHERE FOUND Sandy and recently disturbed soils; roadsides and abandoned fields.

ORIGIN Native.

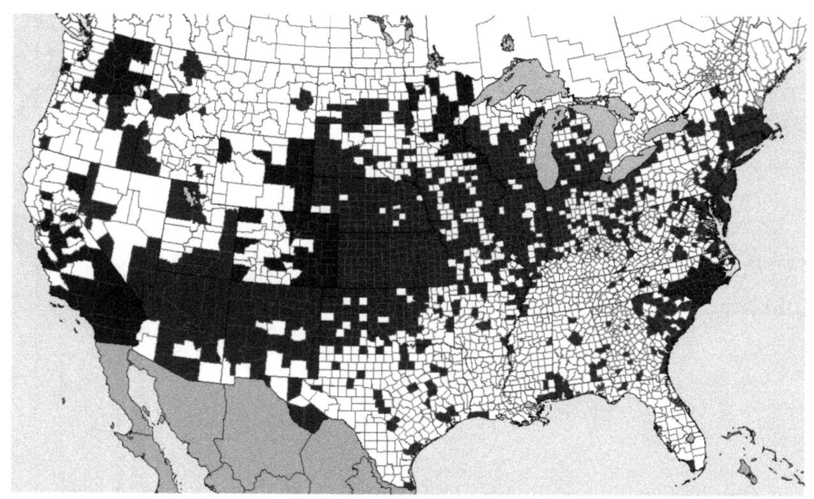

Distribution of **Cenchrus longispinus**, LONG-SPINE SANDBUR

Cenchrus longispinus, LONG-SPINE SANDBUR. A, habit. B, ligules. C, bur.

Cynodon dactylon (L.) Pers. BERMUDAGRASS

Perennial grass, reproducing by seed (rarely), rootstocks, and stolons. Rootstock hard, scaly, sharp-pointed, forming dense heavy sod.

DESCRIPTION Stolons (stems) flat, glabrous, extensively creeping, bearing at each joint the dead bladeless sheath. **Flowering culms** erect or ascending, 15–45 cm tall. **Leaf blades** 3 mm wide, gray-green, slightly hairy or glabrous, except for a fringe of long hair at the edge just above the collar. **Sheath**, often two, opposite per node, sparsely hairy or hairless, strongly flattened. **Ligule** a conspicuous ring of white hairs. **Inflorescence** finger-like, 3- to 7-parted, 2–7 cm long. **Spikelets** in two rows tightly appressed to one side of the rachis, 2 mm long. **Lemma** boat-shaped, acute, longer than the glume. **Caryopsis** free within the lemma and palea.

FLOWERING Summer.

WHERE FOUND Open places, pastures, and most cultivated areas; may become a serious weed when established but an important pasture grass in Southern States; sometimes used for lawns.

ORIGIN Introduced from Africa.

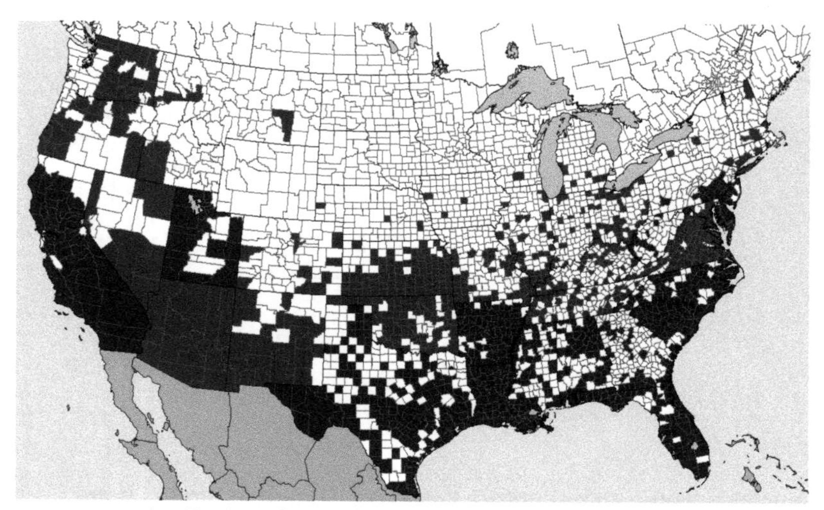

Distribution of **Cynodon dactylon**, BERMUDAGRASS

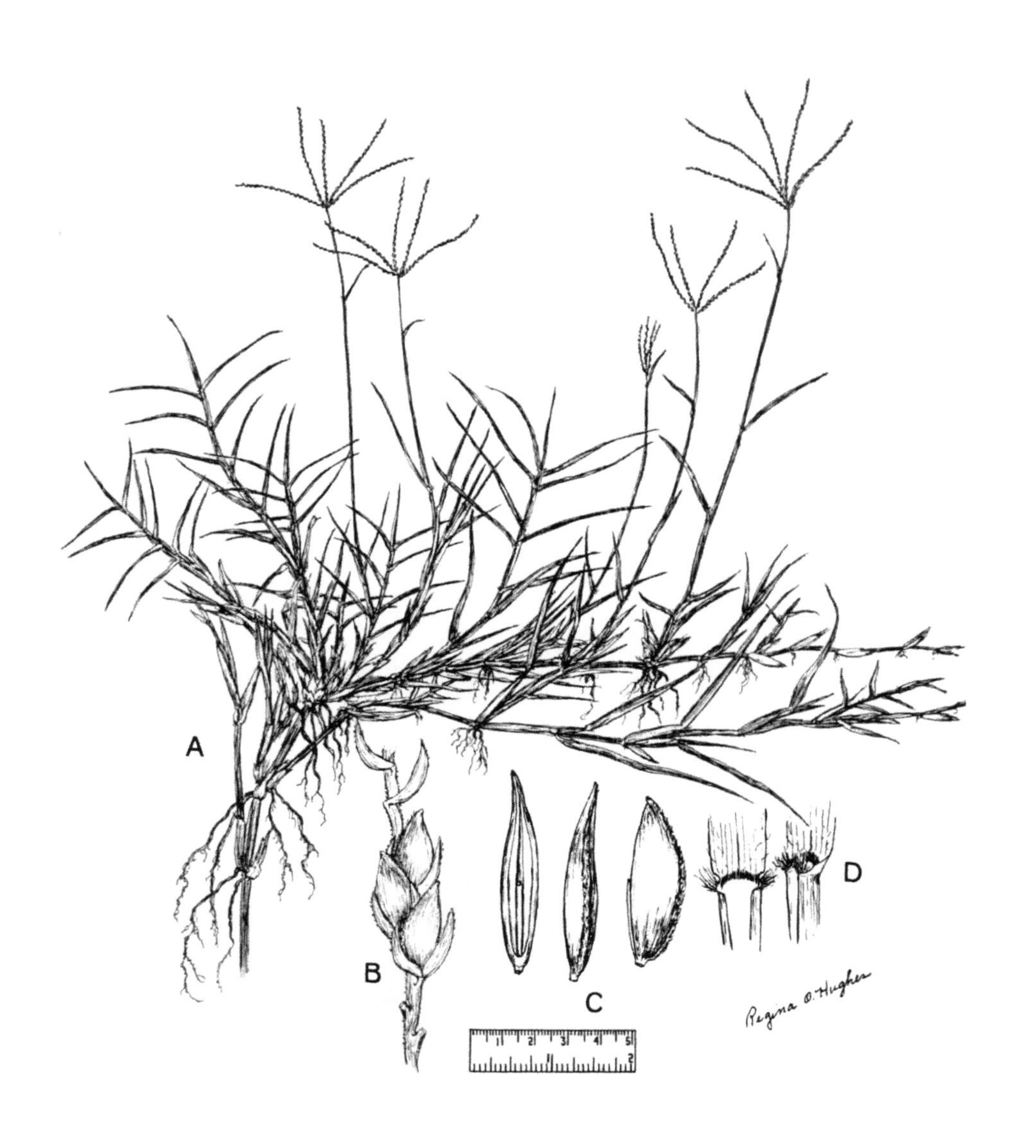

Cynodon dactylon, BERMUDAGRASS. A, habit. B, spike. C, florets. D, ligules.

Digitaria ischaemum (Schreb. ex Schweig.) Schreb. ex Muhl.

SMOOTH CRABGRASS

Annual grass.

DESCRIPTION Culms 0.2-4 dm tall, erect or usually soon decumbent-spreading. **Leaves** 2-10 cm long, 3-6 mm wide, glabrous, bluish to purplish. **Racemes** mostly 2-6, commonly purple, 4-10 cm long, the rachis with thin wings wider than the midrib. **Spikelets** on one side of rachis, solitary or in 2's, about 2 mm long. **First glume** hyaline, obscure. **Second glume** and sterile lemma as long as the dark fertile lemma, pubescent with capitellate hairs.

FLOWERING July-October.

WHERE FOUND Waste places and often troublesome in lawns.

ORIGIN Naturalized from Eurasia.

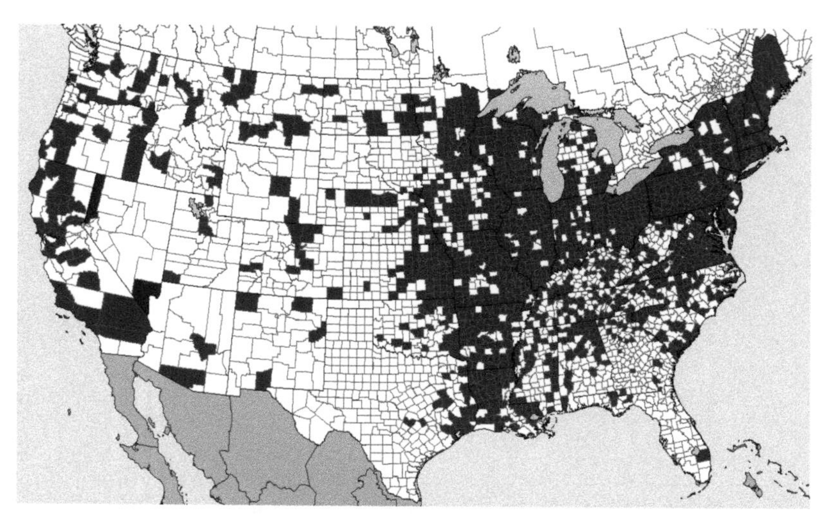

Distribution of **Digitaria ischaemum**, SMOOTH CRABGRASS

Digitaria ischaemum, SMOOTH CRABGRASS. A, habit. B, spike detail, showing arrangement of florets on rachis. C, florets. D, ligules.

Digitaria sanguinalis (L.) Scop. HAIRY CRABGRASS

Annual grass, reproducing by seed, branching and spreading, plants often purplish.

DESCRIPTION **Culms** stout, smooth, up to 3-12 dm long, when prostrate, rooting at the nodes, the flowering snoots ascending. **Leaf blades** lax, 5-15 cm long, 4-10 mm wide, somewhat hairy, sheaths densely long-hairy, especially the lower ones, rough to the touch, often more or less pilose. **Spike** 5-15 cm long, with 3-13 finger-like segments, in whorls at the top of the stem. **Spikelets** along one side of rachis, about 3 mm long. **First glume** minute, but evident. **Second glume** about half as long as the spikelet, narrow, with marginal hairs. **Sterile lemma** strongly nerved, the lateral inter-nerves appressed-pubescent, the hairs somewhat spreading at maturity. Fertile lemma pale or grayish. **Caryopsis** about 2 mm long, alternate on the branches of the inflorescence.

FLOWERING June-October.

WHERE FOUND Lawns, gardens, and fields; serious in lawns and cultivated ground. Starts late when ground is quite warm and grows well under dry, hot conditions.

ORIGIN Native of Europe.

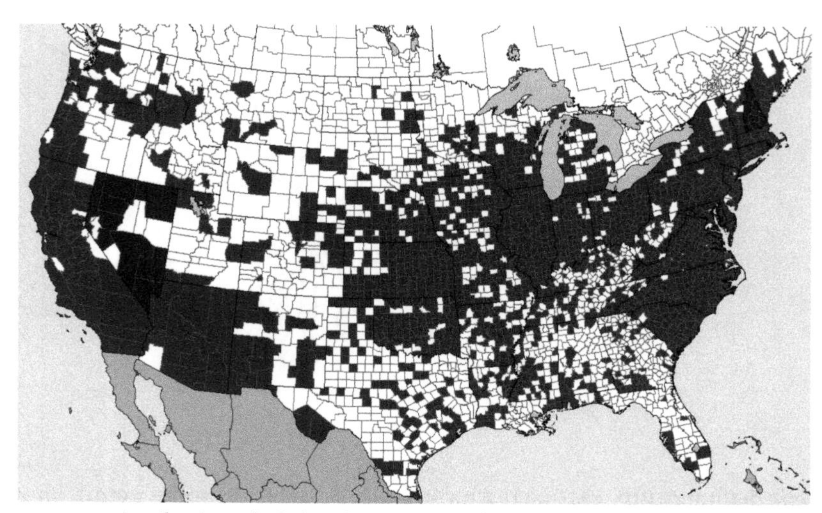

Distribution of **Digitaria sanguinalis**, HAIRY CRABGRASS

Digitaria sanguinalis, HAIRY CRABGRASS. A, habit. B, florets, front and back views. C, caryopsis.

Echinochloa crus-galli (L.) Beauv. BARNYARD-GRASS

Annual grass, with fibrous, rather shallow roots.

DESCRIPTION Culms stout, erect to decumbent, often branching from the base, usually less than 1 m (but occasionally up to 1.5 m) tall. **Sheaths** glabrous. **Ligule** absent. **Leaf blades** glabrous, elongate, 5–15 mm wide, light-green. **Panicle** erect or nodding, green or purple-tinged, 10–20 cm long. **Racemes** numerous, 2–4 cm (but occasionally up to 10 cm) long, spreading, ascending, sometimes branched. **Spikelets** crowded, about 3 mm long, excluding the awns. **First glume** nearly half as long as the spikelet. **Second glume** and sterile lemma with short bristly hairs on the nerves, typically awnless. Awns variable, mostly 5–10 mm (but occasionally up to 3 cm), long. **Caryopsis** ovate, obtuse, usually 2.5–3.5 mm long, the short beak withering, somewhat inflexed and sharply differentiated from the obtuse body. **Seed** tan to brown, and with longitudinal ridges on the convex surface.

FLOWERING June–October.

WHERE FOUND Cultivated areas, waste ground, ditches, and fields, especially soybeans, clover, and alfalfa fields, in late summer and fall.

ORIGIN Native of Europe.

NOTE A very variable species with several similar species in the genus:

Echinochloa crus-pavonis (Kunth) J.A. Schultes—COCK'S-SPUR GRASS. Racemes more or less appressed, the spikelets less strongly hispid but papillose, usually green, the culms less succulent. Most often in alkaline places.

Echinochloa frumentacea Link.—JAPANESE MILLET. Racemes thick, appressed, incurved, with turgid, awnless, mostly purple spikelets, the nerves hispid but not tuberculate. Occasionally cultivated as a forage plant and escaped.

Echinochloa muricata (Beauv.) Fern.—ROUGH BARNYARD-GRASS. Racemes dense, mostly spreading-flexuous, with spikelets awnless or nearly so (awns less than 3 mm long) and the sheaths occasionally pubescent with stiff hairs.

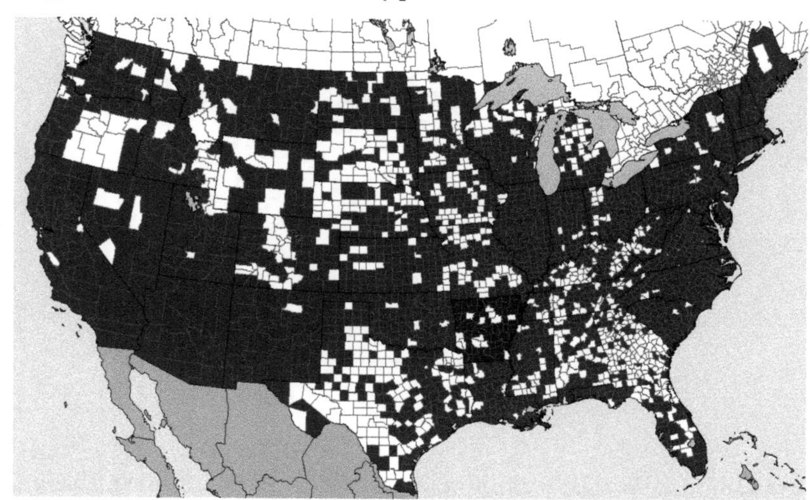

Distribution of **Echinochloa crus-galli**, BARNYARD-GRASS

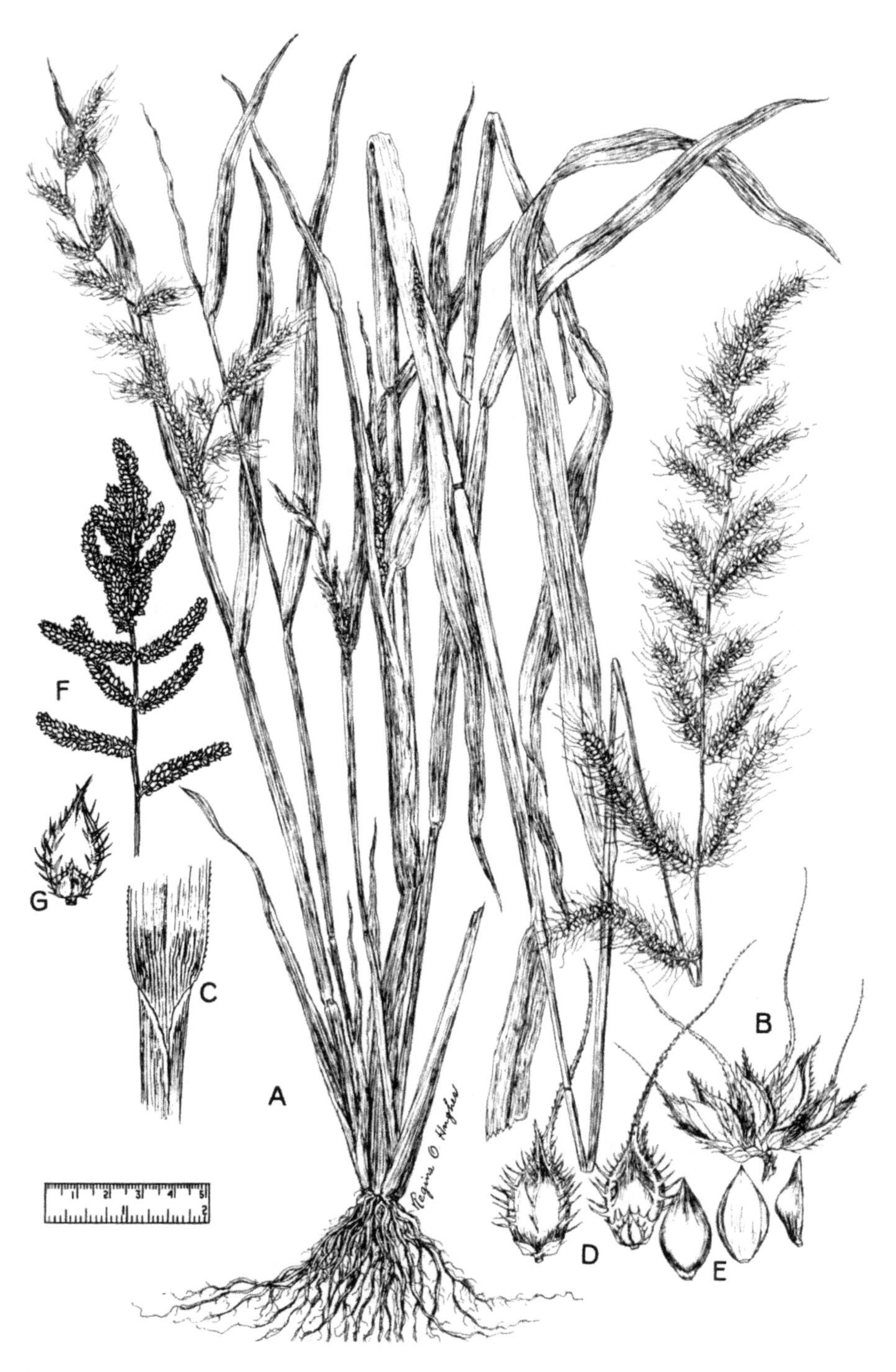

Echinochloa crus-galli, BARNYARD-GRASS. A, habit. B, spikelet. C, ligule. D, florets. E, caryopses. F, spike. G, floret of awnless form.

Eleusine indica (L.) Gaertn. INDIAN GOOSE GRASS

Annual grass, reproducing by seeds; plants coarsely tufted, branching at the base, ascending to prostrate, very smooth. Roots fibrous.

DESCRIPTION Culms 15-60 cm (up to 1 m) long, erect or decumbent at the base, flattened, glabrous. **Leaf blades** flat or folded, smooth but sometimes slightly roughened, 3-8 mm wide, 5-25 cm long. **Sheaths** flattened and keeled, glabrous or somewhat scabrous, margin sometimes pilose. **Ligules** with short marginal hairs, 1 mm long or less. **Spikes** 4-15 cm long, finger-like with 2-6 finger-like segments (2.5-7.5 cm long), crowded and whorled or 1 inserted lower. **Spikelets** sessile on one side of the rachis, 3-5 mm long, 3-6 florets along the edges of the rachilla, crowded. **First glume** 2-2.5 mm long, 1-nerved. **Second glume** about 3 mm long, 3- to 9-nerved, both glumes acute, rough to the touch on the keel, dry and thin. **Lemmas** 3-4 mm long, with 3 strong nerves close together forming a keel and another pair near the margins, obtuse or nearly so, rarely abrupt tipped; glabrous except rough to the touch on the keel. Paleas shorter than the lemmas. **Caryopsis** very small, 1-1.5 mm long, reddish-brown, granular, ridged, loosely inclosed within the pericarp.

FLOWERING July-October.

WHERE FOUND Waste places, fields, open ground, lawns, gardens, and roadsides; a common weed in the warmer regions; often confused with crabgrass, but plants darker green and growing only in tufts.

ORIGIN Naturalized from the Old World.

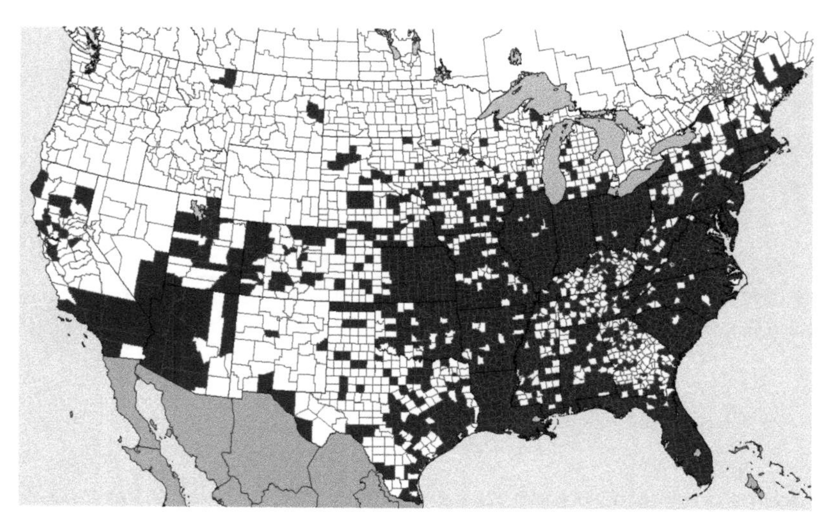

Distribution of **Eleusine indica,** INDIAN GOOSE GRASS

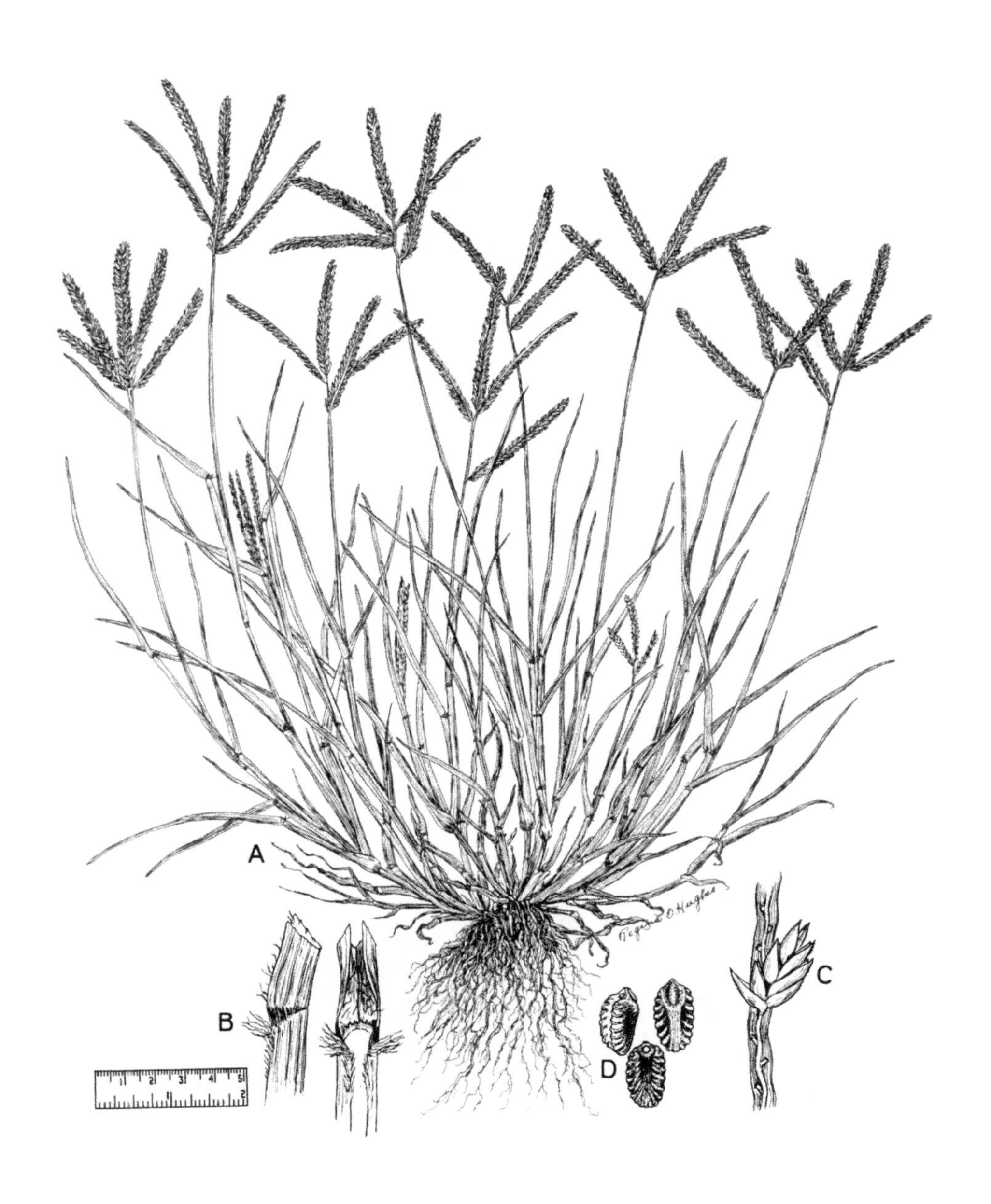

Eleusine indica, INDIAN GOOSE GRASS. A, habit. B, ligules. C, spikelet. D, caryopses.

Elymus repens (L.) Gould QUACKGRASS, COUCHGRASS

Perennial grass, reproducing by seeds and extensively creeping, underground slender rhizomes; roots arising only at the nodes.

DESCRIPTION Culms 3-12 dm tall, smooth, with 3-6 joints, hollow at the tip; the cartilaginous bands of upper nodes longer than thick. **Leaves** with auricles, soft, flat, with crowded fine ribs, scabrous or sparsely pilose above. **Ligules** 0.5 mm long. **Sheaths:** the lower ones hairy, the upper ones glabrous or slightly pilose. **Spike** dense or lax, 0.5-2.5 dm long, 2-9 short-awned florets in the compressed spikelet (0.6-2.2 cm long). **Glume** herbaceous, oblong to lanceolate, narrowed from above the middle, strongly 5- to 7-nerved. **Lemmas** obtuse, acute or awned. A highly variable species.

FLOWERING Late May-September.

SYNONYMS *Agropyron repens* (L.) Beauv.

WHERE FOUND Open waste places, in most cropped areas and in pastures; on gravelly and sandy shores. Formerly used for pasture and as a grass hay plant but very aggressive and difficult to eradicate.

ORIGIN Introduced from Europe.

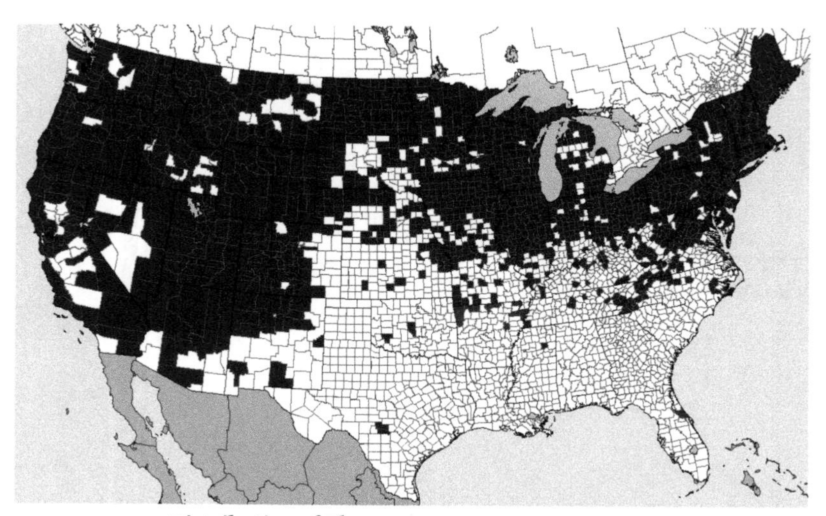

Distribution of **Elymus repens,** QUACKGRASS

Elymus repens, QUACKGRASS. A, habit. B, spikelet. C, ligule. D, florets.

Holcus mollis L. — CREEPING VELVET GRASS

Perennial grass. Rhizomes slender, vigorous.

DESCRIPTION Culms glabrous, 5-10 dm tall. **Leaves** with the blades with long soft hairs or velvety, 4-10 mm wide and the sheaths except the lower ones glabrous. **Panicle** ovate or oblong, rather loose, 6-10 cm long. **Spikelets** 2-flowered, the pedicel disarticulating below the glumes, 4-5 mm long. **Glumes** about equal, longer than the 2 florets, glabrous; the first floret perfect, the lemma awnless; the second floret staminate, its lemma bearing a short awn (about 3 mm long) on the back.

FLOWERING Summer.

WHERE FOUND Damp places.

ORIGIN Introduced from Europe and spreading.

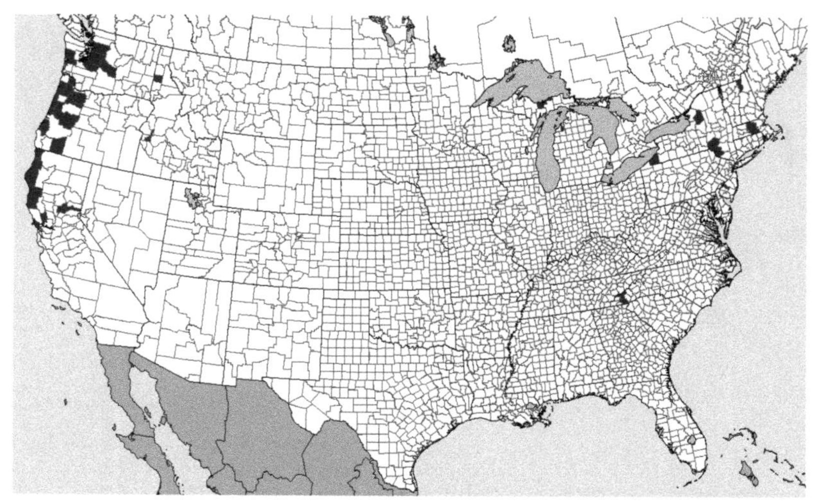

Distribution of **Holcus mollis**, CREEPING VELVET GRASS

Holcus mollis, CREEPING VELVET GRASS. A, habit. B, inflorescence. C, spikelet (lower floret perfect, awnless; upper floret staminate, awned). D, glumes. E, ligule. F, leaf sheath (retrorse hairs of sheath at joint).

Hordeum jubatum L. FOXTAIL BARLEY

Perennial grass; plants in clumps or tufted. Roots densely fibrous.

DESCRIPTION Culms usually erect, 30-60 cm tall, or decumbent at the base. **Leaves** alternate, sheaths smooth, the blades 3-6 mm wide, rough on the upper surface. **Spike** 5-12.6 cm long, about as wide, nodding with rather soft, yellowish-green or purplish bristles about 5 cm long. **Lateral spikelets** reduced to 1-3 spreading awns. **Glumes** of central perfect spikelet awn-like, 2.5-6 cm long, spreading. **Lemma** 6-8 mm long, with an awn as long as the glumes. **Caryopsis** about 3 mm long, yellow, hairy.

FLOWERING June-September.

WHERE FOUND Open wastes, meadows, and pastures, where it may be troublesome as the bristles may injure the mouths of livestock; other uncultivated areas and waste places.

ORIGIN Native, but considered adventive in northeastern states.

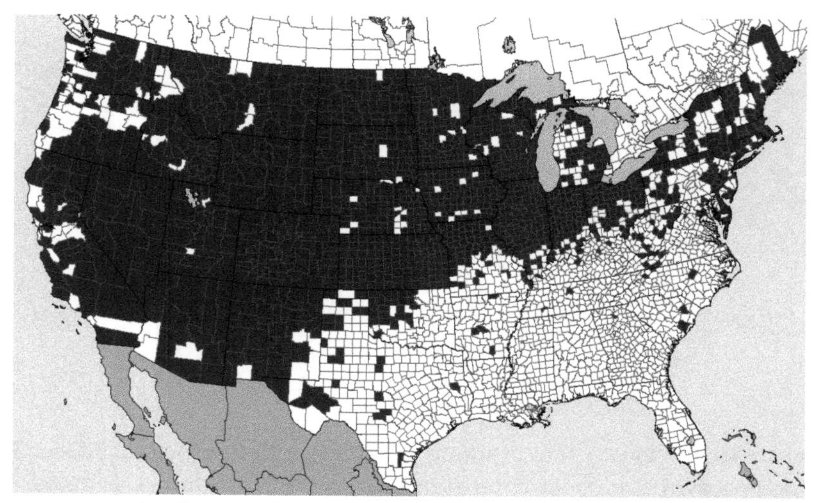

Distribution of **Hordeum jubatum**, FOXTAIL BARLEY

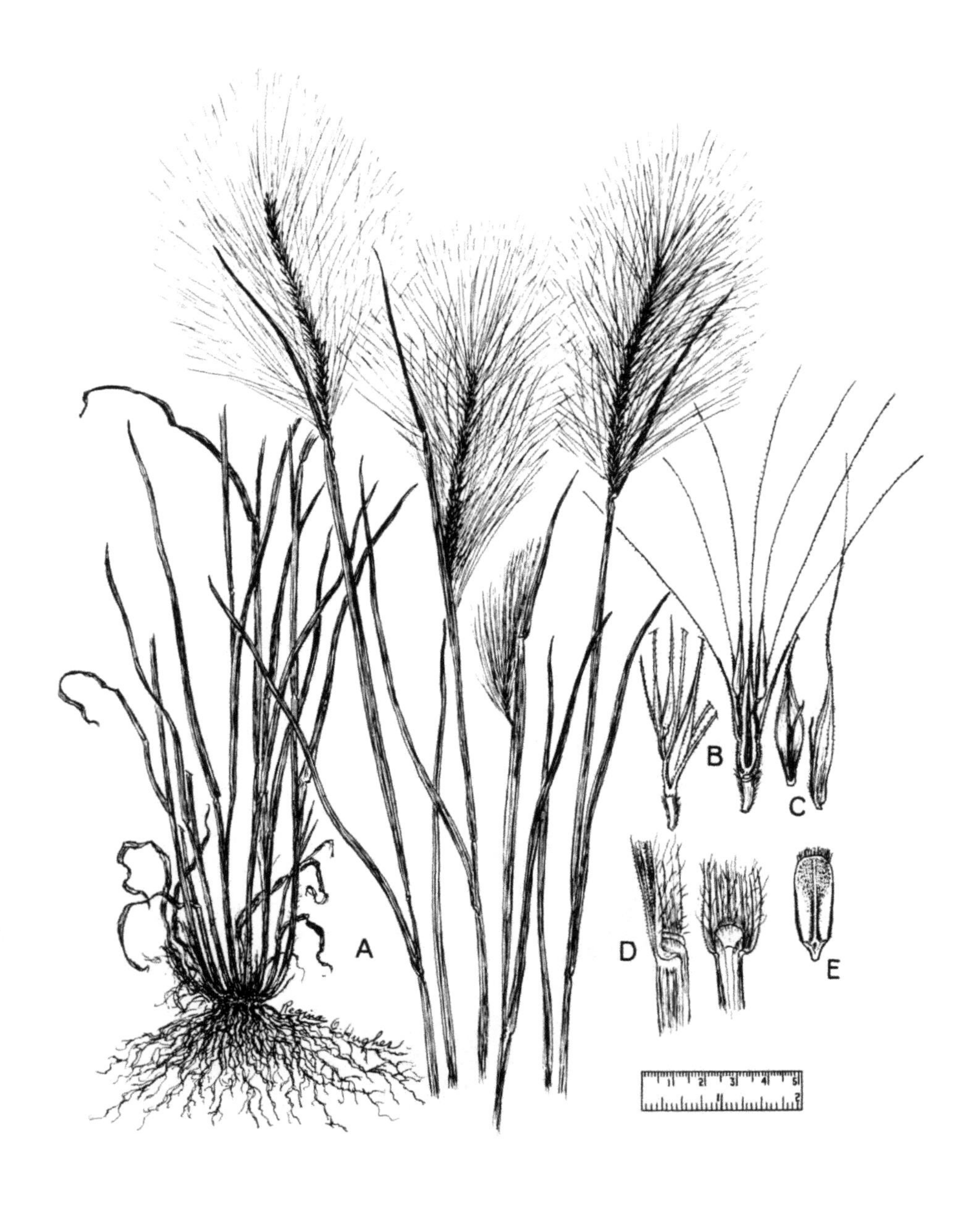

Hordeum jubatum, FOXTAIL BARLEY. A, habit. B, spikelets, side and face views. C, florets. D, ligules. E, caryopsis.

Hordeum pusillum Nutt. LITTLE BARLEY

Annual or winter-annual grass. Roots shallow, fibrous.

DESCRIPTION **Culms** smooth, 1-4 dm tall, bent slightly at each node. **Leaves** erect, 2-7 cm long, 10-14 mm wide, rough on the upper surfaces. **Spike** resembling small dense head of rye, erect, 3.3-8.3 cm long, 1-1.5 cm wide, with short (1.2-1.5 cm) stiff bristles. **First glume** of the lateral spikelet and both glumes of the fertile spikelet dilated at the base, attenuate into a slender awn, 8-15 mm long, the glumes very rough to the touch. **Lemma** of central spikelet awned, of the lateral spikelets awn-pointed. **Caryopsis** 3-8 mm long, hairy at the apex, yellow.

FLOWERING May-June.

WHERE FOUND Plains, pastures, other uncultivated areas, and open places; especially in alkaline ground.

ORIGIN Native; throughout most of the United States; South America.

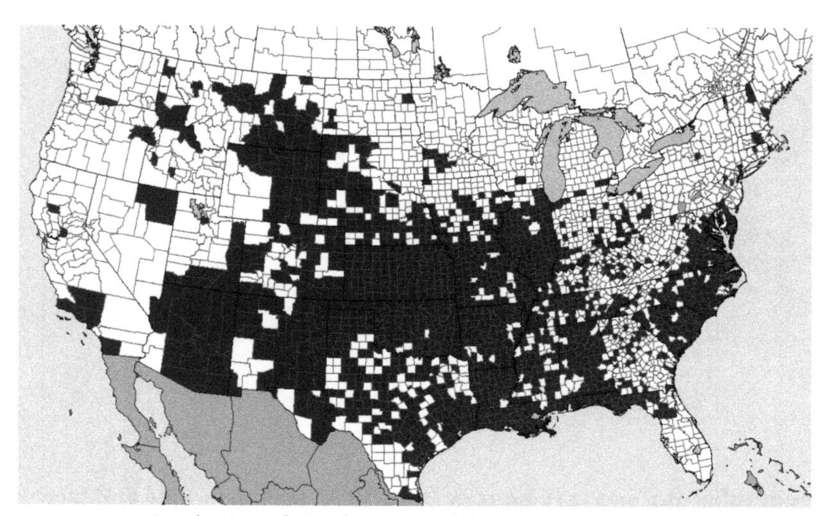

Distribution of **Hordeum pusillum**, LITTLE BARLEY

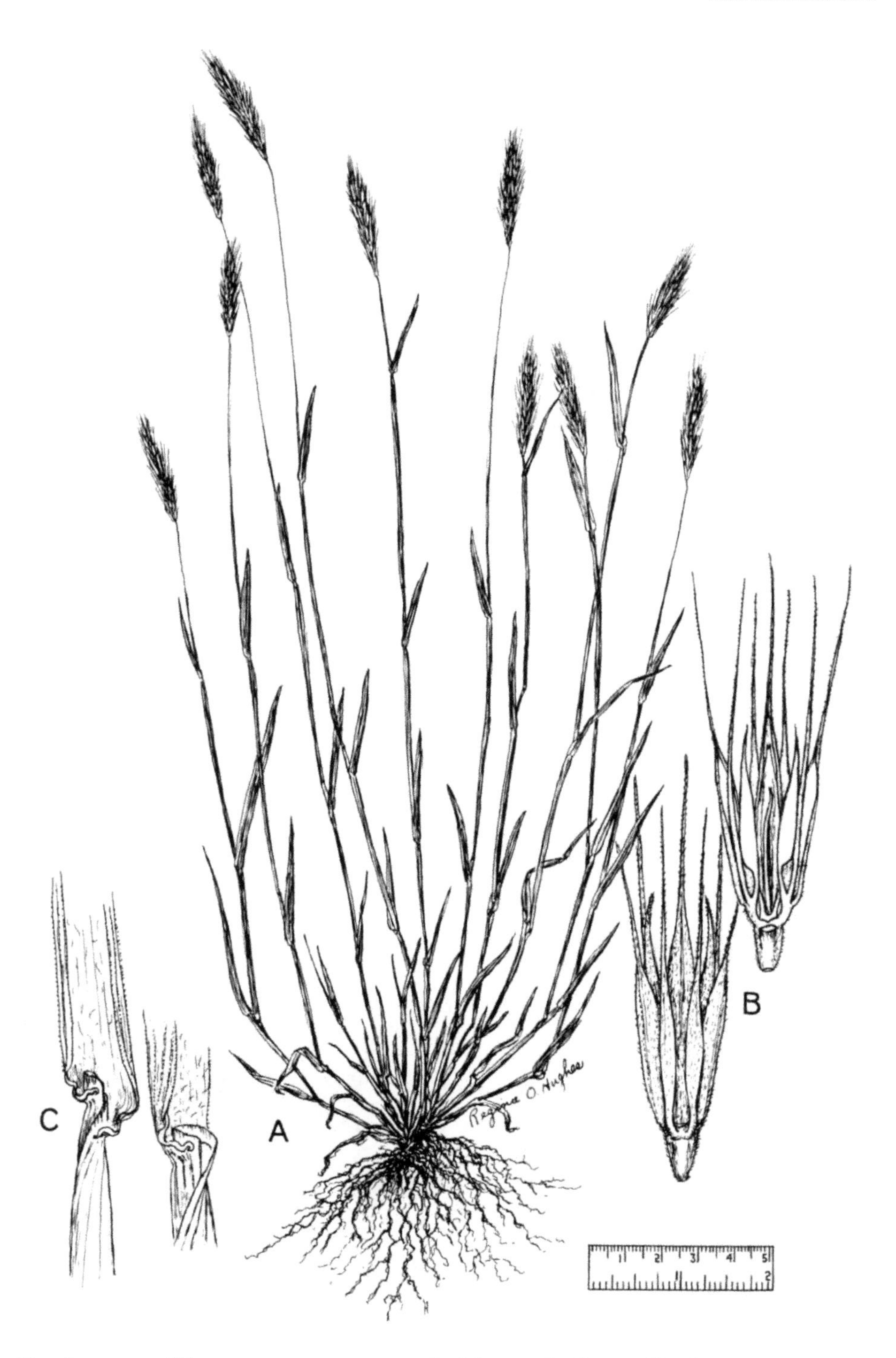

Hordeum pusillum, LITTLE BARLEY. A, habit. B, spikelets. C, ligules.

Muhlenbergia schreberi J. F. Gmel. NIMBLEWILL

Perennial grass.

DESCRIPTION Culms slender, diffuse, branching, spreading and decumbent at the base, often rooting at the lower nodes, but not forming definite creeping stolons, 1.5–6 dm long, freely forking into capillary ascending branches. **Leaf blades** flat, mostly less than 5 cm long, 2–4 mm wide, spreading or loosely ascending. **Panicles** terminal and axillary, threadlike to linear-cylindric, slender, loosely flowered, lax, nodding, 6–18 cm long. **Glumes** minute, the first often obsolete, the second rounded, 0.1–0.2 mm long. **Lemma** green or purple, narrow, somewhat pubescent around the base, strongly 3-nerved, the body about 2 mm long, the slender awn 2–5 mm long. **Palea** about equaling the blade of the lemma. **Caryopsis** loosely infolded by the thin lemma, linear-cylindric, reddish-brown, 1–1.4 mm long.

FLOWERING In spring and early summer the culms are short and erect with spreading blades, the plants being very different in appearance from the flowering phase in the fall.

WHERE FOUND Damp places, woodlands, thickets, roadsides, and dooryards; often a troublesome weed, especially in lawns.

ORIGIN Native.

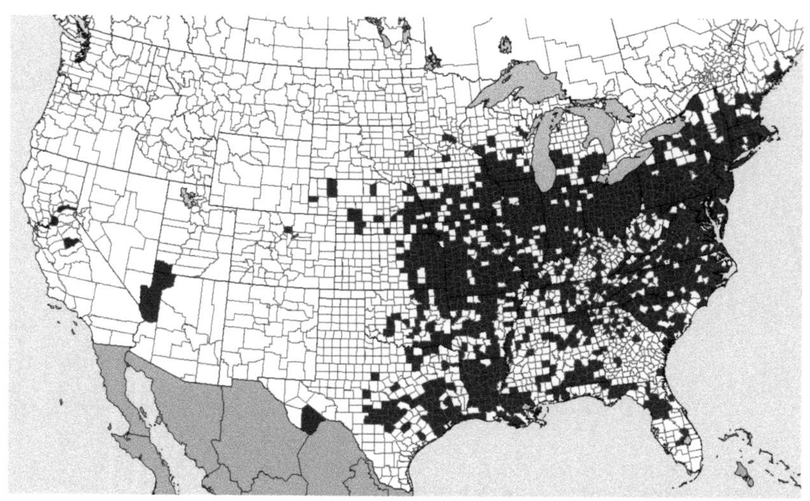

Distribution of **Muhlenbergia schreberi**, NIMBLEWILL

Muhlenbergia schreberi, NIMBLEWILL. A, habit. B, ligules. C, part of spikelet to show glumes. D, florets.

Panicum dichotomiflorum Michx. FALL PANICUM

Annual grass; plants simple or commonly divergently branching from the base and the nodes.

DESCRIPTION Culms ascending or spreading, bent abruptly at the joints at the base, 5-10 dm (but occasionally up to 20 dm) long, flattened, succulent, lower nodes enlarged. **Ligule** a dense ring of white hairs, 1-2 mm long. **Leaf blades** narrowly lanceolate, rough to the touch and sometimes sparsely pilose on the upper surface, 10-50 cm long, 3-25 mm wide, the white midrib usually prominent. **Panicles** terminal and axillary, mostly included in the sheath at the base, 10-40 cm long or more, the main branches ascending. **Spikelets** narrowly oblong-ovate, 2-4 mm (usually 2.5 mm) long, acute, mostly longer than their rough-angled pedicels and on one side only toward the tips of the branchlets. **First glume** rounded-triangular, one-fifth to one-fourth as long as the second and the sterile lemma. A variable species.

FLOWERING June-October.

WHERE FOUND Moist ground and along streams; a weed in waste places and in cultivated soil.

ORIGIN Native, but considered adventive in the western states.

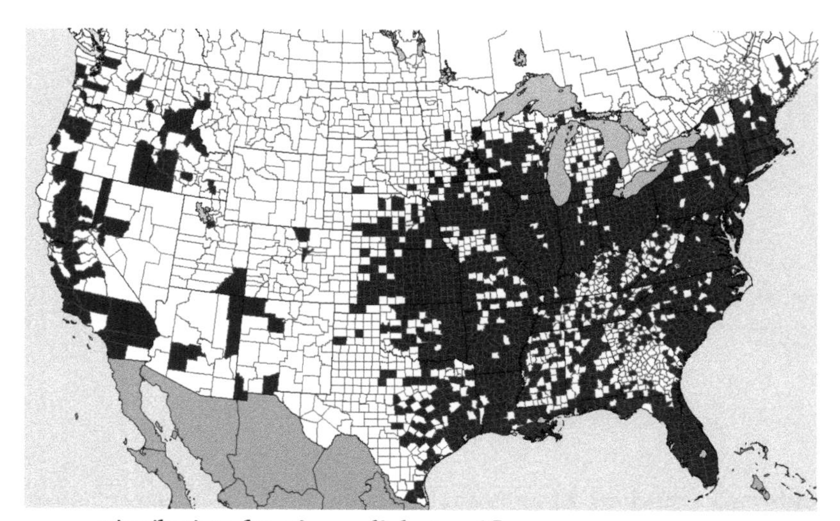

Distribution of **Panicum dichotomiflorum**, FALL PANICUM

Panicum dichotomiflorum, FALL PANICUM. A, habit. B, spikelet, showing the dichotomous florets. C, ligule. D, caryopses.

Paspalum dilatatum Poir. DALLISGRASS

Perennial grass, from a short rhizome.

DESCRIPTION Culms tufted, stoutish, 4.5–17 dm tall, glabrous except the ligules and crowded spikelets. **Leaves** elongated, 10–25 cm long, 4–12 mm wide. **Racemes** 3–5 (occasionally 2–10), 5–19 cm long, loosely ascending and spreading. **Spikelets** egg-shaped, tapering to a point, 3–4 mm long, 2–2.5 mm broad. **Glume and sterile lemma** silky, long soft hairs, over-topping the caryopsis.

FLOWERING May–October.

WHERE FOUND Low ground, from rather dry prairie to marshy meadows, roadsides, and borders of ditches; occasionally on ballast northward; a valuable pasture grass.

ORIGIN Introduced and naturalized from Uruguay and Argentina.

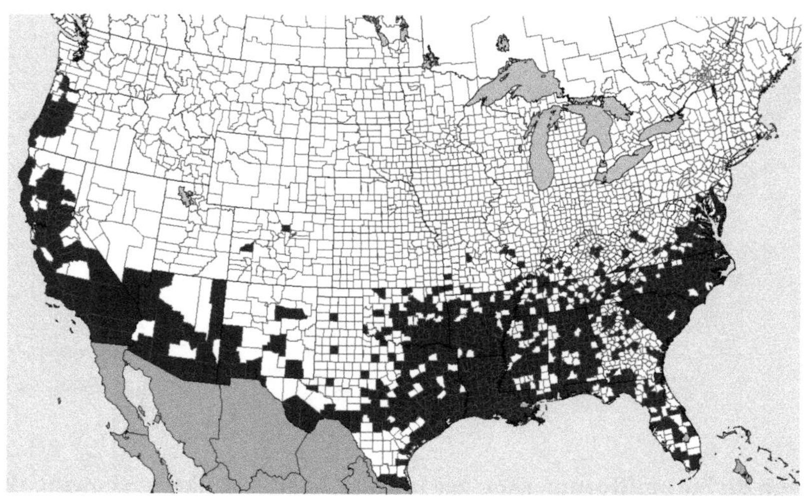

Distribution of **Paspalum dilatatum**, DALLISGRASS

Paspalum dilatatum, DALLISGRASS. A, habit. B, ligule. C, spikelet. D, florets. E, caryopses.

Paspalum distichum L. KNOTGRASS

Perennial grass, stoloniferous.

DESCRIPTION **Culms** creeping, rooting at the pubescent nodes, with ascending culms 1-6 dm long. **Leaves** short, 2-8 cm long, 2-5 mm wide, usually crowded, sometimes sparsely hairy on the margins, ciliate at the base. **Racemes** 2 (rarely 3 or 4), strictly terminal, 1.5-7 cm long, ascending and often incurved. **Spikelets** solitary, 2.5-4 mm long, 1.3-1.5 mm wide, ovate, abruptly acute, sparsely pubescent, pale-green. **First glume** occasionally present, triangular, to 1 mm long. **Second glume** appressed-pubescent, 5-nerved. **Sterile lemma** 5-nerved, glabrous.

FLOWERING July-October.

WHERE FOUND Swamps, ditches, muddy and sandy shores, and, rarely, in brackish areas.

ORIGIN Native.

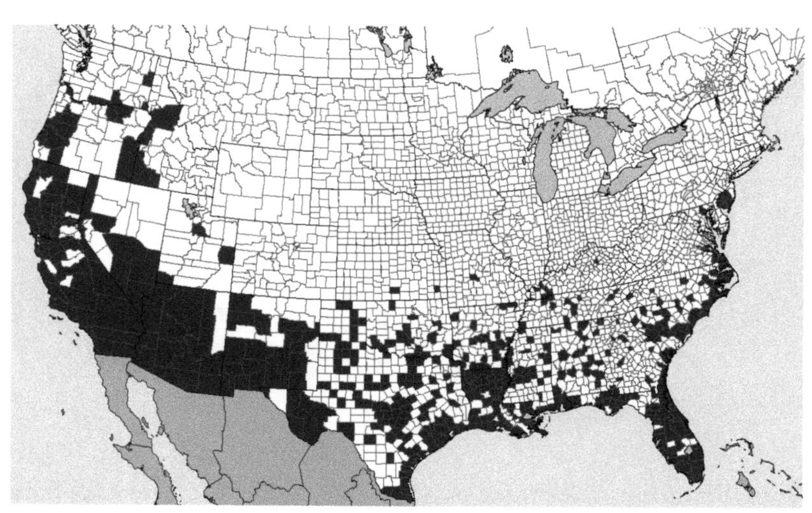

Distribution of **Paspalum distichum**, KNOTGRASS

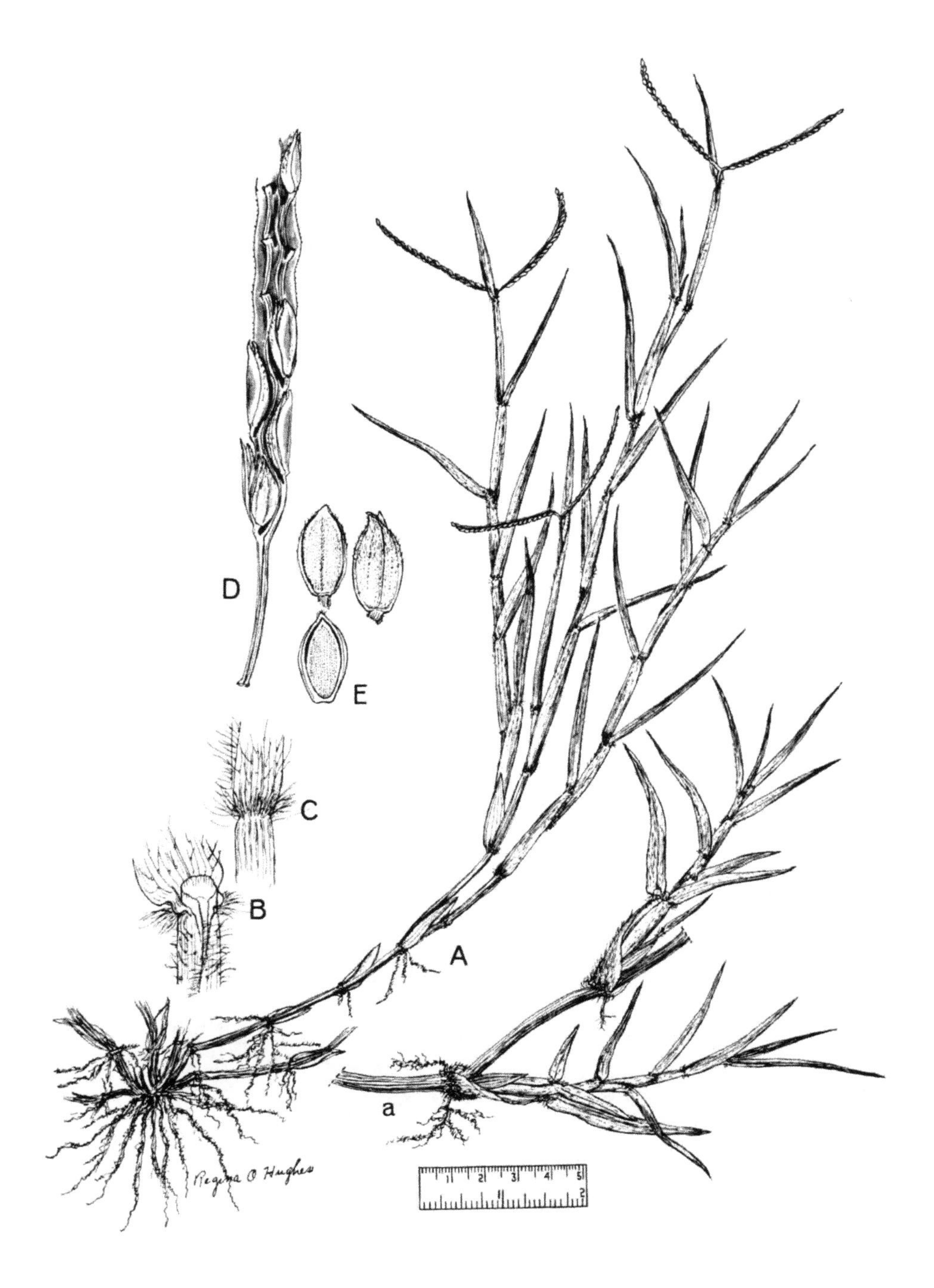

Paspalum distichum, KNOTGRASS. A, habit, with a, young shoot. B, ligule. C, back of leaf, showing fringe of long hairs at base of blade. D, spike. E, florets.

Phalaris arundinacea L. REED CANARY GRASS

Perennial grass, with creeping rhizomes; plants glaucous.

DESCRIPTION Culms erect, 6–20 dm tall. **Leaves** flat, elongate, 0.6–2 cm wide. **Panicles** 5–20 cm (but occasionally up to 30 cm) long, narrow, the branches spreading during anthesis, the lower as much as 5 cm long, tightly contracted in fruit. **Spikelets** lanceolate, 4–6 mm long, pale. **Glumes** about 5 mm long, narrow, acute, the keel rough to the touch, very narrowly winged, or wingless. **Fertile lemma** lanceolate, 4 mm long, with a few flat-lying hairs. **Sterile lemma** with long soft hairs, 1 mm long. **Caryopsis** 3–4.2 mm long, 0.7–1.5 mm broad.

FLOWERING June–September.

WHERE FOUND Shores, swales, and meadows; an important constituent of lowland hay from Montana to Wisconsin.

ORIGIN Native.

NOTE Form with white-striped leaves grown for ornament in gardens, called ribbongrass or gardener's garters.

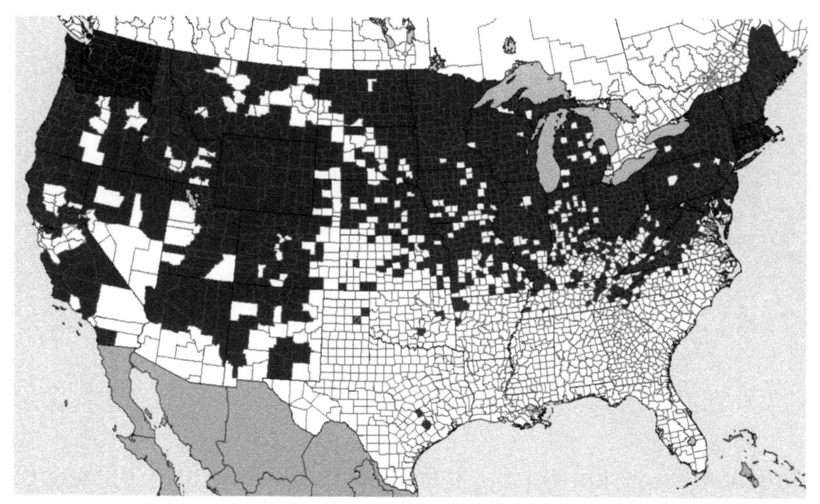

Distribution of **Phalaris arundinacea**, REED CANARY GRASS

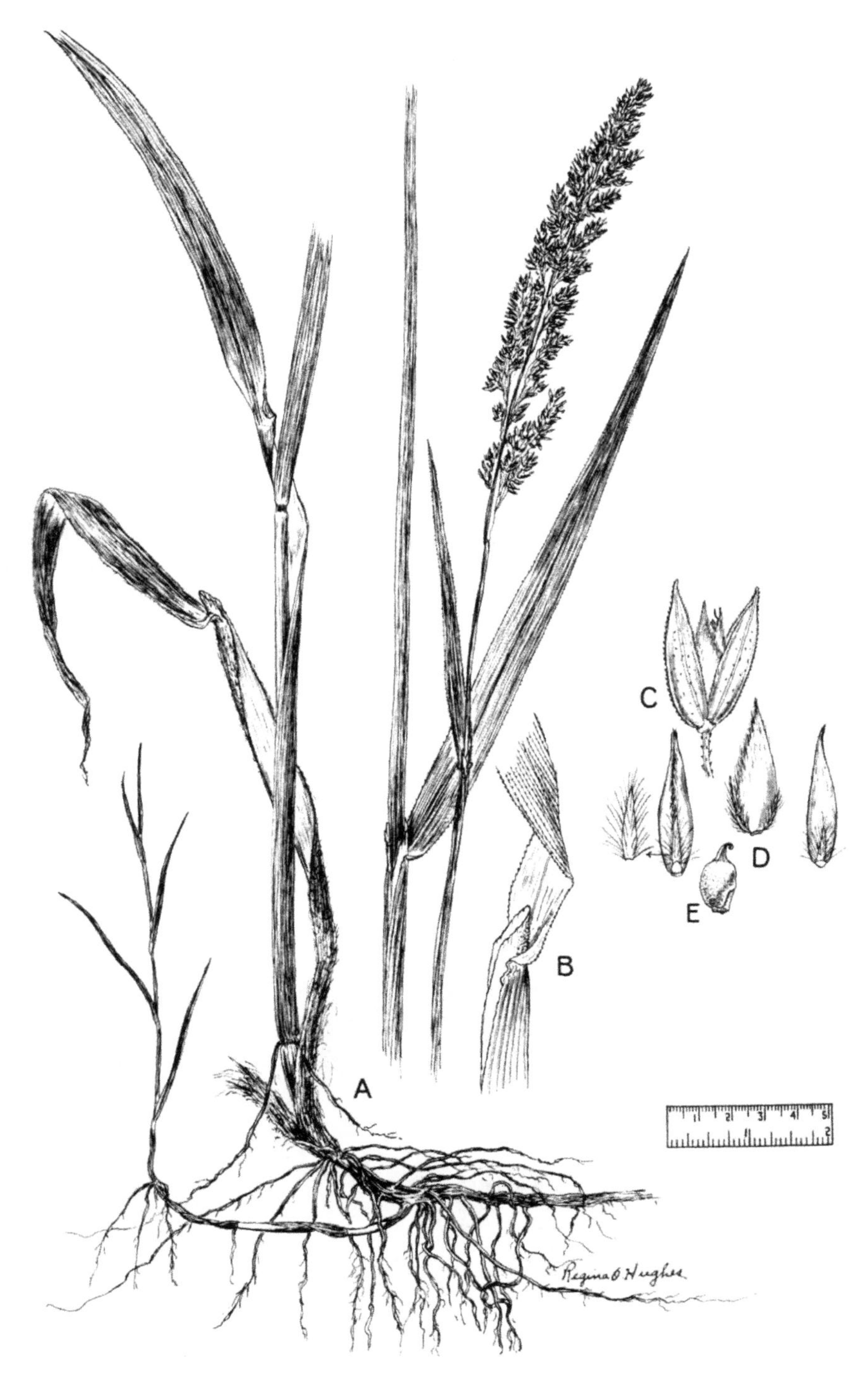

Phalaris arundinacea, REED CANARY GRASS. A, habit. B, ligule. C, spikelet. D, florets. E, caryopsis.

Phragmites australis (Cav.) Trin. ex Steud. COMMON REED

Large perennial reedgrass, with broad flat, linear blades and large terminal panicles.

DESCRIPTION Culms erect, 2-4 m (but occasionally up to 6 m) tall, with stout creeping rhizomes, often also with stolons. **Leaf blades** flat, 1.5-6 dm long, 1-6 cm wide, glabrous, the sheaths overlapping. **Panicle** tawny or purplish, 15-40 cm long, the branches ascending, rather densely flowered. **Spikelets** 10-17 mm long, the florets exceeded by the hairs of the rachilla. **First glume** 2.5-5 mm long; **second glume** 5.7 mm long.

FLOWERING Late July-September.

SYNONYMS *Phragmites communis* Trin.

WHERE FOUND Marshes, banks of lakes and streams, and wet waste places.

ORIGIN Introduced.

NOTE American reed [*Phragmites americanus* (Saltonstall, P.M. Peterson, & Soreng) A. Haines] now separated from introduced and invasive common reed. The most apparent distinguishing field character is that the middle and upper stem internodes of American reed are smooth, shiny and red-brown to dark red-brown during the growing season. In common reed, the middle to upper stem internodes are dull, ridged, and tan-colored during the growing season.

In the Southwest, used for lattices in the construction of adobe huts; stems used by Native Americans for shafts or arrows; in Arizona and Mexico, used for mats, screens, thatching, cordage, and for carrying nets.

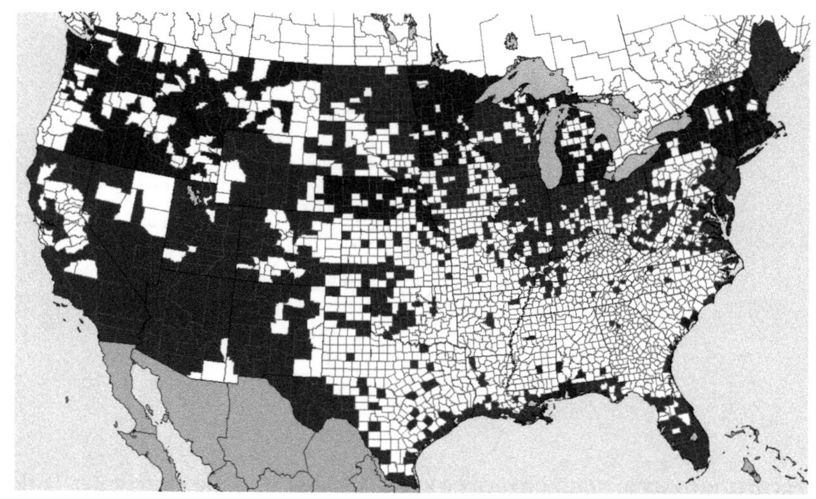

Distribution of **Phragmites australis,** COMMON REED

Phragmites australis, COMMON REED. A, habit. B, ligule. C, spikelet.

Setaria faberi Herrm. GIANT FOXTAIL

Annual grass, branching at the base, reproducing by seed.

DESCRIPTION Culms 8–18 dm tall, plants often so weak as to fall over unless supported by each other or by other vegetation. **Leaf blades** usually softly pubescent beneath to becoming glabrous, and with flattened straight stiff hairs on the upper surface, 8–17 mm broad. **Panicle** dense, 7.5–20 cm long, flexuous to conspicuously nodding, 1–1.7 dm long, 2.3 cm thick. **Spikelets** about 3 mm long, with 3–6 bristles extending from the base of each spikelet, the second glume shorter than the wrinkled seed. **Caryopsis** usually greenish, about 1.5 mm long, transversely and abundantly cross-wrinkled.

FLOWERING June–September.

WHERE FOUND Fields, waste places, disturbed soils, roadsides, and cultivated ground; a serious weed in Illinois and Missouri.

ORIGIN Introduced from China, probably in seed of Chinese millet (1931).

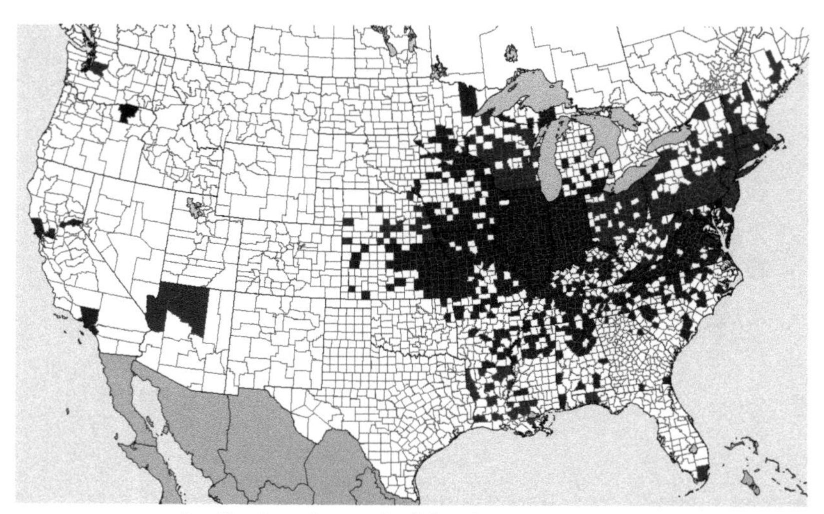

Distribution of **Setaria faberi,** GIANT FOXTAIL

A, **Setaria faberi**, GIANT FOXTAIL, a, habit. b, spikelet, showing subtending bristles. c, ligule. d, caryopses. B, **Setaria viridis**, GREEN BRISTLEGRASS, a, habit. b, spikelet. c, ligule. d, caryopses. C, **Setaria pumila**, YELLOW BRISTLEGRASS, a. habit. b, spikelet. c, ligule. d, caryopses.

Setaria pumila (Poir.) Roemer & J.A. Schultes YELLOW BRISTLEGRASS

Annual grass, branching at the base.

DESCRIPTION Culms flattened, often tufted, erect to prostrate, mostly 5-12 dm tall. **Leaves** with keeled sheaths. Blades as much as 25 cm long and 3-10 mm wide, flat, twisted in a loose spiral, with hairs long and soft toward the base above. **Panicle** dense, evenly cylindric, spike-like, yellow at maturity, mostly 1.5-12 cm long, 0.9-1.4 cm thick, the axis densely pubescent. Bristles 5-20 in a cluster, 3-8 mm long, the longer ones 2 to 3 times as long as the spikelet. **Spikelets** 3 mm long, with undulate-wrinkled fertile lemma. **Caryopsis** strongly wrinkled.

FLOWERING June-September.

SYNONYMS *Setaria glauca* auct. non (L.) Beauv., *Setaria lutescens* (Weigel) Hubb.

WHERE FOUND Cultivated soils, waste areas, and other disturbed soil.

ORIGIN Introduced from Europe.

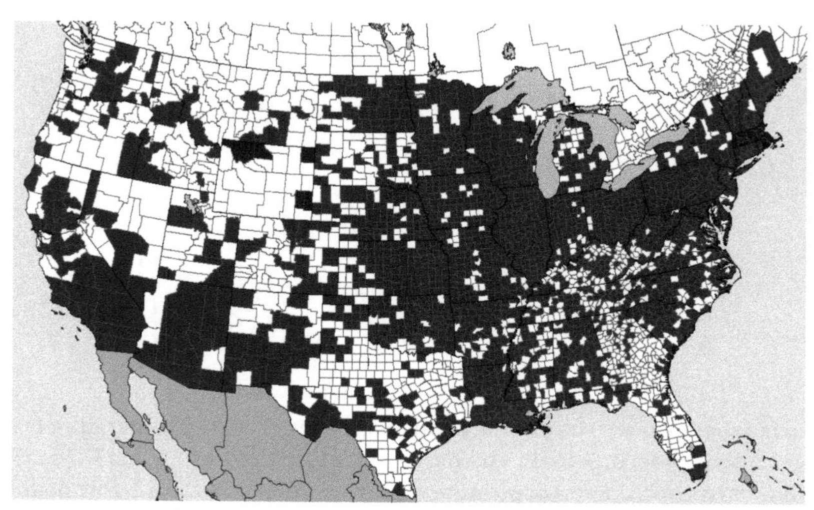

Distribution of **Setari pumila**, YELLOW BRISTLEGRASS

A, **Setaria faberi**, GIANT FOXTAIL, a, habit. b, spikelet, showing subtending bristles. c, ligule. d, caryopses. B, **Setaria viridis**, GREEN BRISTLEGRASS, a, habit. b, spikelet. c, ligule. d, caryopses. C, **Setaria pumila**, YELLOW BRISTLEGRASS, a. habit. b, spikelet. c, ligule. d, caryopses.

Setaria viridis (L.) Beauv. GREEN BRISTLEGRASS

Annual grass; plants simple or tufted, branching at the base, sometimes spreading.

DESCRIPTION Culms 20–40 cm (but occasionally up to 1 m) tall. **Leaf blades** flat, usually less than 15 cm long, 5–15 mm wide, glabrous. **Panicle** erect or somewhat nodding, densely flowered, green or purple, cylindric but tapering a little at the summit, usually 1.5–15 cm long, 1–2.3 cm wide. **Bristles** 1–3 below each spikelet, mostly 3–4 times their length. **Spikelets** 1.8–2.5 mm long, green, the green to purplish bristles upwardly rough to the touch, spreading-ascending and 3–4 times as long. **Caryopsis** very finely rugose.

FLOWERING June–September.

WHERE FOUND A widespread weed of cultivated soils and waste places.

ORIGIN Introduced from Europe. Common throughout all the cooler parts of the United States.

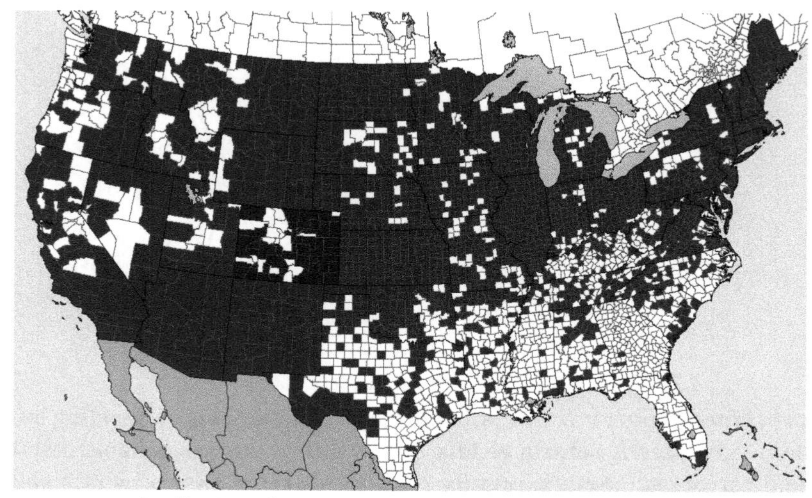

Distribution of **Setaria viridis**, GREEN BRISTLEGRASS

A, **Setaria faberi**, GIANT FOXTAIL, a, habit. b, spikelet, showing subtending bristles. c, ligule. d, caryopses. B, **Setaria viridis**, GREEN BRISTLEGRASS, a, habit. b, spikelet. c, ligule. d, caryopses. C, **Setaria pumila**, YELLOW BRISTLEGRASS, a. habit. b, spikelet. c, ligule. d, caryopses.

Sorghum halepense (L.) Pers. JOHNSON GRASS

Perennial grass, reproducing by large rhizomes and by seeds. Root system freely branching, fibrous, the rhizomes stout, creeping, with purple spots, usually with scales at the nodes.

DESCRIPTION Culms erect, stout, 5-15 dm tall. **Leaves** alternate, simple, smooth, 20-50 cm long, less than 2 cm wide. **Panicles** large, purplish, hairy, 15-50 cm long. **Sessile spikelet** 4.5-5.5 mm long, ovate, with flat-lying straight hairs, the readily deciduous awn 1-1.5 cm long, bent abruptly, twisted below. **Pedicillate spikelet** 5-7 mm long, lanceolate. **Caryopsis** nearly 3 mm long, oval, reddish-brown, marked with fine lines on the surface.

FLOWERING June-October.

WHERE FOUND Open ground, fields, and waste places; cultivated for forage, but becomes a troublesome weed.

ORIGIN Native of the Mediterranean region.

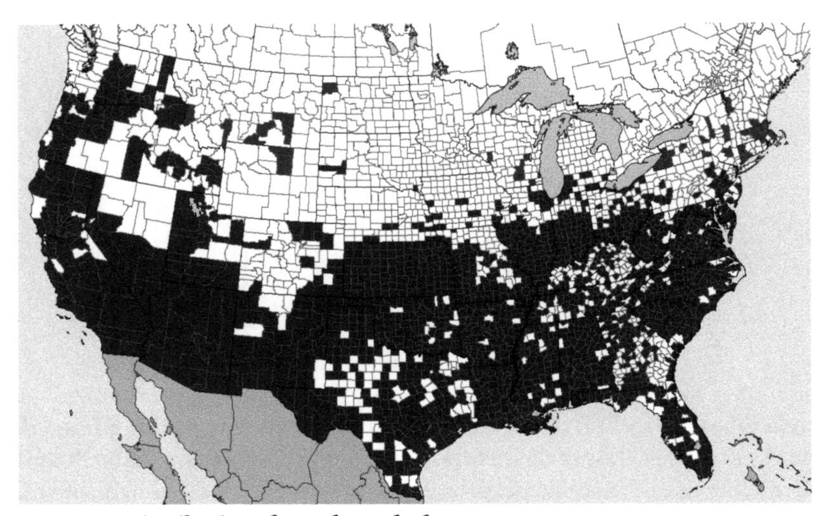

Distribution of **Sorghum halepense**, JOHNSON GRASS

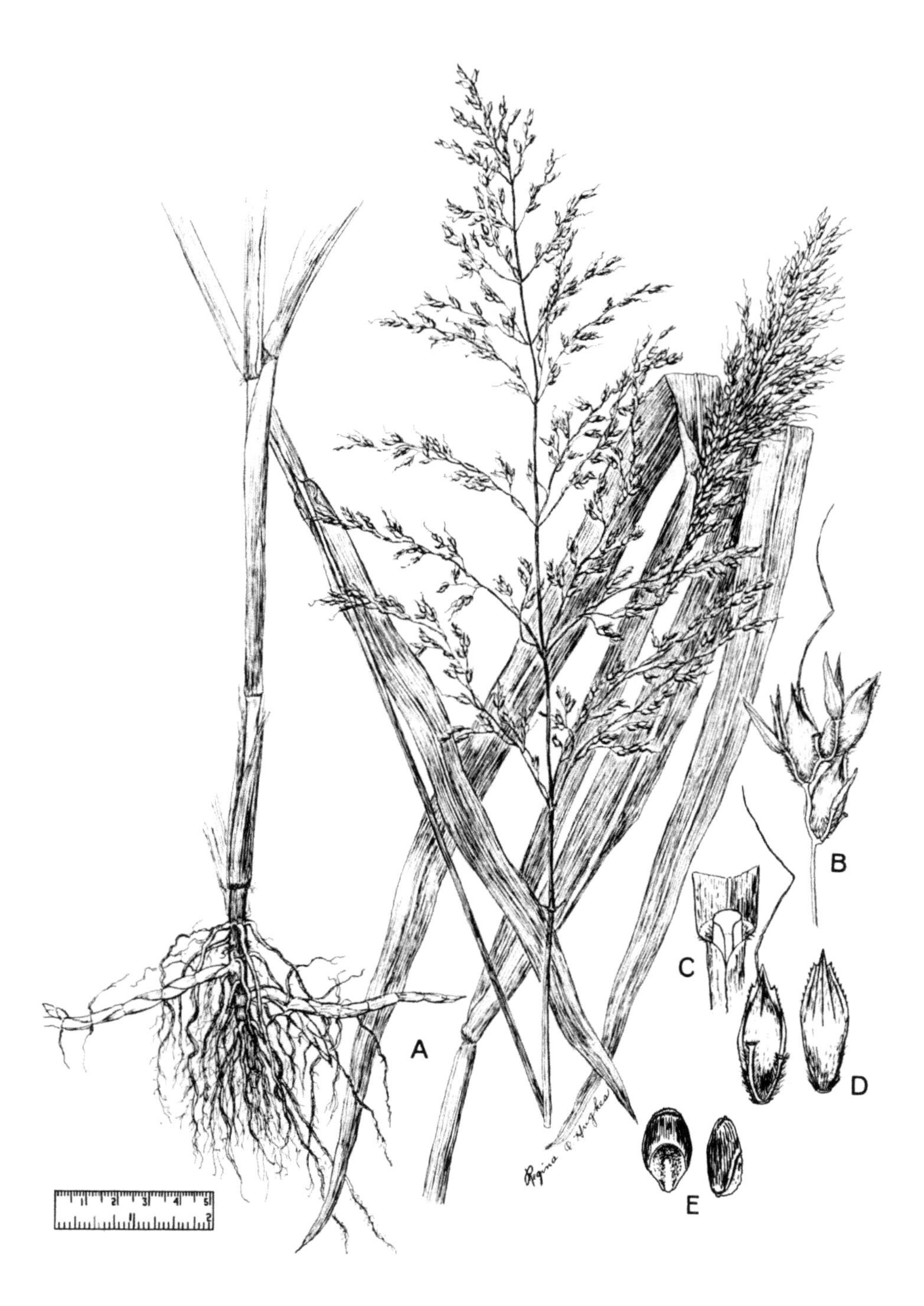

Sorghum halepense, JOHNSON GRASS. A, habit. B, spikelet. C, ligule. D, florets. E, caryopses.

Sporobolus indicus (L.) R. Br. SMUT GRASS

Perennial grass.

DESCRIPTION Culms solitary or tufted, 3–10 dm tall, erect, wiry, with 2 or 3 leaves. **Leaves** flat to slightly rolled inward, rather firm, 10–30 cm long, 2–5 mm wide at base, slenderly tapering. **Panicle** linear-cylindric, stiff, one-fourth to one-half the entire length of the plant, 10–40 cm long, the branches close to the central axis or ascending. **Spikelets** 1.5–2 mm long, shining, crowded on the slender erect branches. **Glumes** obtuse, unequal, the second half as long as the tapering pointed **lemma**, which is slightly longer than the blunt palea.

FLOWERING May–October.

SYNONYMS *Sporobolus poiretii* (Roem. & Schult.) Hitchc.

WHERE FOUND Dry sandy soils, open ground, and waste places; on ballast in New Jersey and Oregon.

ORIGIN Naturalized from tropical America.

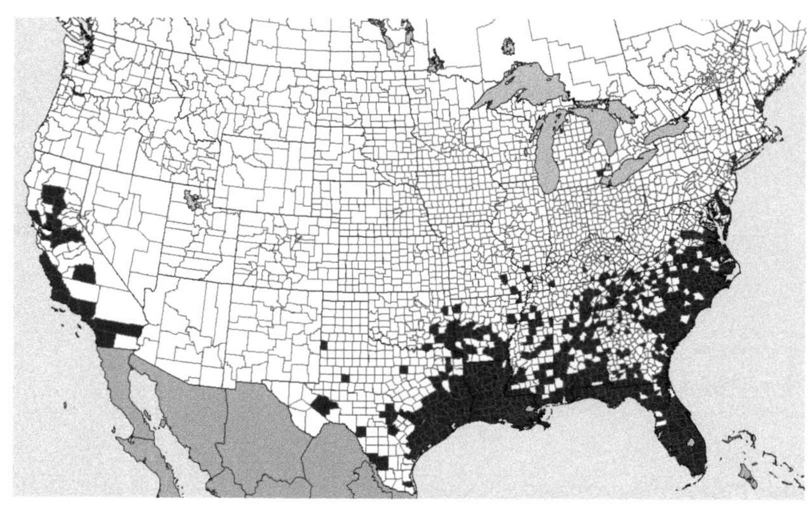

Distribution of **Sporobolus indicus**, SMUT GRASS

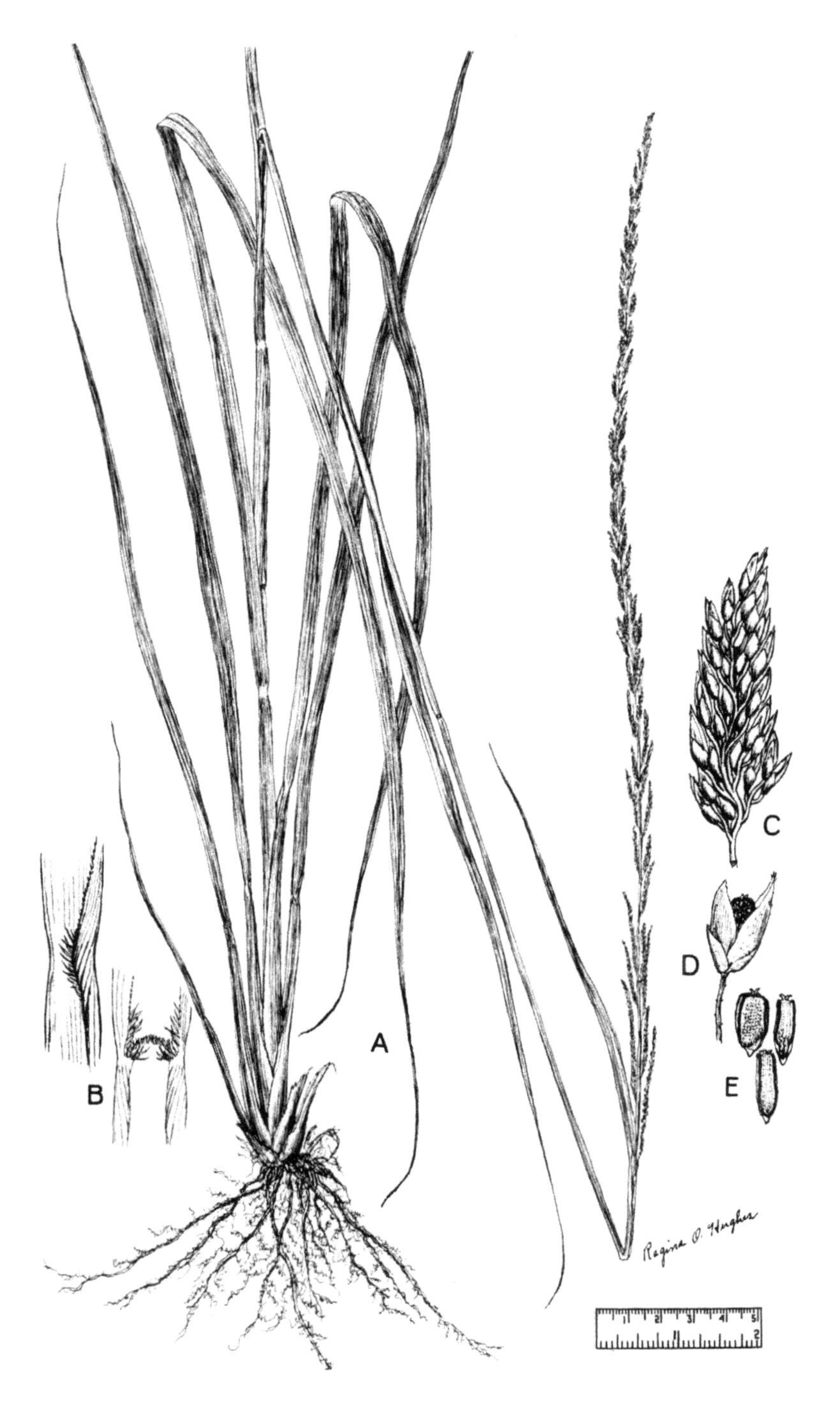

Sporobolus indicus, SMUT GRASS. A, habit. B, ligules (left, as in nature; right, spread open). C, spikelet. D, floret. E, caryopses.

Taeniatherum caput-medusae (L.) Nevski MEDUSAHEAD

Annual grass.

DESCRIPTION Culms ascending from a decumbent, branching base, slender, 2.6 dm tall. **Leaf blades** narrow, short. **Spike** very bristly, 2-5 cm long, excluding the long spreading awns. **Spikelets** 2 (occasionally 3), each containing 1 seed located on each node of the rachis. **Glumes** awl-shaped, smooth, hardened below, tapering into a slender awn, 1-2.5 cm long. **Lemmas** lanceolate, 3-nerved, 6 mm. long, very rough to the touch, tapering into a flat awn 5-10 cm long, containing upward pointing barbs.

FLOWERING Summer.

SYNONYMS *Elymus caput-medusae* L.

WHERE FOUND Open ground; a bad weed, spreading on the ranges in northern California, Oregon, Washington, and Idaho.

ORIGIN Introduced from Europe.

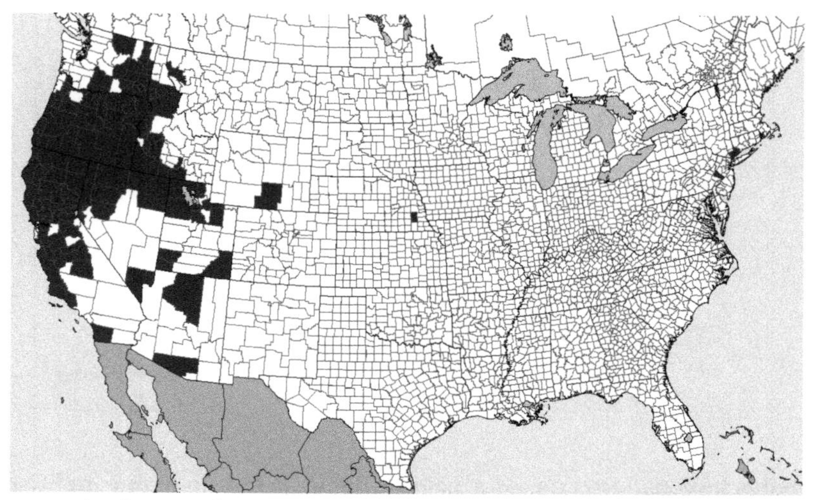

Distribution of **Taeniatherum caput-medusae**, MEDUSAHEAD

Taeniatherum caput-medusae, MEDUSAHEAD. A, habit. B, spikelet, showing arrangement of florets in pairs. C, florets, front and back. D, ligule.

Brunnichia ovata (Walt.) Shinners REDVINE, BUCKWHEAT-VINE

Perennial, woody, tendril-bearing vine.

DESCRIPTION Stems high-climbing, much-branched, sometimes 2 cm thick. **Leaves** alternate, deciduous, the **ocreae** (tubular stipules) obsolete, the blades ovate to ovate-lanceolate, 3–15 cm long. **Flowers** in branching panicles. Perianth becoming 3–3.5 cm long. Sepals 5, oblong, about 5 mm long, surmounting the hypanthium. Hypanthium developing into an indehiscent winged fruit, 2.5–3.5 mm wide, extending the length of the elongate pedicel-like base. Stamens 8. Stigmas 3, each 2-cleft. **Achene** 7–10 mm long, 3-angled, included in the dry hypanthium.

FLOWERING May–September.

SYNONYMS *Brunnichia cirrhosa* Gaertn.

WHERE FOUND Mostly on river banks. In the extreme southeastern coastal area of the United States, along the Gulf coast, north paralleling each side of the Mississippi River to Illinois.

ORIGIN Native.

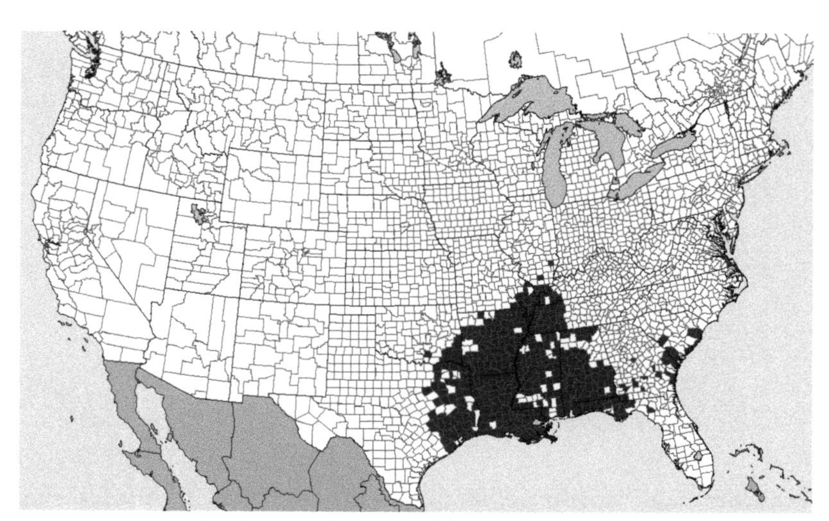

Distribution of **Brunnichia ovata**, REDVINE

Brunnichia ovata, REDVINE. A, habit. B, branch showing fruiting calyces. C, flowers. D, achenes.

Fallopia convolvulus (L.) A. Löve — WILD BUCKWHEAT

Annual herb, reproducing by seeds.

DESCRIPTION Stems glabrous to slightly roughish, scurfy, branched at the base, twining or procumbent, 20–100 cm long, the internodes long. **Leaves** alternate, 2–6 cm long, heart-shaped with basal lobes directed backward, pointed with smooth margins, dull-green, long-petioled; the upper leaves lanceolate, with basal lobes directed backward, tapering to a gradual point. **Stipule sheaths** (ocreae) with entire margins. **Flowers** small, greenish-white, borne in short axillary clusters, or axillary and terminal interrupted or spike-like racemes. Pedicels mostly shorter than the minutely pubescent green or purple-tipped calyx. Fruiting calyx nearly oblique-angled to egg-shaped, 4–5 mm long, 3-angled, with scarcely developed keels. **Achene** 3-angled, black, often covered with a dull-brown minutely roughened hull.

FLOWERING May–November.

SYNONYMS *Polygonum convolvulus* L.

WHERE FOUND Most cultivated areas, waste grounds, gardens, grainfields, and thickets, and along fences; a serious weed.

ORIGIN Naturalized from Eurasia.

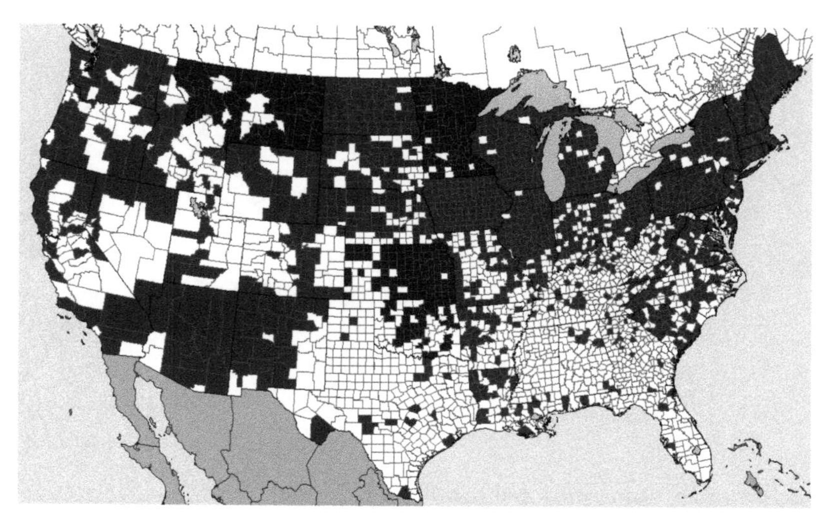

Distribution of **Fallopia convolvulus**, WILD BUCKWHEAT

Fallopia convolvulus, WILD BUCKWHEAT. A, habit. B, branchlet with fruiting calyx. C, flower. D, achenes.

Persicaria amphibia (L.) S. F. Gray p. p. WATER SMARTWEED

Perennial herb, with aquatic and terrestrial forms, reproducing by seeds and by slender and tough, forking rhizomes, stolons, and rooting stems.

DESCRIPTION Branches elongate, simple, leafy to the summit, the young stems, ocreae (tubular stipules) and leaves glabrous to pubescent. **Leaves** alternate, floating or spreading to ascending, elliptic-oval to lanceolate, petioled, gradually tapering or rounded at the base, rarely subcordate. **Ocreae** cylindric, with or without spreading and hairy summit margins. **Spikes** 1–4, dense, straight, erect on erect or terminal glabrous peduncles, cylindric, oval or egg-shaped, the leading spike 1–4 cm long, at maturity 1–2 cm thick. Calyx pink or rose-colored, the sepals spreading and blunt in anthesis, closed in fruit. **Achene** lens-shaped, broadly egg-shaped to circular, 2.5–3 mm long. Highly variable in appearance as influenced by its habitat.

FLOWERING July–October.

SYNONYMS *Polygonum amphibium* L., *Polygonum coccineum* Muhl.

WHERE FOUND Gravelly shores, slopes, prairies, open and wooded swamps, and meadows; also in aquatic situations as margins of lakes and ponds, in quiet waters, and in floating and stranded situations. Throughout the northern areas of the world and in south Africa.

ORIGIN Native.

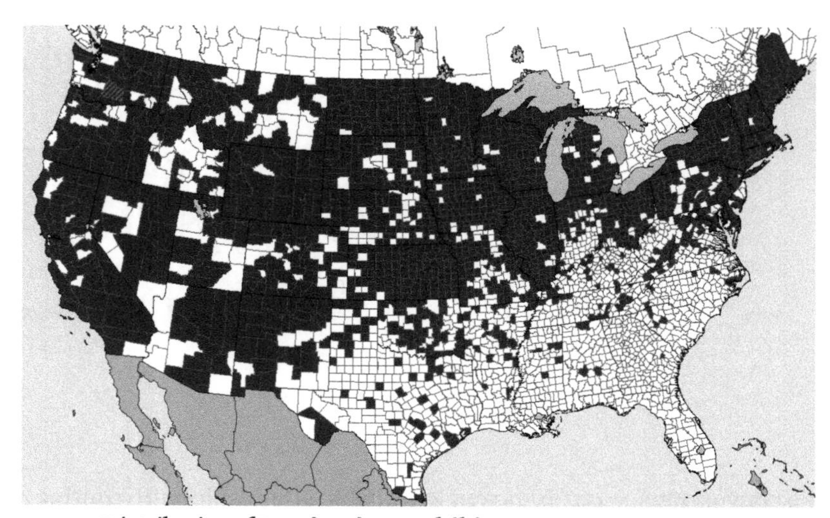

Distribution of **Persicaria amphibia**, WATER SMARTWEED

Persicaria amphibia, WATER SMARTWEED. A, habit. B, ocrea. C, flower cluster, showing sheath and bilobed sheathing bracts. D, flowers. E, achene.

Persicaria hydropiper (L.) Delarbre MARSH-PEPPER SMARTWEED

Annual herb, reproducing by seeds.

DESCRIPTION Stems jointed, somewhat flattened, ascending or erect, 3-6 dm tall, glabrous, green or more or less reddened, intensely acrid or peppery. **Leaves** alternate, simple, narrowly lanceolate to oblanceolate, acute or tapering to a gradual point, entire, 9 cm long, sessile or extending downward into short petioles, glabrous, with wavy margins. **Ocreae** (tubular stipules) thin, dry, membranaceous, brown, truncate, bristly, hairy-margined, the hairs 1-2 mm long. **Spikes** slender and lax, arching at the tips of the branches, with distant cleistogamous fascicles extending down to and often inclosed in the upper sheaths. Ocreolae (small secondary sheaths) inversely top-shaped, greenish, sometimes with red tips, without hairs or with short hairs. Pedicels included or exserted. Calyx greenish or red-tipped, 2-4.5 mm long in fruit, covered with dark sessile glands. Stamens 6. Pistil with 1-celled ovary, the style 2- to 3-parted. **Achene** lens-shaped to 3-angled, dull, minutely dotted with depressions in lines, 2-3.5 mm long, dark-brown to black, the tip often slightly exserted, with the remains of the perianth usually attached.

FLOWERING June-November.

SYNONYMS *Polygonum hydropiper* L.

WHERE FOUND Damp soils, low meadows, pastures, cultivated ground, and waste places.

ORIGIN Native to North America and Europe, widespread.

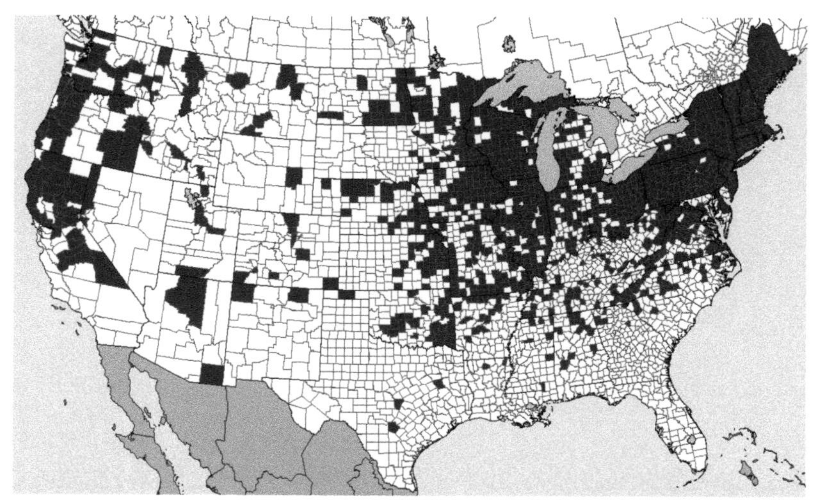

Distribution of **Persicaria hydropiper,** MARSH-PEPPER SMARTWEED

Persicaria hydropiper, MARSH-PEPPER SMARTWEED. A, habit. B, node showing ocrea. C, spike. D, flower. E, achenes.

Persicaria maculosa S. F. Gray LADY'S-THUMB

Annual herb, reproducing by seeds.

DESCRIPTION Stems smooth, sometimes hairy, simple to much-branched, ascending or decumbent, 0.2–1 m tall. Leaves alternate, narrowly to broadly lanceolate, pointed at both ends, 2.5–15 cm long, 0.5–3 cm broad, with smooth edges and usually with a purplish blotch in the middle. **Sheath** (ocreae) at base of the leaf fringed with short bristles. **Flowering spikes** 1 to several in a panicle, glabrous-peduncled or the secondary spikes short-peduncled to nearly sessile, oblong to thick-cylindric, dense, 7–11 mm thick, the leading spikes 1.5–4.5 cm long. Ocreolae (secondary sheaths) usually with hairy margins. Calyx pink, purplish, or green and pink, rarely white, 2–3 mm long (4 mm in fruit), the mature sepals with prominently netted-veined bases. **Achene** almost circular, flattened or 3-angled, smooth, black, shiny, 2.5–3 mm long.

FLOWERING June–October.

SYNONYMS *Polygonum persicaria* L.

WHERE FOUND Cultivated ground, waste places, and roadsides and along ditches, damp clearings, and shores.

ORIGIN Naturalized from Europe.

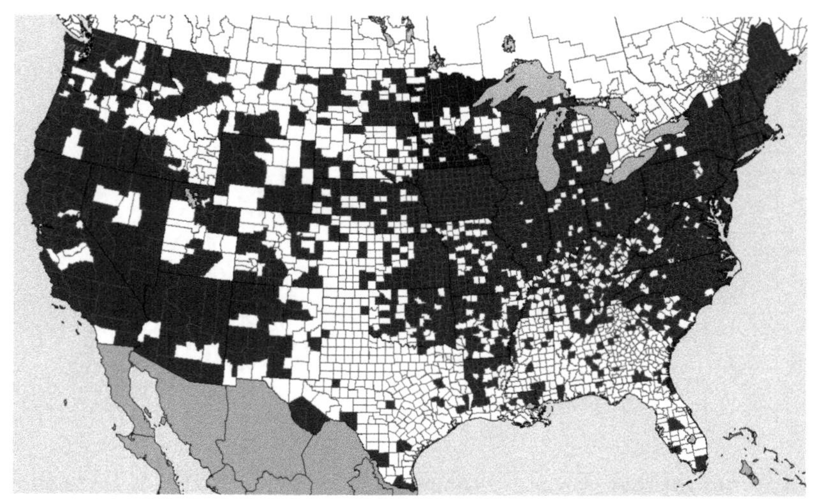

Distribution of **Persicaria maculosa**, LADY'S-THUMB

Persicaria maculosa, LADY'S-THUMB. A, habit. B, spike. C, ocrea. D, achenes.

Persicaria pensylvanica (L.) M. Gómez PENNSYLVANIA SMARTWEED

Annual herb, reproducing by seeds.

DESCRIPTION Stems ascending to erect, up to 1.2 m tall, swollen at the nodes, branching, glandular to nearly glabrous, often strigose-hispid. **Leaves** alternate, glabrous, often with close-lying stiff hairs, pointed, lanceolate to elliptic or oval, 5–15 cm long. **Sheath** (ocrea) at the base extending around the stem, thin, membranaceous, cylindric, soon falling away, without hairs, the upper sheaths often glandular or strigose. **Flowers** in dense erect spikes, bright-pink or rose, nearly round to thick-cylindric, 1–6 cm long, 1–1.5 cm thick, the larger ones peduncled; the flowers all alike or if occasionally of two types (sexes) borne in the same inflorescence. **Achene** shiny, black, smooth, lens-shaped, round-ovoid to circular, flattened on one face, concave on the other, 2.2–3.5 mm broad. Several varieties may be distinguished.

FLOWERING Late May–October.

SYNONYMS *Polygonum pensylvanicum* L.

WHERE FOUND Cultivated ground and waste places and along ditches, damp shores, and thickets.

ORIGIN Native.

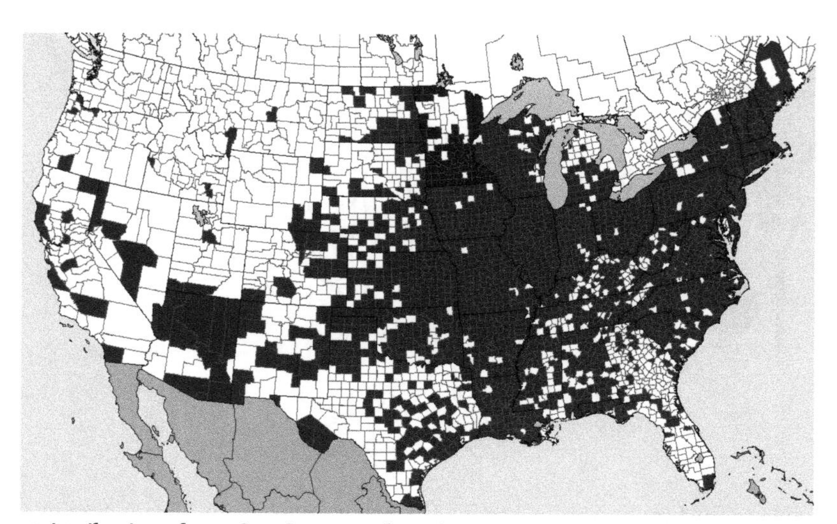

Distribution of **Persicaria pensylvanica**, PENNSYLVANIA SMARTWEED

Persicaria pensylvanica, PENNSYLVANIA SMARTWEED. A, habit. B, spike. C, achenes.

Polygonum aviculare L. PROSTRATE KNOTWEED

Annual herb, reproducing by seeds.

DESCRIPTION Stems 10–100 cm long, prostrate or loosely ascending to erect, the main stem corrugated, much-branched, mostly forming mats from thin taproots. **Leaves** alternate, entire, sharp-pointed to rounded at the end, narrowed at the base, blue-green, lanceolate, linear to oblong or elliptic, 5–30 mm long, 1–8 mm wide, scattered to close together on the stem, petioles very short, united to the short sheath formed by the stipules, veinless or inconspicuously veined. **Flowers** in axillary clusters, perfect, small. Pedicels soon exserted beyond the hyaline and flaccid soon torn ocreae (tubular sheaths). **Perianth** (fruiting calyx) 2–3 mm long, with 5–6 lobes, green with pinkish or whitish margins, appressed at maturity. Stamens mostly 8. Pistil solitary, ovary 1-celled, styles 3. **Achene** egg-shaped, 2–2.5 mm long, 3-angled, dark reddish-brown to black, dull to nearly lustrous, granular-striate to minutely punctuate, the remains of the perianth usually attached. Variable species with many forms and varieties.

FLOWERING July–November.

WHERE FOUND Disturbed soils, lawns, along streets, gardens, roadsides, and wastes; a ubiquitous weed.

ORIGIN Native to North America but also naturalized from Europe. Throughout all the United States and southern Canada; north to Alaska and Newfoundland.

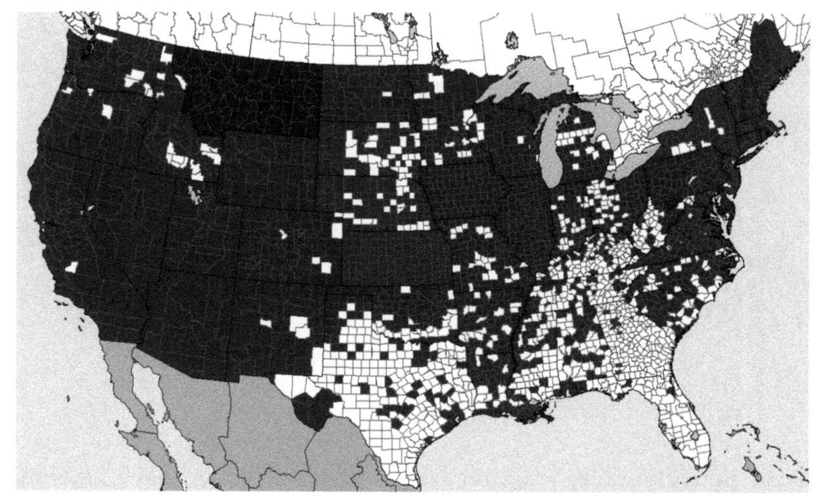

Distribution of **Polygonum aviculare**, PROSTRATE KNOTWEED

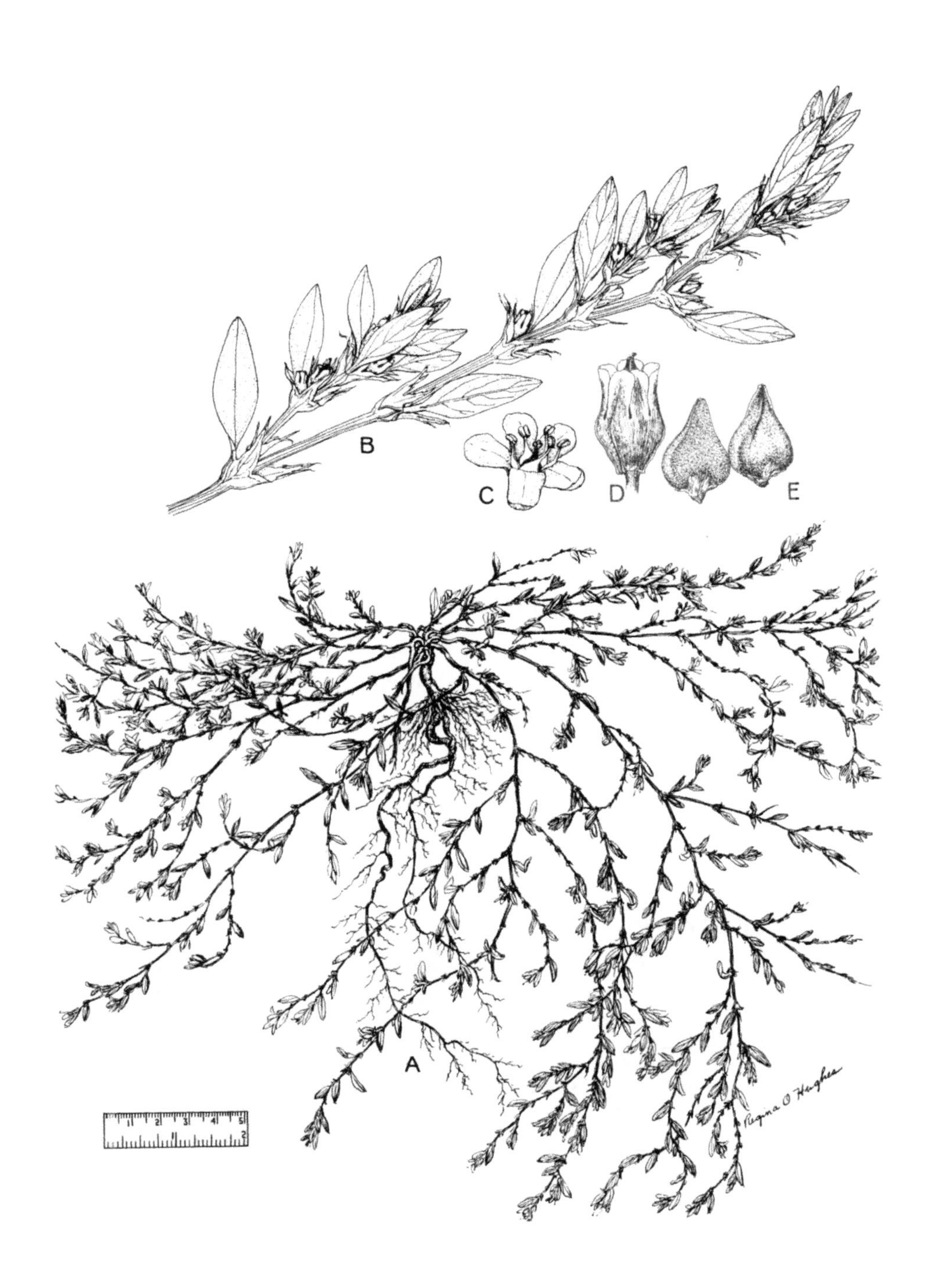

Polygonum aviculare, PROSTRATE KNOTWEED. A, habit. B, flowering branch, enlarged. C, flower D, fruiting calyx. E, achenes.

Rumex acetosella L. COMMON SHEEP SORREL

Perennial, reproducing by seeds and creeping rhizomes. Roots and rhizomes extensive but rather shallow.

DESCRIPTION **Stems** slender, erect, 1.5–4.5 dm tall, branched at the top, several stems from one crown or from the rhizomes. **Leaves** alternate, simple and entire, a rosette of basal leaves in early growth, the stem leaves arrow-shaped with the 2 basal lobes somewhat divergent, thick, glabrous, narrowly lanceolate to almost linear upward, 2.5–7 cm long, acid to the taste. **Inflorescence** of slender racemes near the top of the plant, erect in panicles. **Flowers** yellow to red, male and female flowers on different plants, nodding on short-jointed pedicels, the outer sepals lanceolate, the inner sepals in staminate flower 1.5–2 mm long, obovate, in pistillate flower broadly ovate. **Hull** reddish-brown, rough, often adhering to the seed. **Achene** 3-sided, about 1.5 mm long, reddish-brown to golden-brown, shiny, exserted from and divested of the calyx. A highly variable species in color and size of plant.

FLOWERING June–October.

WHERE FOUND A ubiquitous weed in gardens, pastures, meadows, and lawns; persists in areas of poor drainage and low soil fertility; in gravelly sterile fields; very difficult to eradicate.

ORIGIN Naturalized from Eurasia; throughout the United States.

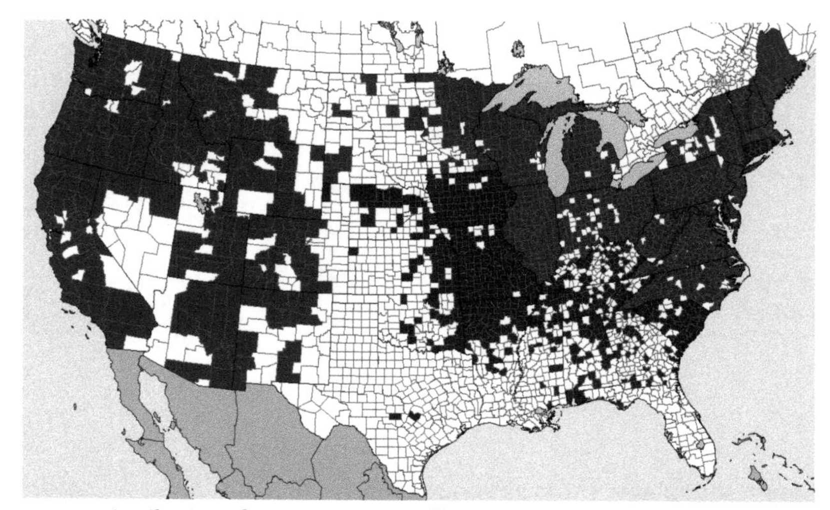

Distribution of **Rumex acetosella**, COMMON SHEEP SORREL

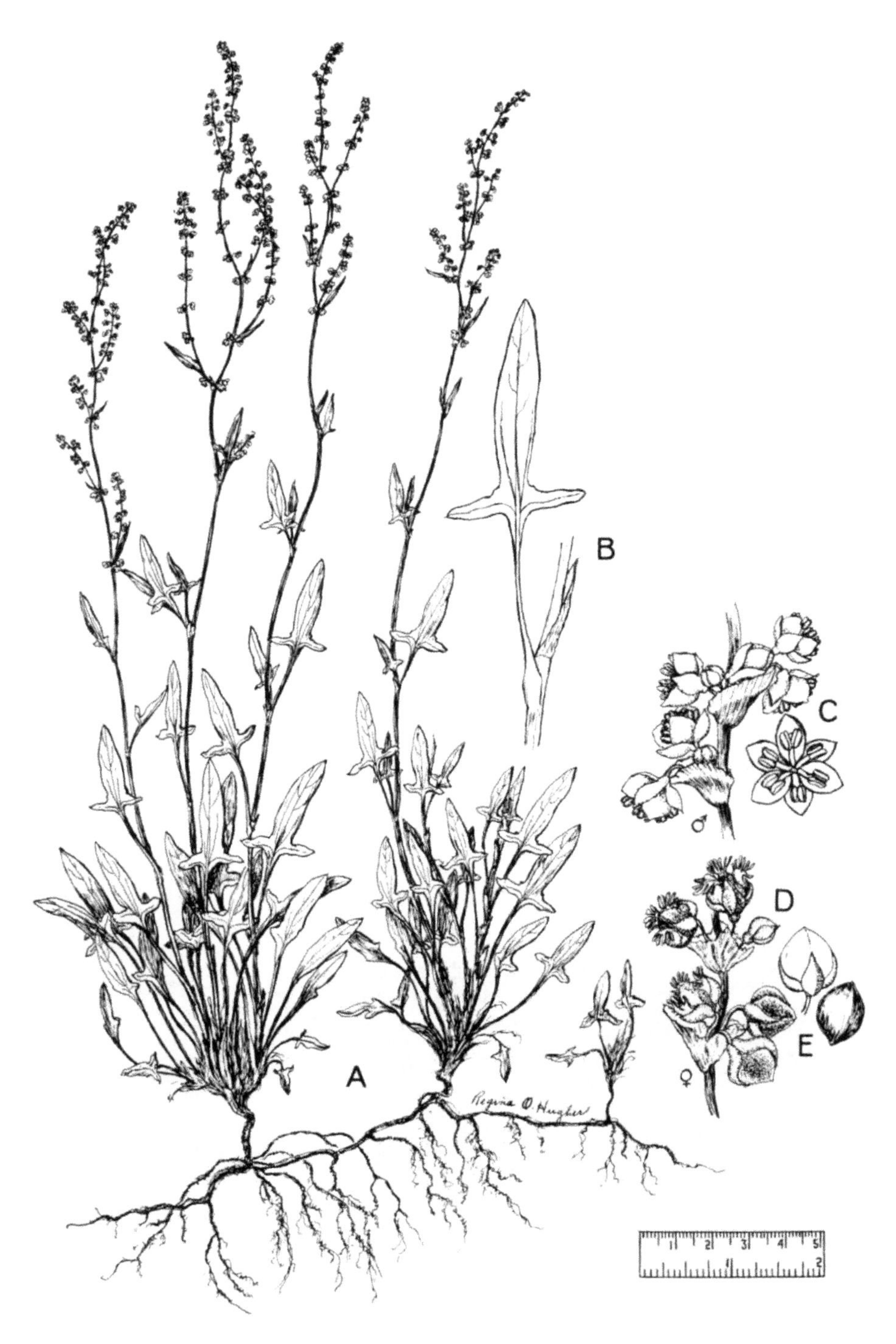

Rumex acetosella, COMMON SHEEP SORREL. A, habit. B, leaf detail. C, staminate flowers D, pistillate flowers E, achenes, in and out of calyx.

Rumex crispus L. CURLY DOCK

Perennial, reproducing by seeds. Taproot large, yellow somewhat branched.

DESCRIPTION Stems glabrous, erect, single or in groups from the root crown, simple to the inflorescence, up to 1 m tall. **Leaves** simple, mostly basal, glabrous, 15–30 cm long, lanceolate, the larger leaves rounded to nearly heart-shaped at the base, with wavy-curled or crisped margins; upper leaves alternate, the base of the short petiole with a papery sheath surrounding the stem. **Inflorescence** large with many erect or ascending branches, with few to many linear leaves intermingled. **Flowers** small, greenish, becoming reddish-brown at maturity, in dense clusters of ascending racemes in branches at the ends of the stems, on long-slender pedicels 5–10 mm long. Calyx of 6 greenish sepals more or less persistent, the 3 inner enlarged (in fruit called the valves), heart-shaped, and nearly entire, 4–6 mm wide, and each bearing a rounded plump grain (tubercule). **Grains** 3, often unequal, the larger ovoid, very turgid, rounded at both ends, about half as long as the valves. **Achene** brown, shiny, triangular, and sharp-edged, about 2 mm long, surrounded with 3 heart-shaped valves with smooth edges.

FLOWERING June–September.

WHERE FOUND Pastures, hayfields, meadows, waste areas, and gardens and along roadsides; often a pernicious weed.

ORIGIN Introduced and naturalized from Eurasia; throughout the United States.

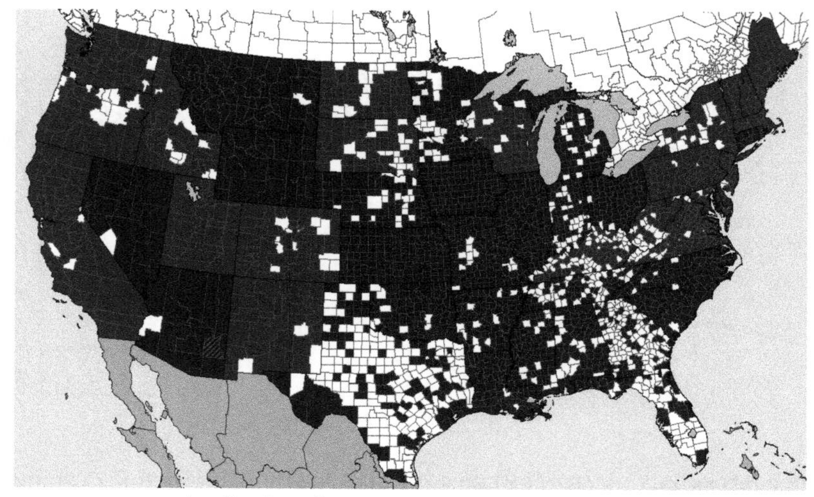

Distribution of **Rumex crispus**, CURLY DOCK

Rumex crispus, CURLY DOCK. A, habit. B, Fruit; a, surrounded by persistent calyx. B, showing 3 valves. C, achene.

Eichhornia crassipes (Mart.) Solms WATER-HYACINTH

Perennial aquatic herb. Rhizome and stems normally floating, rooting at the nodes, with long black pendent roots.

DESCRIPTION Leaves usually with greatly inflated spongy petioles, the leaf blades circular to kidney-shaped, 4–12 cm wide. **Inflorescence** a contracted panicle, 4–15 cm long, with several flowers. **Perianth** lilac, bluish-purple, or white, the upper lobe bearing a violet blotch with a yellow center. Stamens 6. Stalk of the inflorescence soon becoming goose-necked, forcing the dead flowers under the water. **Capsule** dehiscent, surrounded by the perianth, membranous, many-seeded.

FLOWERING July–October, longer in the warmer regions.

SYNONYMS Genus also sometimes spelled *Eichornia.*

WHERE FOUND Ditches, quiet streams, rivers and waterways, lakes, and ponds; usually very prolific and troublesome, often clogging and obstructing waterways. Widely naturalized in all tropical regions; sometimes cultivated in the northern regions of the range as the 'water orchid.'

ORIGIN Introduced and naturalized from South America; throughout the southeastern United States. California, scattered locations inland; introduced in Europe and Asia.

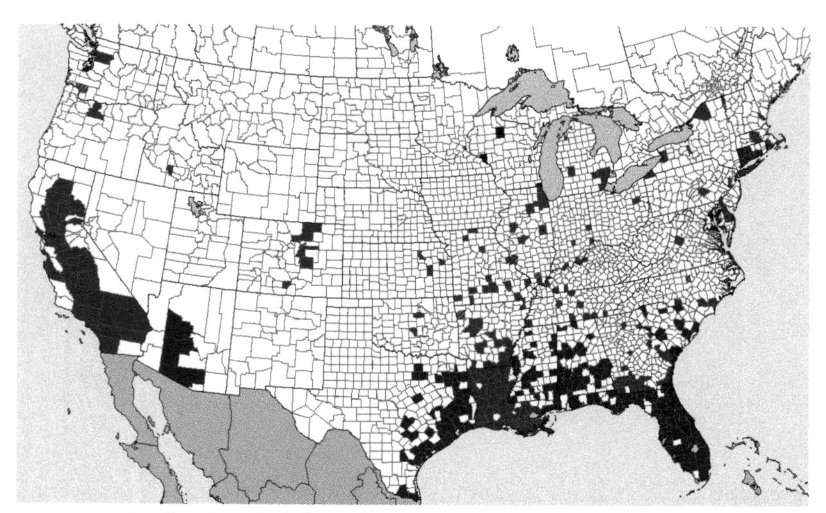

Distribution of **Eichornia crassipes,** WATER-HYACINTH

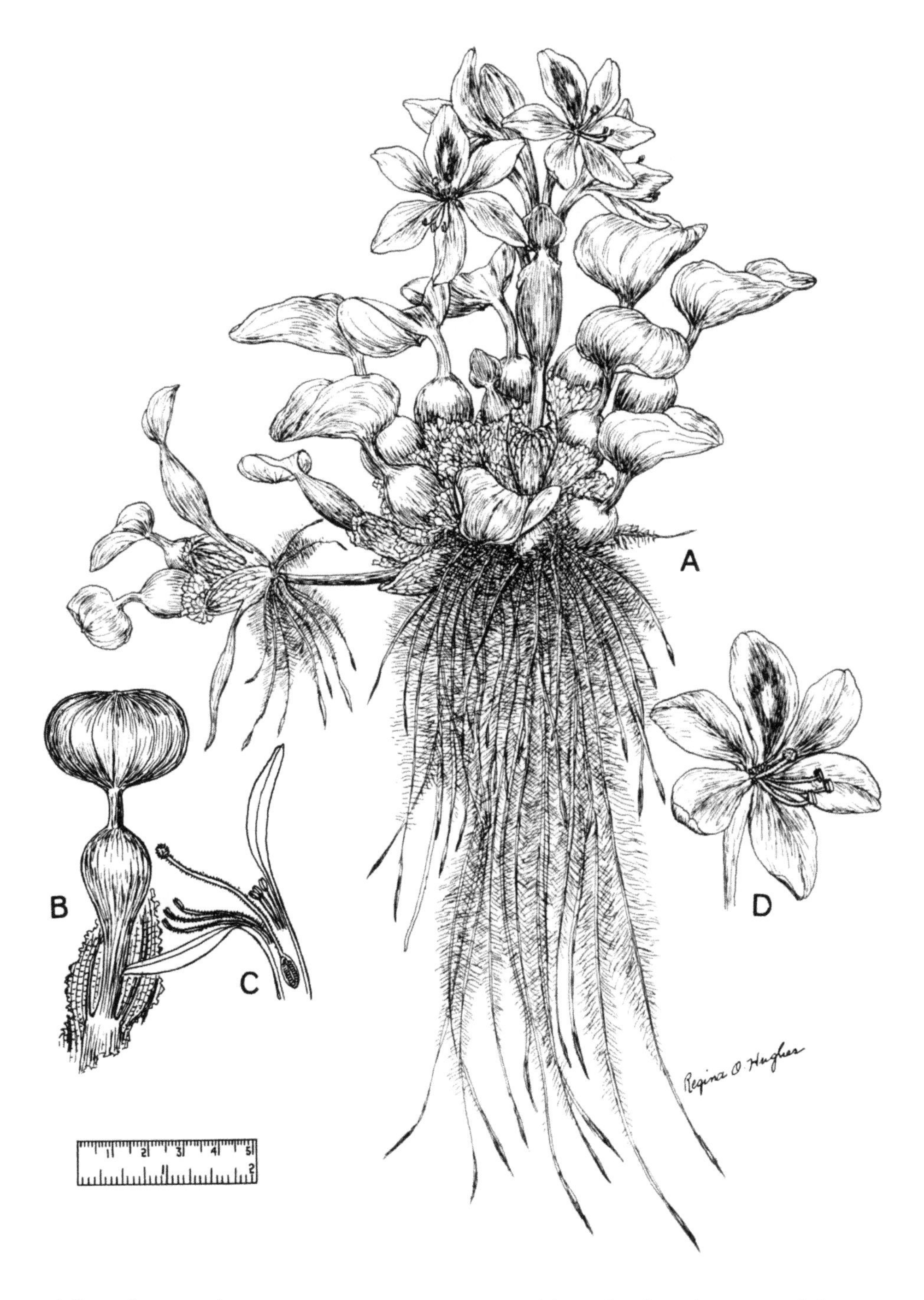

Eichornia crassipes, WATER-HYACINTH. A, Habit. B, leaf. C, diagram of flower. D, flower.

Heteranthera dubia (Jacq.) MacM. WATER-STARGRASS

Aquatic herb; plants submersed, floating, or sometimes stranded.

DESCRIPTION Stems elongate, branched, slender, submersed. **Leaves** linear, grass-like, sessile, translucent, 2-6 mm wide, up to 15 cm long. **Inflorescence** 1-flowered, nearly sessile, completely enclosed by the 1-leaved spathe. **Spathe** 2-5 cm long, slender, sessile in the axils, the small flowers expanded on the water surface, the spathe abruptly short-caudate at the spreading tip. **Flowers** often fertilized in unopened buds, or perianth tube 2-7 cm long, the limb spreading, pale-yellow, the outer segments linear, the inner linear-lanceolate. Stamens 3, all alike, with filaments dilated near the middle, tapering at both ends. **Fruit** an indehiscent 1-locular, few-seeded capsule, narrowly egg-shaped, about 1 cm long.

FLOWERING June–September.

SYNONYMS Zosterella dubia (Jacq.) Small

WHERE FOUND Quiet water, streams, meadows, and shores of quiet water areas, as lakes and ponds.

ORIGIN Tropical America.

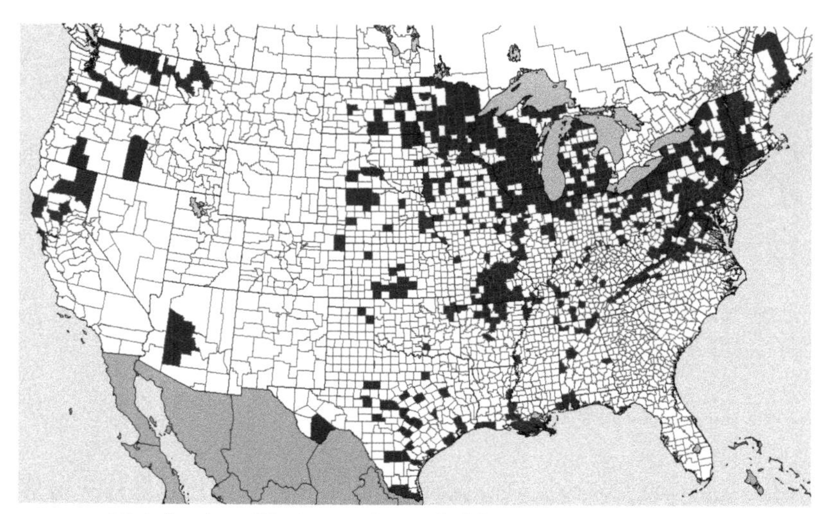

Distribution of **Heteranthera dubia**, WATER-STARGRASS

Heteranthera dubia, WATER-STARGRASS. A, habit. B, enlarged habit, showing flower. C, sheath at base of leaf, showing the stipule-like appendages. D, capsule. E, seeds.

Portulaca oleracea L. COMMON PURSLANE

Annual herb, reproducing by seeds and stem fragments on moist soil.

DESCRIPTION Stems succulent, smooth, fleshy, commonly prostrate, arising from a taproot, usually purplish-red and often forming mats, freely branched, 10–56 cm long, glabrous. **Leaves** alternate or nearly opposite, often in clusters at the ends of the branches, thickened, sessile, margins smooth and broad rounded tips, 0.4–2.8 cm long. **Flowers** yellow, sessile, solitary in the leaf axils or several together in the leaf clusters at the ends of the branches, 3–10 mm broad, including the 5 pale-yellow petals (which open only on sunny mornings). Styles 4–6. Calyx with the lower portion fused with the ovary, the upper part with 2 free sepals, pointed at the tip and 3–4 mm long. Petals and the 6–12 stamens appearing to be inserted on the calyx. **Fruit** a globular, many-seeded capsule, 4–8 mm long, splitting open around the middle, the upper half (with the 2 sepals on top) falling away like a lid. **Seed** nearly oval, tiny, only about 0.5 mm in diameter and length, the surface covered with curved rows of minute wrinkles, black with a whitish scar at one end.

FLOWERING and fruiting from June or July until frost; in hot regions from April to June, disappearing in the hottest period, reappearing in late summer, and continuing until frost.

WHERE FOUND Common locally in cultivated lands; a serious pest in vegetable crops, ornamental crops, citrus and other fruit and nut orchards, small fruits and berries, cotton, lawns, and peanuts in some Southern States; also waste places, barren driveways, eroded slopes, and bluffs, from sea level to about 8,500 feet.

ORIGIN Native of western Asia, introduced into the United States from southern Europe, but native status in North America uncertain as the species was widely spread by humans.

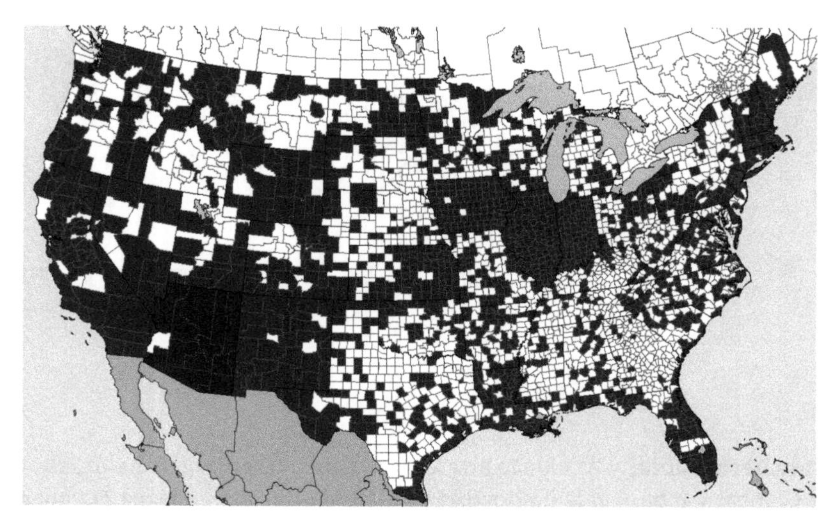

Distribution of **Portulaca oleracea**, COMMON PURSLANE

Portulaca oleracea, COMMON PURSLANE. A, habit. B, flowers and capsules. C, flower open. D, seeds.

Potamogeton crispus L. CURLY-LEAF PONDWEED

Aquatic perennial herb.

DESCRIPTION Stem flattened, the broader flattened sides channeled, branched above, 3–8 dm long. **Leaves** alternate, all submersed, sessile, linear-oblong, 3–8 cm long, 5–12 mm wide, rounded, obtuse or minutely tipped with a sharp and firm point at the apex, often reddish, sharply and finely serrate, at maturity often crisped and wavy-margined 3- to 5-nerved. **Stipules** about 4 mm long, thin, dry and membranaceous soon disintegrating and deciduous. **Peduncles** curved, 2–5 cm long. **Spikes** loosely to densely flowered, slightly tipped with a sharp and firm point, up to 1.8 cm long. **Achene** obliquely ovate to ovoid, 3–5 mm long, flattened, shallowly pitted, with 3 rounded dorsal keels, the central one prolonged at the base, shaped somewhat like an arrowhead, into a projecting appendage, the beak erect, conic, 2–2.5 mm long.

FLOWERING May–September.

WHERE FOUND Fresh calcareous and brackish ponds and streams.

ORIGIN Naturalized from Europe, locally introduced. Throughout Northeastern United States into southern Canada; also in three distinct areas in Western United States.

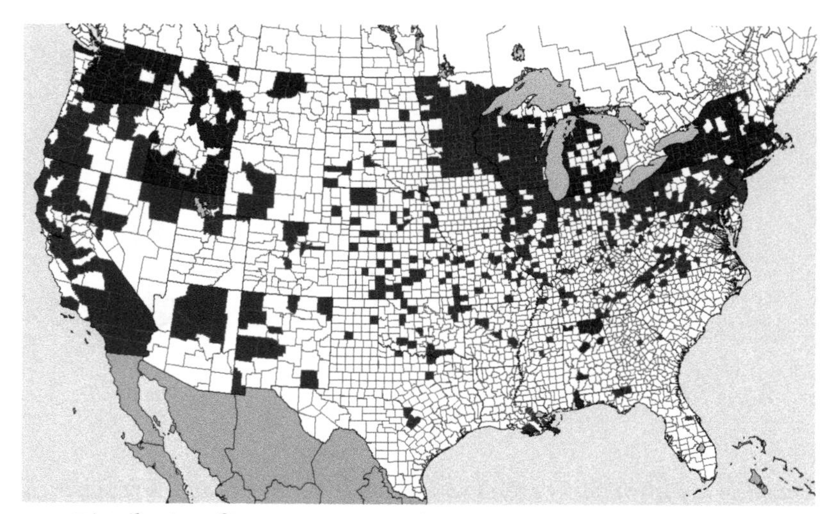

Distribution of **Potamogeton crispus**, CURLY-LEAF PONDWEED

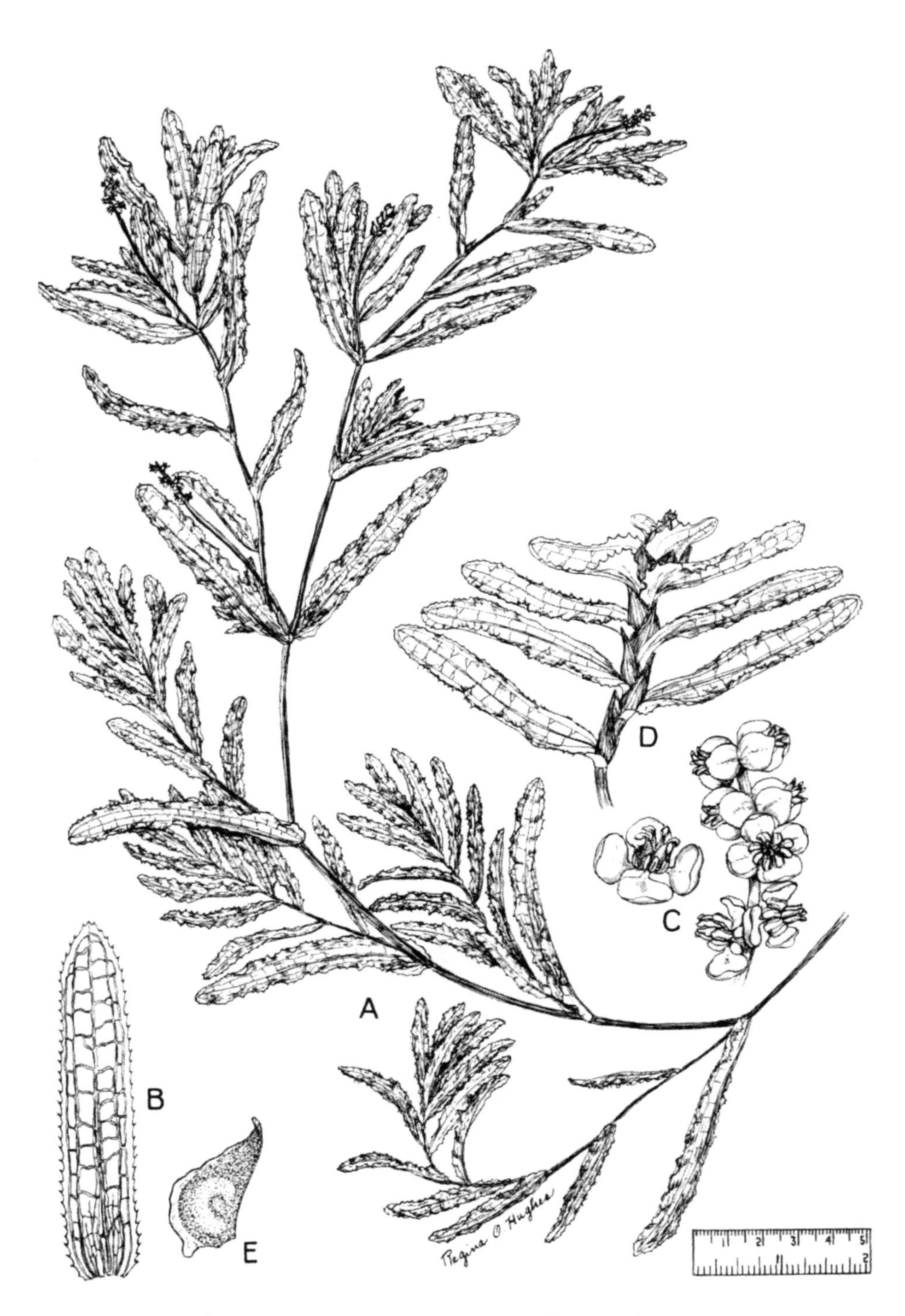

Potamogeton crispus, CURLY-LEAF PONDWEED. A, habit. B, leaf venation. C, flowers. D, winter bud. E, achene.

Potamogeton foliosus Raf. LEAFY PONDWEED

Aquatic perennial herb. Rootstock long and slender, freely branching, rooting at the nodes.

DESCRIPTION Stems compressed-filiform, simple to loosely branched, up to 1 m long. **Leaves** alternate, all submersed, dark-green to bronze, narrowly linear, somewhat tapering at the base, usually without basal glands; the primary leaves 4-10 cm long, 1.4-2.7 mm wide, 3- to 5-nerved, the midrib prominent, compound below the middle, with 1-3 rows of air cells on each side at the base. **Stipules** at first with united margins, forming tubular fibrous blunt sheaths 0.7-1.8 cm long, soon splitting and deciduous. **Peduncles** few, arising in the upper forks, slightly thickened upward, 3-10 mm (rarely 30) long. **Spike** nearly head-shaped or thick-cylindric, 2-5 mm long, about 4 mm thick, with 2 or 3 whorls of 2 flowers each. Sepal-like connectives 0.6-1 mm long, brownish. **Achene** brownish to olive-green, unequally circular, strongly compressed, 2-2.5 mm long, the body circular to broadly inverted ovate, the dorsal keel thin and undulate or dentate, the beak 0.2-0.4 mm long. **Winter buds** sessile in the axils or terminating short branches, their hard bodies 1-1.6 cm long, 1-2 mm thick.

FLOWERING July-October.

WHERE FOUND Streams and ponds.

ORIGIN Native. Throughout all the United States excepting southern Florida; north into southern Canada.

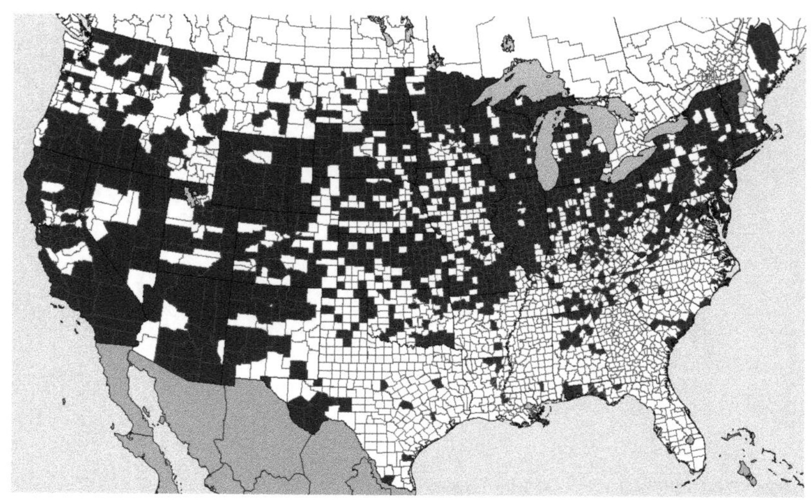

Distribution of **Potamogeton foliosus**, LEAFY PONDWEED

Potamogeton foliosus, LEAFY PONDWEED. A, habit. B, enlarged habit. C, flower diagram. D, achene. E, tip of leaf showing nerves.

Potamogeton nodosus Poir. AMERICAN PONDWEED

Aquatic perennial herb.

DESCRIPTION Stem branched, up to 2 m long. **Submersed leaves** alternate, linear to elliptic-lanceolate, up to 3 dm long, 1-3.5 cm wide, tapering to petiole 2-13 cm long, and to acutish tip, 7- to 15=nerved, with 2-5 rows of air cells each side of midrib, the young blades margined by temporary translucent minute teeth, submersed stipules early decaying, linear, 3-10 cm long. **Floating leaves** long-petioled, lance-oblong to lance-elliptic, rounded to acutish at base and apex, 3-13 cm long, 1.5-4.5 cm broad, with petioles 5-20 cm long, 9- to 21-nerved, emersed stipules similar to the submersed ones but broader. **Spike** dense, becoming loose, of 10-17 whorls, in fruit 3-7 cm long, 0.8-1 cm thick. **Achene** inverted oval to broadly nearly egg-shaped, 3.5-4.3 mm long, 2.5-3 mm wide, the dorsal keel sharp but narrow, often tuberculate, the lateral keels often rough with spines, in maturity brownish or reddish, the facial beak short.

FLOWERING August-September.

WHERE FOUND Deep or shallow water; ponds and streams.

ORIGIN Native. Throughout all the United States excepting most of New England, northern New York, and southern Florida.

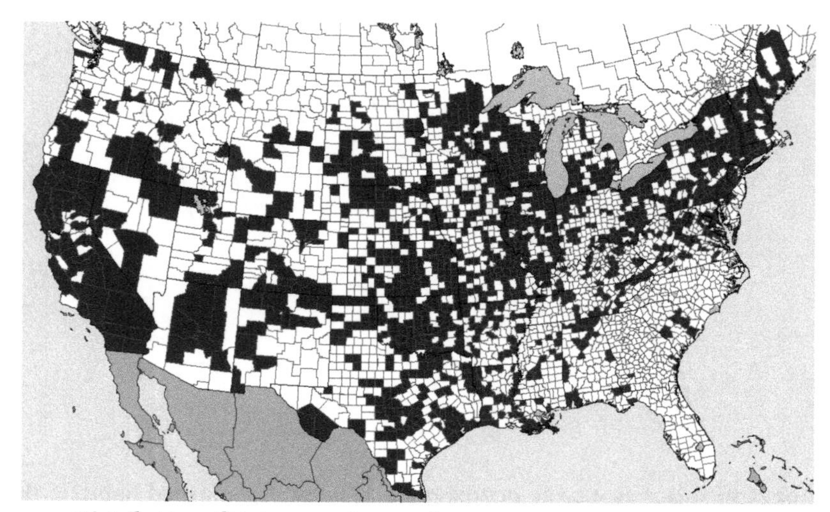

Distribution of **Potamogeton nodosus,** AMERICAN PONDWEED

Potamogeton nodosus, AMERICAN PONDWEED. A, habit. B, flower spike. C, flower diagram, showing stamens. D, achenes.

Potamogeton pusillus L. SMALL PONDWEED

Aquatic perennial herb.

DESCRIPTION Stems very slender, slightly flattened, usually much branched toward the summit. **Leaves** alternate, all submersed, light-green, narrowly linear, firm, usually with 2 small translucent basal glands; the primary leaves 1-7 cm long, 0.5-3 mm wide, acute to obtuse, 3-nerved, the midrib prominent, the lateral veins delicate and often indistinct, usually not bordered by air cells, or with 1-4 rows of air cells at least in the lower half. **Stipules** thin, dry, membranaceous, slender tubular, with margins united to above the middle, finally rupturing and soon deciduous, 0.6-1.7 cm long. **Peduncles** axillary, long and slender, 1.5-8 cm long. **Spikes** elongate to short cylindric dense clusters, strongly interrupted, 6-15 mm long, of 1 or 3-5 distant whorls. **Achene** olive-green, unevenly egg-shaped, often somewhat S-shaped, 1.9-2.8 mm long, 1-1.8 mm broad, smooth but often deeply impressed on the somewhat flattened sides, the back rounded, the ventral face arching to the prominent marginal beak. **Winter buds** axillary along the branches and terminal, their bodies 9-17 mm long, 0.5-1.5 mm broad.

FLOWERING July-September.

WHERE FOUND In basic or alkaline lakes and ponds.

ORIGIN Native. Throughout all the United States excepting the southeastern area; north to Alaska, Canada, and Greenland and south to central Mexico; West Indies; Azores. Eurasia.

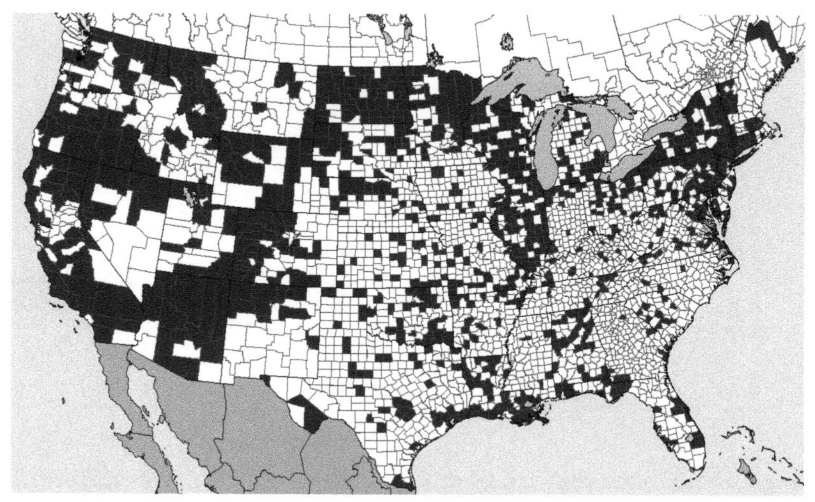

Distribution of **Potamogetan pusillus**, SMALL PONDWEED

Potamogetan pusillus, SMALL PONDWEED. A, habit. B, enlarged habit. C, flower diagram. D, achene. E, tip of leaf showing nerves. F, winter bud.

Potamogeton richardsonii (A. Benn.) Rydb.

RICHARDSON PONDWEED

Aquatic perennial herb.

DESCRIPTION Stems freely branched, densely leafy, 1-2.5 mm thick. **Leaves** alternate, all submersed, the lowest ones ovate to ovate-lanceolate, to narrowly lanceolate, 1.5-12 cm long, 5-20 mm broad, prominently 7- to 33-nerved, with 3-7 stronger than the others, cordate-based, closely sessile and cordate clasping one-half to three-fourths the circumference of the stem, somewhat acute to blunt, the margin minutely toothed on young leaves. **Stipules** whitish, coarsely nerved, ovate to lanceolate, obtuse, keel-less, 1-2 cm long, soon disintegrating into stringy white fibers. **Peduncles** often upwardly thickened, about as thick as the stem, 0.15-2.5 dm long. **Spikes** dense, with 6-12 whorls, cylindric, in fruit 1.5-4 cm long, 1 cm thick. **Achene** grayish-green to olive-green, inverted ovate, rounded on back and at the base, 2.5-4 mm long, 2.3 mm wide, the prominent beak up to 1 mm long.

FLOWERING July-September.

WHERE FOUND Ponds, lakes, rivers, and streams; frequently in brackish or alkaline waters.

ORIGIN Native.

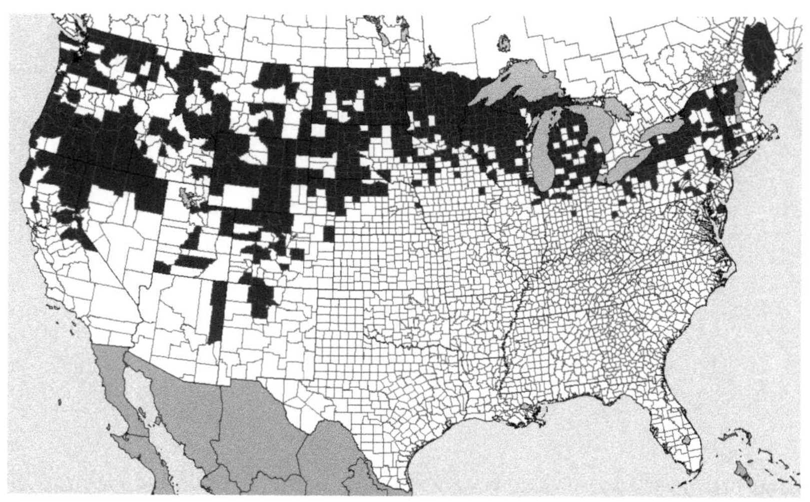

Distribution of **Potamogeton richardsonii**, RICHARDSON PONDWEED

Potamogeton richardsonii, RICHARDSON PONDWEED. A, habit. B, enlarged leaves with young stipules. C, enlarged bases of leaves. D, flowers. E, achenes.

Delphinium barbeyi Huth — TALL LARKSPUR

Perennial herb, reproducing by seeds. Rootstocks stout, woody.

DESCRIPTION Stems 5–20 dm tall, hollow, more or less glandular stiff-haired, densely so in the inflorescence, often nearly glabrous near the base. **Leaves** 5–15 cm broad, pubescent, a little reduced above, broader than long, divided or cleft into 3 primary segments, segments entire in the lower half and cleft or coarsely toothed in the upper half. Racemes rather compact, short-oblong. **Flowers** scented and showy with conspicuous bracts below. Sepals 14–16 mm long, narrowly ovate, slenderly tapering or acute, rich dark-purple, the spur about 10 mm long, curved near the tip. Petals, the lower ones dark-purple, the clefts between lobes about 3 mm deep, the upper petals little exserted, edged with white. **Follicle** 14–17 mm long, oblong-ovate, with purplish veins, nearly glabrous. **Seed** smoky-brown, wing-angled.

FLOWERING July–August.

SYNONYMS *Delphinium cockerellii* A. Nels., *Delphinium subalpinum* (Gray) A. Nels.

WHERE FOUND Meadows and open woods.

ORIGIN Native; central and southern Rocky Mountains.

NOTE Poisonous to cattle.

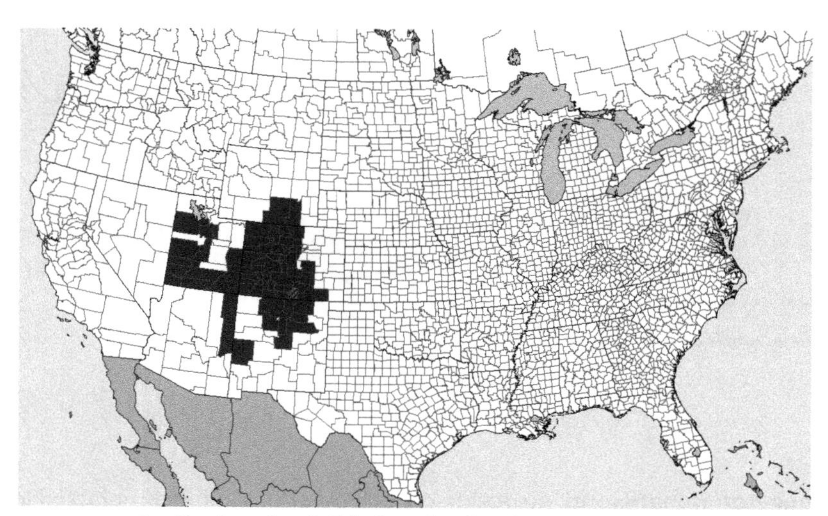

Distribution of **Delphinium barbeyi**, TALL LARKSPUR

Delphinium barbeyi, TALL LARKSPUR. A, habit. B, flower diagram. C, capsules. D, seeds.

Delphinium geyeri Greene GEYER LARKSPUR, PLAINS LARKSPUR

Perennial herb, reproducing by seeds. Rootstocks tough, woody, fibrous, usually vertical.

DESCRIPTION **Stems** 2-7 dm tall, usually several strict, ashy-puberulent especially below. **Leaves** several, usually toward the base, 5-8 cm broad, divided, the divisions repeatedly dissected, the ultimate divisions straight and linear to linear-filiform. **Racemes** narrow, strict and rather dense, flowers showy. Sepals 10-15 mm long, oblong-ovate, acute to obtuse, the spur up to 1-1/2 times as long as the sepal, straight or nearly so, deep-blue. Petals, the lower ones hairy near the middle, the lower sinuses obsolete or short, closed, the upper petals whitish. **Follicle** 12-15 mm long, finely hirsute. **Seed** 2-3.5 mm long, flat, the body roughened.

FLOWERING July-August.

WHERE FOUND Open plains, slopes, dry mesas, and rocky hills; often growing among shrubs.

ORIGIN Native.

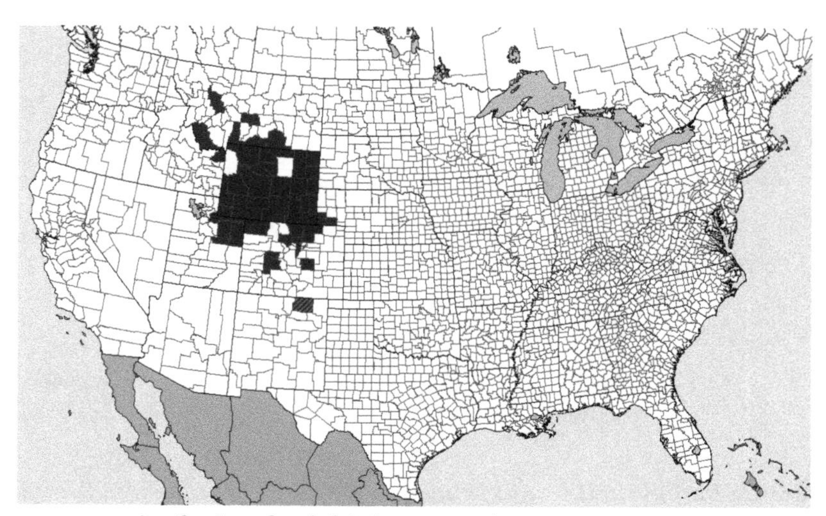

Distribution of **Delphinium geyeri**, GEYER LARKSPUR

Delphinium geyeri, GEYER LARKSPUR. A, habit. B, enlarged leaf. C, flower. D, capsule. E, seeds.

Delphinium menziesii DC. PUGET SOUND LARKSPUR

Perennial herb, reproducing by seeds, with tubers in small shallow globose clusters.

DESCRIPTION Stems 2-7 dm high, soft-pubescent, with spreading white hairs. **Leaves** both on stems and basal, reduced upward to leafy bracts, the blades round 5-sided, 3-6 cm wide, pubescent, palmately arranged into broad approximate wedge-shaped divisions, these shallowly to deeply lobed, the ultimate segments blunt to short-acute, the lower petioles 5-10 cm long. **Racemes** 3- to 10-flowered, short, hairy. Pedicels spreading. Sepals deep rich blue, oblong-ovate, 10-15 mm long, 9-11 mm wide, ending in an abrupt short-pointed tip, hairy. Upper petals pale, rhombic, lower petals rounded, dark-blue, sometimes white-lined, 7-9 mm wide, shallowly notched at rounded apex to 2-lobed. **Follicle** somewhat spreading, 10-14 mm long, white-hairy. **Seed** few, oblong-prismatic, brownish, 1.5 mm long, narrowly wing-margined.

FLOWERING March-May.

WHERE FOUND Open places above the Pacific Ocean.

ORIGIN Native.

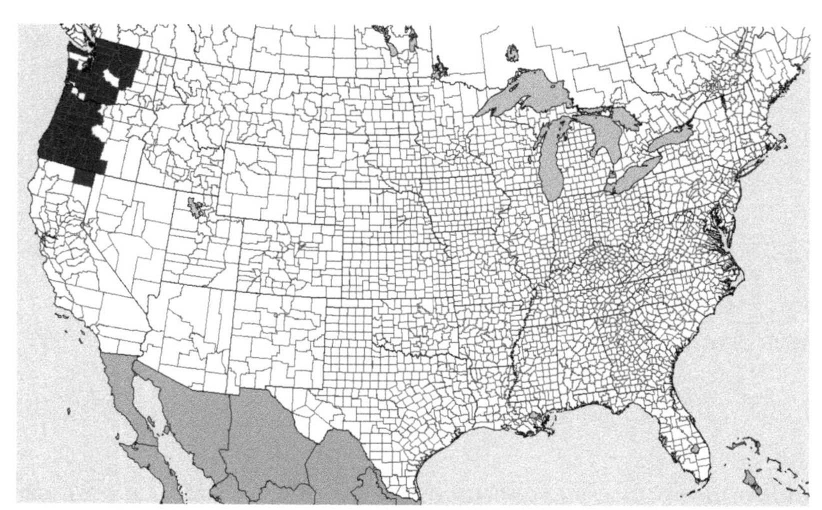

Distribution of **Delphinium menziesii**, PUGET SOUND LARKSPUR

Delphinium menziesii, PUGET SOUND LARKSPUR. A, habit. B, leaf. C, flowers. D, capsule. E, seeds.

Delphinium occidentale (S. Wats.) S. Wats. DUNCECAP LARKSPUR

Perennial herb, reproducing by seeds. Rootstock woody, deep, vertical.

DESCRIPTION Stems 6-20 dm tall, somewhat straw-colored, especially at the base, somewhat glaucous and glabrous below. **Leaves** 1-1.5 dm wide, divided into 3-7 rhombic divisions, these cleft below the middle, pubescent on both sides. **Racemes** usually over 15 cm long, dense, resembling a spike or loosely panicle-like, rachis thinly to densely glandular-hairy, bracts small. Sepals 6-12 mm long, narrowly egg-shaped to oblong, rounded or acute at apex, usually paler or hoary gray, with pubescence on the back, otherwise blue-purple, spur 9-12 mm long, horizontal, straight or curved somewhat near the tip; lower petals with sinuses 1-2 mm long; upper petals small and included. **Follicle** 9-15 mm long, short-oblong, glabrous to glandular-pubescent. **Seed** wing-angled.

FLOWERING July-August.

WHERE FOUND Meadows, thickets, and open woods.

ORIGIN Native.

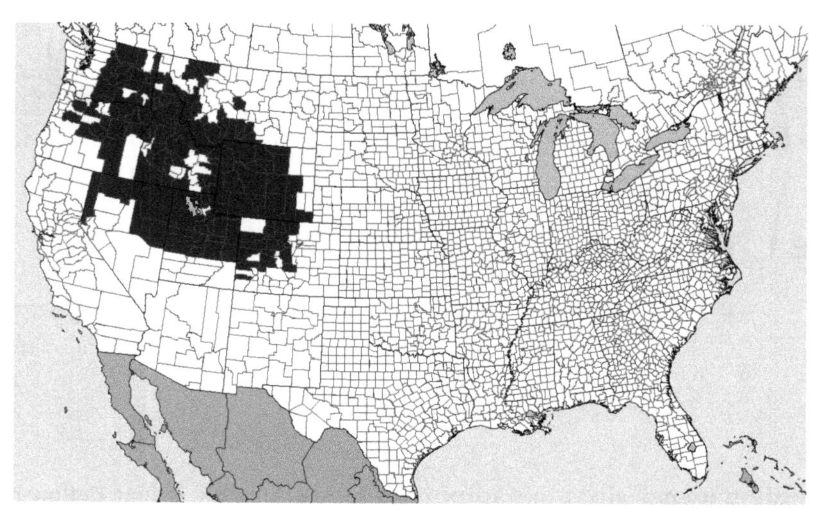

Distribution of **Delphinium occidentale,** DUNCECAP LARKSPUR

Delphinium occidentale, DUNCECAP LARKSPUR. A, habit. B, flower diagram. C, capsules. D, seeds.

Ranunculus acris L. TALL BUTTERCUP

Perennial herb, reproducing by seeds. Roots thick, fibrous.

DESCRIPTION Stems erect, often in clusters, branched, hairy, often 5-10 dm tall, leafy mainly below the middle. **Leaves** alternate, hairy, kidney-shaped in general outline, palmately 3-divided, the divisions sessile and parted into narrow segments, the dissection of the leaves variable. **Flowers** perfect, regular, solitary or in cymose clusters. Calyx mostly of 5 separate spreading sepals. Petals 5-7, broadly obovate, often notched at the apex 8-16 mm long, bright-yellow, sometimes cream-colored, about twice as long as the sepals. Stamens numerous. Carpels numerous, separate, with short, recurved, persistent styles. **Achenes** numerous, in nearly globose heads, broadly obliquely obovate, 2-3 mm long, flattened, with a prominent, nearly straight beak (0.4-1 mm long), dull, minutely pitted, dark-brown, margin often lighter.

FLOWERING May-September.

WHERE FOUND Fields, pastures, and meadows; introduced in lawns; heavy moist soils; does not persist under cultivation.

ORIGIN Introduced from Eurasia.

NOTE Plant contains an acrid juice that is somewhat poisonous if eaten by livestock, blistering the mouth and intestinal tract.

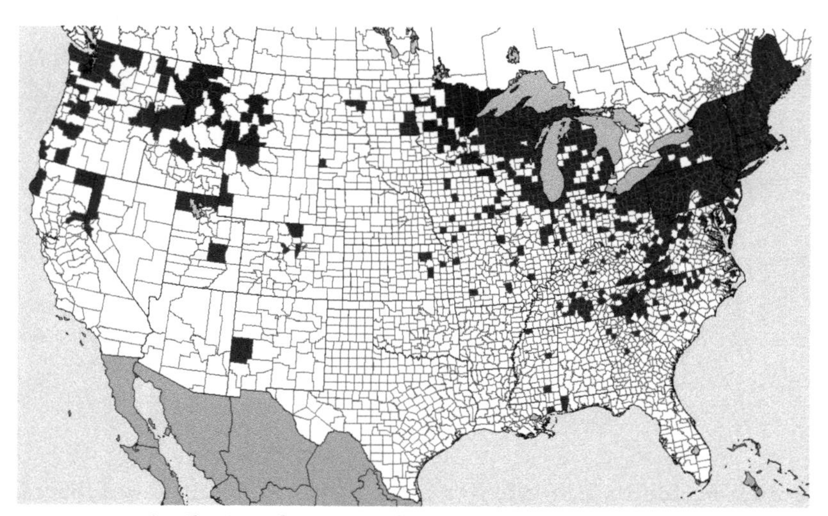

Distribution of **Ranunculus acris**, TALL BUTTERCUP

Ranunculus acris, TALL BUTTERCUP. A, habit. B, fruiting head. C, achene.

Ranunculus repens L. CREEPING BUTTERCUP

Perennial herb, reproducing by seeds and runners.

DESCRIPTION Stems low, normally creeping and rooting at the nodes, rarely ascending or erect, hairy, or rarely glabrous. **Leaves** alternate, long-petioled, 3-divided or 3-lobed, the segments broadly obovate to nearly rounded in general outline, sharply toothed, hairy, dark green or sometimes with light spots. **Flowers** perfect, regular, solitary or in corymb-like clusters. Sepals 5, separate, green, not reflexed. Petals 5-7, bright glossy yellow, 8-15 mm long, about two-thirds as wide. Stamens numerous, the anthers 1-2 mm long. Carpels numerous, separate, with short, recurved styles. **Achenes** numerous in globular heads, broadly obliquely obovate, 2.5-3.5 mm long, sharply but narrowly margined, blackish-brown; achene beak triangular, usually somewhat curved, 0.8-1.5 mm long.

FLOWERING June-August.

WHERE FOUND Fields, moist meadows, pastures, mucklands and lawns, and along ditches and roadsides.

ORIGIN Introduced and naturalized from Europe.

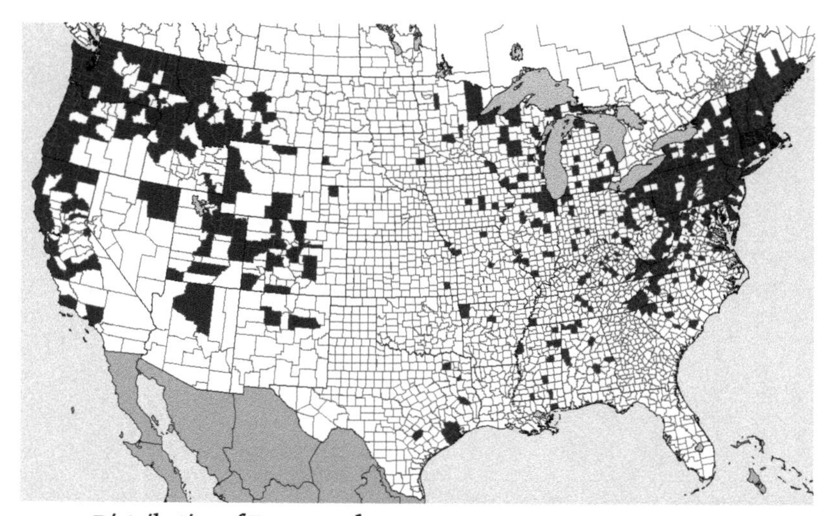

Distribution of **Ranunculus repens**, CREEPING BUTTERCUP

Ranunculus repens, CREEPING BUTTERCUP. A, habit. B, flower. C, head of achenes. D, achene.

Potentilla canadensis L. COMMON CINQUEFOIL

Perennial herb, reproducing by seeds and runners. Rhizome short, to 2 cm long, 4–8 mm thick.

DESCRIPTION Stems at flowering time, 0.1–1.5 dm tall, soon becoming prostrate, runner-like or threadlike, 0.3–1 mm thick, not bearing tuberous enlargments at the tips. Stems, petioles, lower leaf surfaces and calyx silky soft hairs with appressed or loosely ascending soft pubescence. **Leaves** with 5 leaflets, the leaflets narrowly wedge-shaped obovate, the margins coarsely and deeply 5–15 toothed around the rounded summit, entire below the middle, the middle leaflet largest, 1.5–4 cm long. Stipules of basal leaves with oblong-lanceolate, flat auricles, of the mature primary stem leaves mostly 3-cleft and 4–12 mm long. **Flowers** 11.5 cm broad, on long, slender peduncles 1–9 cm long. Bracteoles linear or linear-lanceolate. Petals deep-yellow or cream-colored, 10–15 mm wide, rounded at the summit or retuse.

FLOWERING March–June.

SYNONYMS Potentilla pumila Poir.

WHERE FOUND Dry sandy and gravelly fields and in dry woods.

ORIGIN Native.

Distribution of **Potentilla canadensis,** COMMON CINQUEFOIL

Potentilla canadensis, COMMON CINQUEFOIL. A, habit. B, flower and calyx. C, achenes.

Potentilla norvegica L. ROUGH CINQUEFOIL

Biennial (rarely a short-lived perennial) herb, reproducing by seeds.

DESCRIPTION Stems stout and leafy, commonly branched, hairy with stiff mostly spreading hairs, semi-erect or spreading, 3-9 dm tall. **Leaves** alternate, the lower leaves long-petioled, palmately divided, with 3 leaflets obovate to tapering toward the tip or elliptic, the margins coarsely toothed, the upper leaves sessile, with narrow leaflets, hairy, green. **Flowers** a leafy cyme at the tips of the branches. Calyx in fruit small, inconspicuous, enlarging to 0.8-1.7 cm high, its bracteoles acutish, ovate-lanceolate, about equal at time of pollen shedding, the sepals enlarging with age in the fruit and up to 16 mm long. Petals yellow, obovate, mostly shorter than the calyx lobes. Stamens 15-20. Style slenderly conical at base, subterminal, about equaling the mature carpel. **Achene** 0.8-1.3 mm long, light-brown, with longitudinally curved ridges.

FLOWERING June-October.

SYNONYMS *Potentilla monspeliensis* L.

WHERE FOUND Fields, meadows, pastures, roadsides, waste places, and thickets.

ORIGIN Introduced and naturalized from Eurasia; some varieties appear to be native of North America.

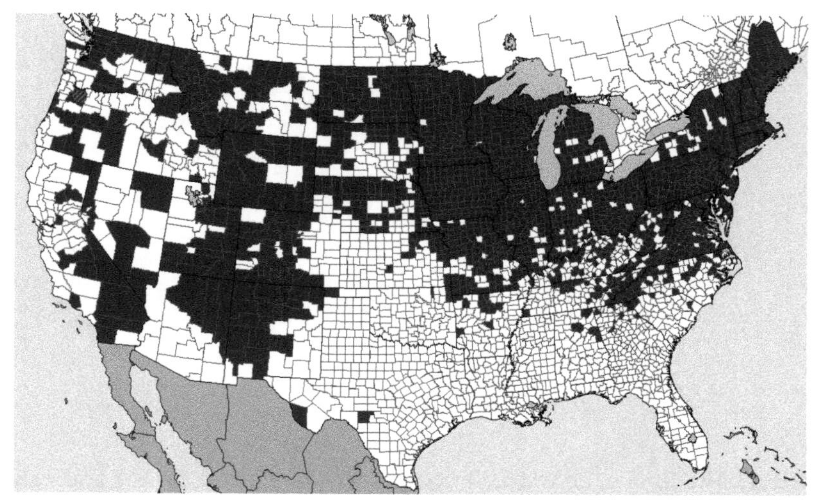

Distribution of **Potentilla norvegica**, ROUGH CINQUEFOIL

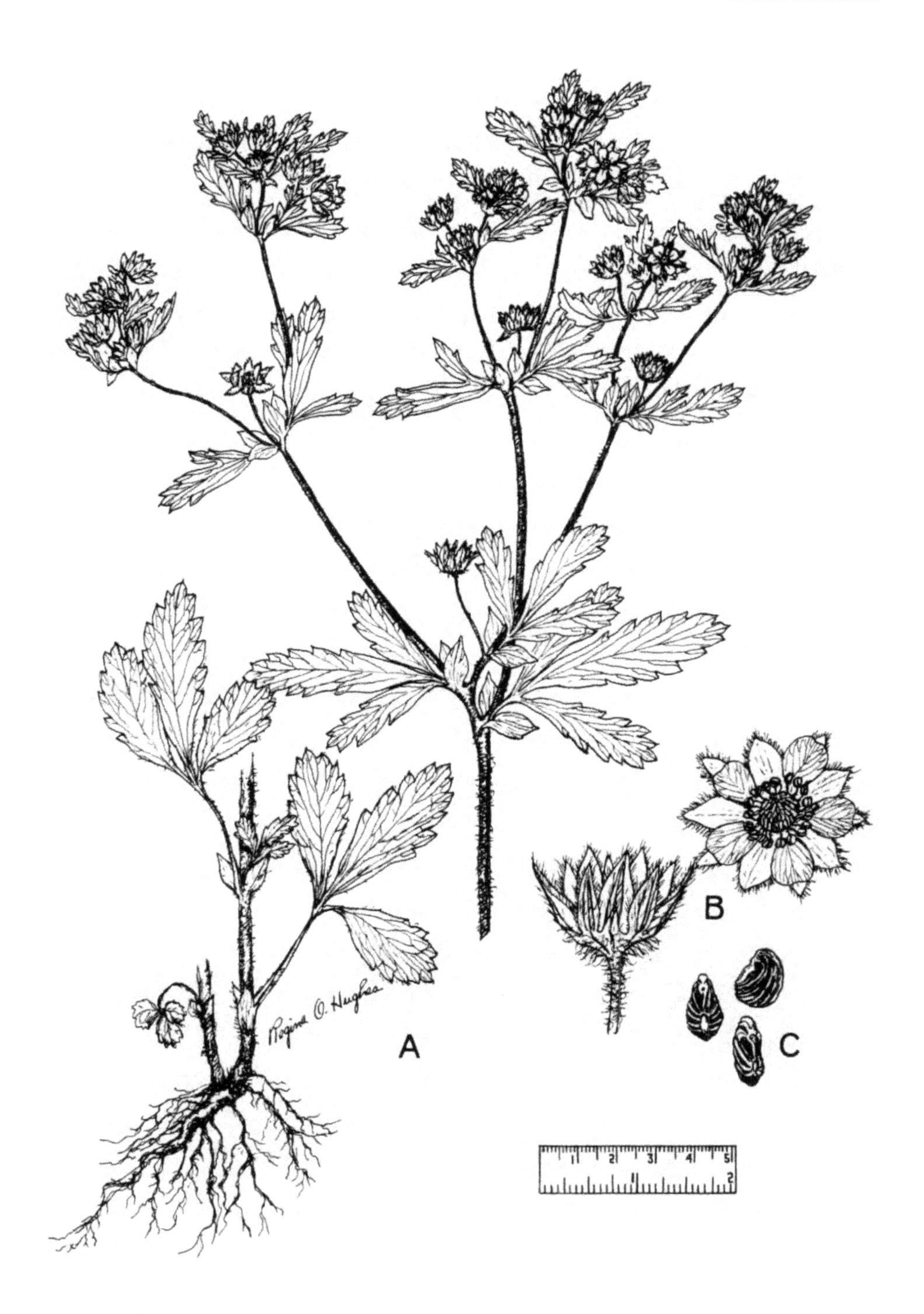

Potentilla norvegica, ROUGH CINQUEFOIL. A, habit. B, flower and calyx. C, achenes.

Potentilla recta L. SULFUR CINQUEFOIL

Perennial herb, reproducing by seeds.

DESCRIPTION Stems stiffly erect, very leafy, 1.5–9 dm tall loosely hairy. **Leaves** alternate, the basal leaves on long hairy petioles, palmately divided with 5-7 leaflets, the leaflets oblanceolate, 3–14 cm long, the margins with 7–17 prolonged triangular teeth, pale beneath, and more or less hirsute on both surfaces. **Flowers** in terminal, stiffly erect cymes. Calyx hairy, becoming 1–1.5 cm high. Corolla conspicuous, sulfur-yellow, the petals inverted heart-shaped, deeply notched at the apex, about 1 cm long. Stamens mostly 30 (rarely 25). Style shorter than the mature carpel, filiform, terminal. **Achene** dark-brown, striate, with minute low curved ridges.

FLOWERING Late May–August.

WHERE FOUND Dry fields, wastelands, pastures, meadows, and roadsides; troublesome in limestone regions.

ORIGIN Naturalized from Europe.

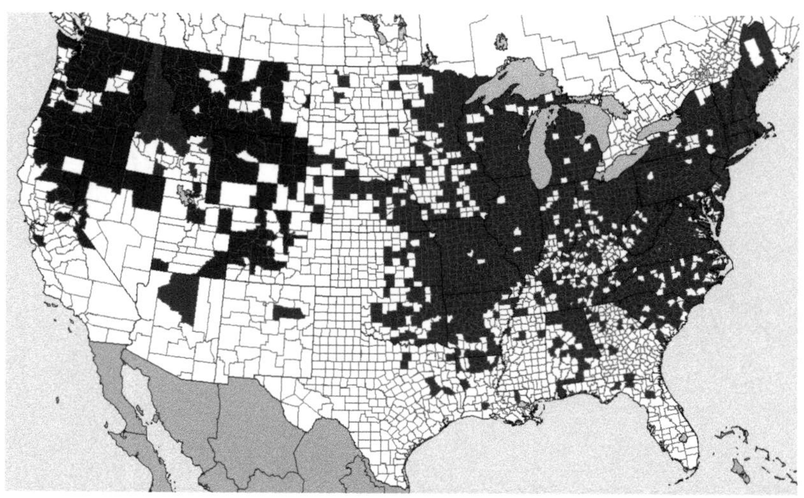

Distribution of **Potentilla recta**, SULFUR CINQUEFOIL

Potentilla recta, SULFUR CINQUEFOIL. A, habit. B, base of leaf showing stipules. C, flower and calyx. D, achenes.

Rosa multiflora Thunb. MULTIFLORA ROSE

Perennial shrub, reproducing by seeds and sometimes rooting at the tips of the drooping side canes.

DESCRIPTION Stems up to 3 m long, in clumps, arching or trailing, usually growing about 2 m erect and then the tips drooping almost to the ground, beset with stiff thorns. **Leaves** pinnately compound, usually with 7 or 9 leaflets, membranaceous, deciduous, 2-4 cm, long, elliptic or obovate, obtuse to acute, nearly glabrous on the upper surface, paler and usually with short soft pubescence beneath, the petioles short-pubescent, the stipules green, membranaceous, short-pubescent and glandular hairs, the free part threadlike at the apex. **Inflorescence** a many-flowered panicle, usually pyramidal, erect, 2-4 cm broad, mostly white, sometimes pinkish, the pedicels often with stalked glands. Sepals 5-8 mm long. Styles glabrous. **Fruits** (hips) bright-red, often lasting until spring, about 0.8 cm in diameter, nearly round. **Achene** enclosed in the fleshy calyx tube (hip).

FLOWERING June-July.

WHERE FOUND Clearings, roadsides, fence rows, waysides, borders of woods, "bumper areas" along superhighways, and curves; introduced as an ornamental and used horticulturally as hardy rootstock on which to bud other ornamental varieties.

ORIGIN Introduced and naturalized from eastern Asia.

Rosa arkansana Porter PRAIRIE ROSE

Perennial shrub, with underground rootstocks.

DESCRIPTION Stems 3-4 dm tall, densely prickly. **Leaves** compound, with 7-11 oval leaflets, about 2.5 cm long. **Flowers** pink, about 4 cm broad, fragrant. **Fruit** (hip) round, 1-1.3 cm in diameter. **Seed** hard, hairy, brown, about 4 mm long.

FLOWERING May-August.

WHERE FOUND Weed in pastures, prairies, fields, roadsides, and fence rows.

ORIGIN Native.

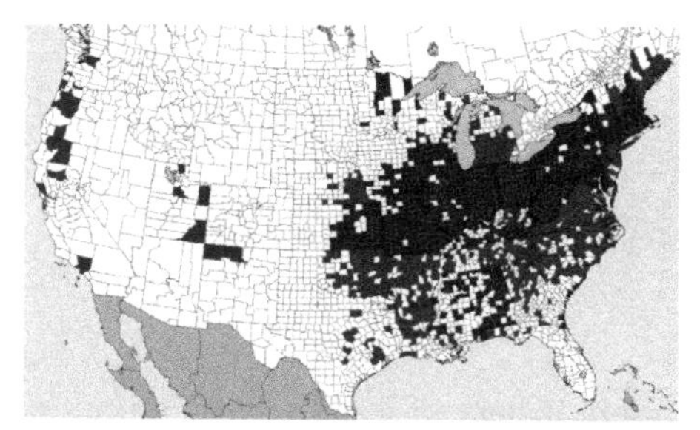

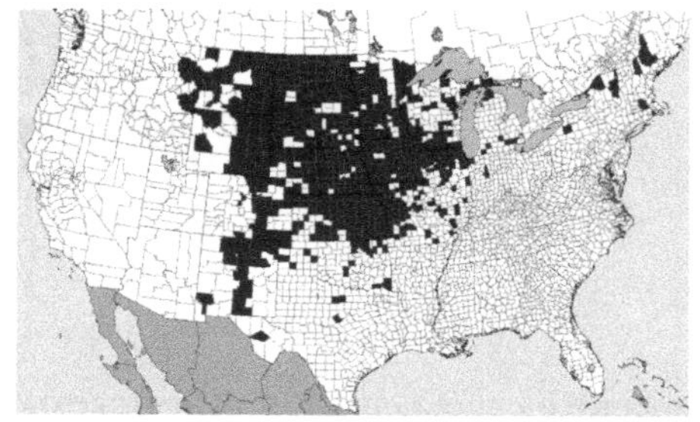

Distribution of **Rosa multiflora** (left), **Roas arkansana** (right)

Rosa multiflora, MULTIFLORA ROSE. A, habit. B, leaf-base. C, bud. D, styles and stamens. E, fruits. F, achenes.

Rubus allegheniensis Porter ALLEGHENY BLACKBERRY

Perennial shrub, reproducing by seeds and underground runners.

DESCRIPTION Stems erect or high-arching, 1-3 m high, armed with scattered broad-based (3-5 mm long) lanceolate prickles, or the prickles absent. **Primocanes** often ridged or angled and finely pubescent. **Primocane leaflets** mostly 5, palmately compound, the upper 3 conspicuously stalked, the petiole and leaflet petioles prickly and pubescent and glandular, the blades typically oblong or elliptic to ovate and gradually tapering, 8-12 cm long, 4-6 cm broad. **Floricanes** mostly retaining strong scattered prickles, normally erect, but sometimes depressed with fruit. **Floricane leaflets** usually 3, the foliage of floral shoots of short-pointed or obtuse small leaflets. **Inflorescences** in elongated racemes, 8-25 cm long, extending beyond the foliage, pubescent and glandular. **Flowers** 12-30 on long clusters, about 2 cm in diameter, widely opening. Petals narrow, white, noticeably separate. Calyx pubescent and glandular, the lobes acute to prolonged-pointed. **Fruit** globose to thimble-shaped, compound, 2 cm or more long, made up of 50-70 drupelets (small one-seeded fruits). **Seed** about 1 mm long, wrinkled. A variable species with several varieties.

FLOWERING May-June; fruiting July.

WHERE FOUND Dry places from lowlands to hills and mountains, open woodlands, roadsides, thickets, fence rows, and clearings, ubiquitous.

ORIGIN Native.

NOTE There are several hundred species of *Rubus* described for North America, varying from blackberries to dewberries and raspberries (but the taxonomy is complicated by hybridization and environmental differences). Perennials often somewhat shrubby with semi-woody stems (canes). **Stems** trailing or upright and arching, frequently rooting at the tips, usually with thorns, or prickles or stiff soft hairs. **Canes** of two types, the **primocanes** that grow the 1st year and only vegetate (rarely flow

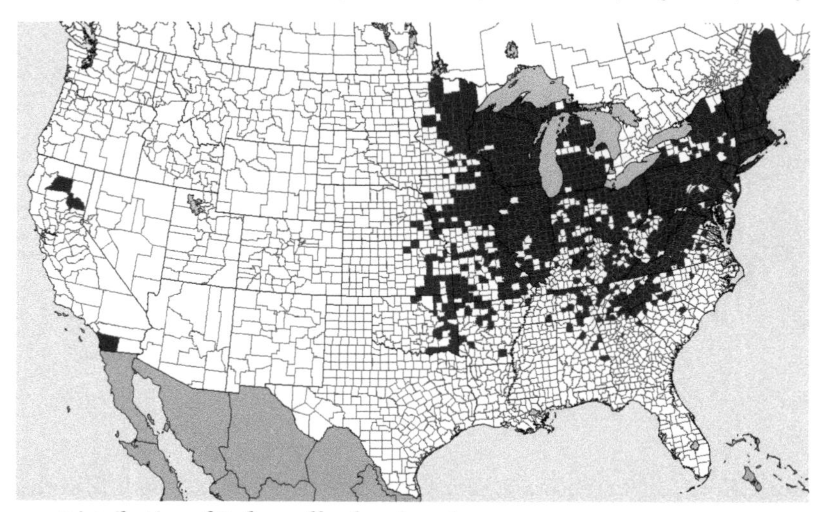

Distribution of **Rubus allegheniensis**, ALLEGHENY BLACKBERRY

ering and fruiting) and these same canes becoming the **floricanes** the 2nd year, bearing flowers and fruits, these canes then dying after the 2nd year. **Leaves** petioled, palmately compound, with 3- or 5-foliolate or lobed leaflets, these varying in size and shape. **Inflorescences** terminal or axial, of few to many flowers in a cluster, the petals usually white, rarely pink or pinkish. **Fruits** compound, made up of numerous small one-seeded fruits varying from dark-purple to red, yellow, or white. Various species throughout the United States, forming thickets, bramble patches, sand dune tangles, etc.

Rubus allegheniensis, ALLEGHENY BLACKBERRY. A, primocane habit. B, floricane habit C, fruit showing drupelets. D, seeds.

Galium aparine L. CATCHWEED BEDSTRAW

Annual herb, reproducing by seeds. Roots branching, short, shallow.

DESCRIPTION Stems 6–15 dm long, glabrous, weak, sprawling, 4-sided with each edge bearing a row of downward-pointing stiff bristles. **Leaves** simple, entire, 1-veined, lanceolate, tapering at the base, 2–7 cm long, rough, bristle-pointed, 6–8 in a whorl at each node. **Flowers** 1–3, perfect, very small, borne on slender branches in the axils of the leaves at the nodes. Calyx obsolete. Corolla with 4 white lobes. **Fruit** of 2 nearly spherical carpels, covered with stiff-hooked bristles, 1.5–4 mm in diameter. **Seed** ball-shaped with a deep pit in 1 side, short sharp spines on the outer surface, 2–3 mm in diameter, gray-brown.

FLOWERING May–July.

WHERE FOUND Rich woods, thickets, seashores, waste ground, and on moist land in meadows, pastures, woodlands, and fencerows.

ORIGIN Native.

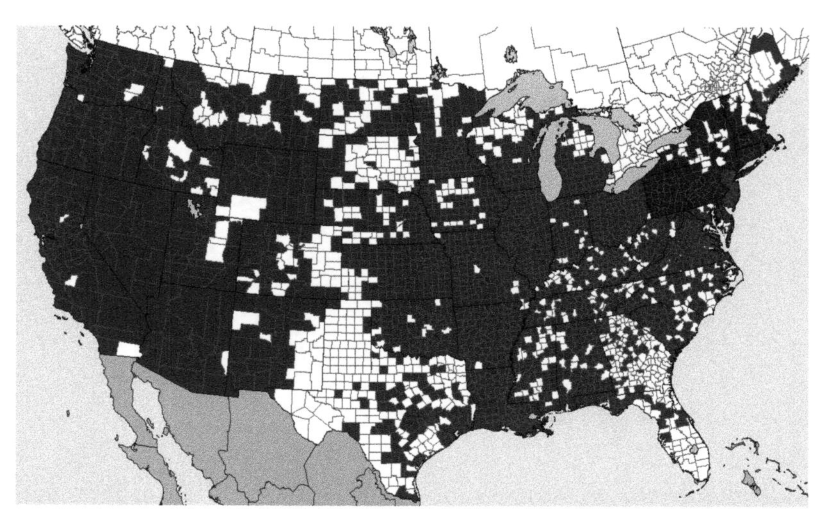

Distribution of **Galium aparine**, CATCHWEED BEDSTRAW

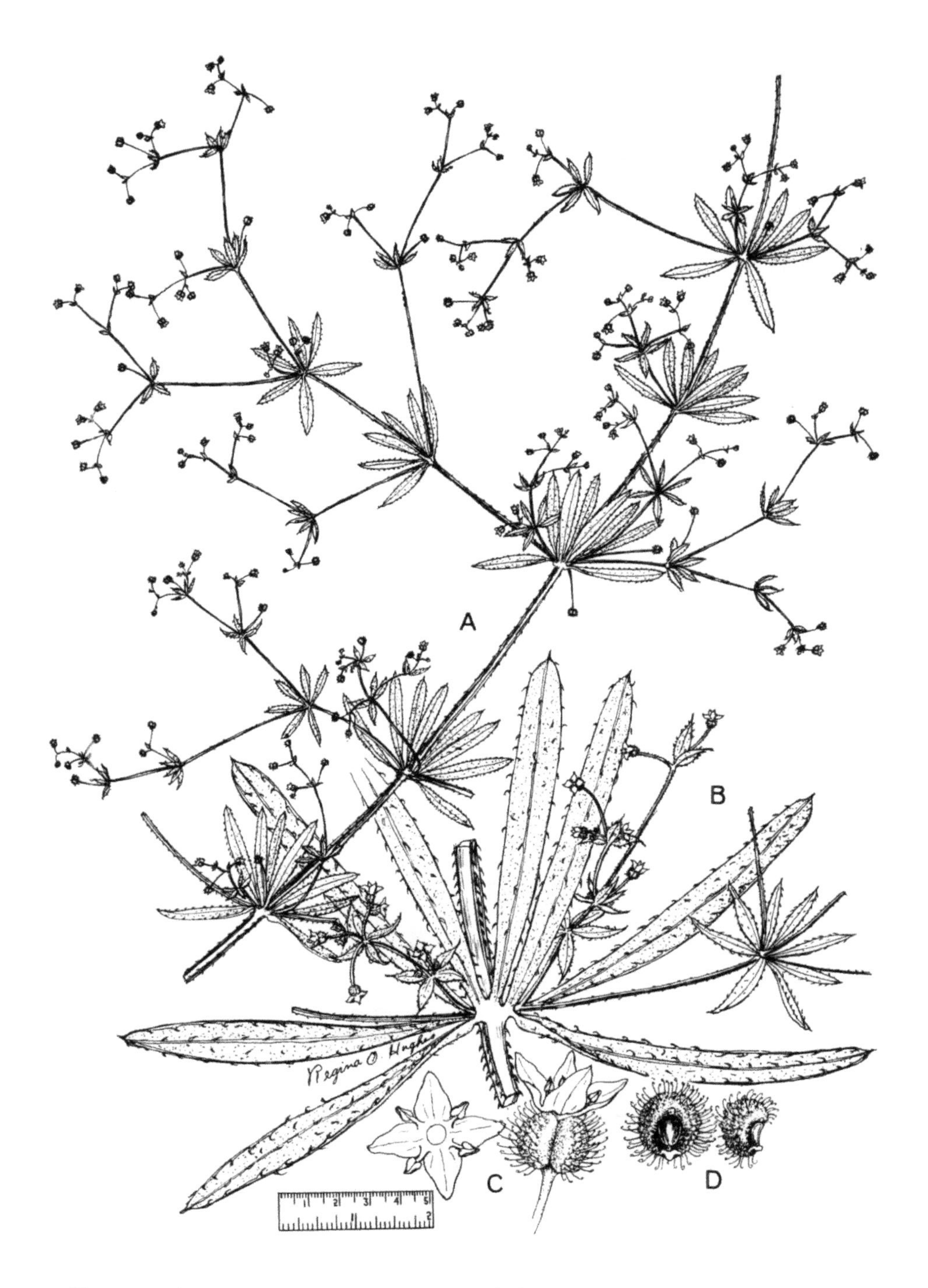

Galium aparine, CATCHWEED BEDSTRAW. A, habit. B, enlarged leaf whorl. C, flowers. D, fruits.

Galium mollugo L. SMOOTH BEDSTRAW

Perennial herb, reproducing by seeds, spreading or ascending, with a rhizome.

DESCRIPTION Stems diffusely branched, glabrous or pubescent, weak, 2-12 dm tall, from a decumbent base; perennial offshoots produced from the base in the summer or fall, these short, slender, ascending, and leafy. **Leaves** whorled, 6-8 at a node, mostly oblanceolate to obovate, or almost linear, sessile, 1-3 cm long (smaller on the branches), rough on the margins with forwardly directed prickles, 1-nerved, with a rigid tip. **Inflorescence** a terminal panicle with spreading branches, almost leafless. **Flowers** numerous, white or nearly so, 2-4 mm wide, corollas 4-lobed, with a rigid tip. **Fruits** glabrous, roughened, 1-1.5 mm long, with appressed segments.

FLOWERING June-August.

WHERE FOUND Meadows, pastures, lawns, roadsides, and waste places; on gravelly or sandy loam soils.

ORIGIN Native of Eurasia.

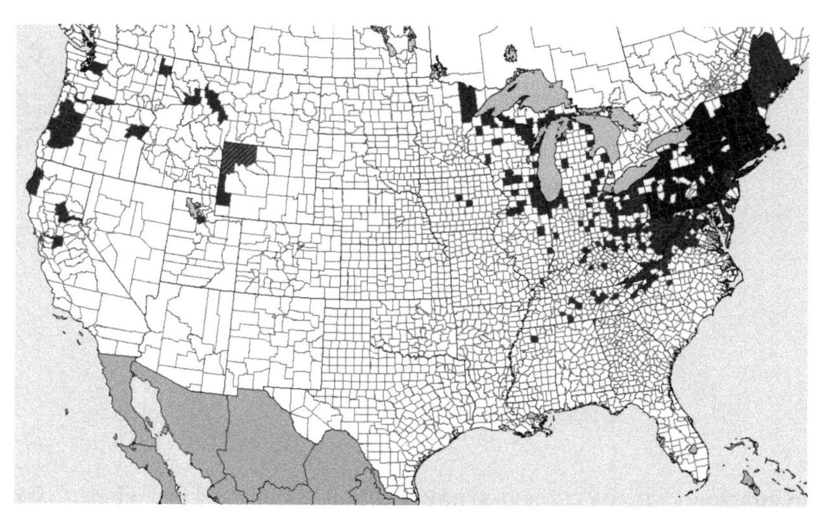

Distribution of **Galium mollugo**, SMOOTH BEDSTRAW

Galium mollugo, SMOOTH BEDSTRAW. A, habit. B, flowers. C, schizocarp. D, mericarps.

Hexasepalum teres (Walt.) J. H. Kirkbr. POOR-JOE

Annual herb. Roots shallow, slender, taprooted.

DESCRIPTION Stems hairy or minutely pubescent, branching, moderately erect to spreading, 1–8 dm tall (or long), nearly circular. **Leaves** opposite, sessile, linear-lanceolate to elliptic, narrow, tapering to long point, edges smooth. Stipules fused, forming several long bristles. **Flowers** perfect, 1–3 in axils of the leaves, small. Corolla whitish-pink to lavender, funnel-shaped, 4–6 mm long, with 4 short lobes. Calyx fused to the ovary, 4-toothed. Stamens 4, attached to the corolla. **Seed pods** hairy, obovoid top-shaped, not furrowed, with the 4 short, green calyx teeth at the top, splitting when ripe into 2 or 3 indehiscent carpels. **Seed** 3–4 mm long, hairy, oval, light-brown to grayish-brown, the inner surface indented with a forked groove.

FLOWERING June–October.

SYNONYMS *Diodia teres* Walt.

WHERE FOUND In abandoned fields, along roadsides, and waste places; mostly on dry or sandy soils.

ORIGIN Native.

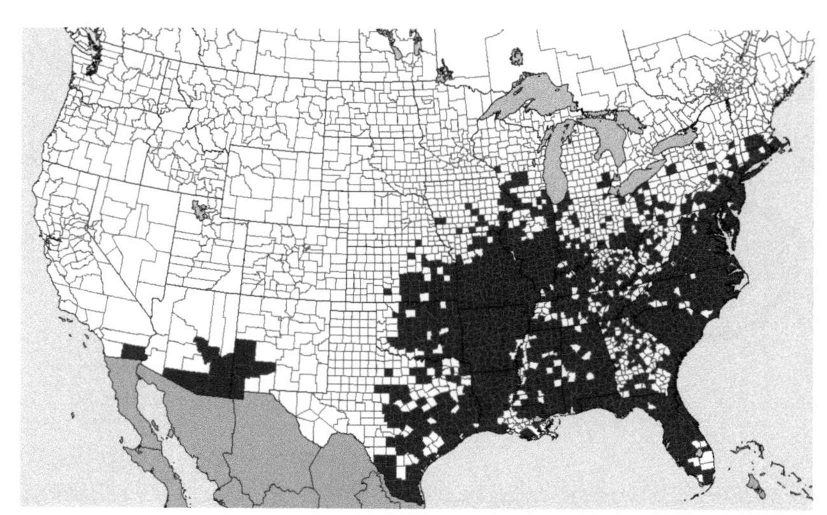

Distribution of **Hexasepalum teres**, POOR-JOE

Hexasepalum teres, POOR-JOE. A, habit. B, flower. C, fruits. D, fruits at node. E, nutlets.

Richardia scabra L. ROUGH MEXICAN-CLOVER

Annual herb, reproducing by seeds.

DESCRIPTION Stems sparingly branched and erect, or copiously branched and diffusely spreading, round, 1-5 dm tall, hairy. **Leaves** opposite, pubescent, the leaf blades flat, entire, oblong or elliptic to lanceolate or ovate, 2-8 cm long, acute or acuminate, wavy-margined, narrowed into short-margined petioles or nearly sessile. Stipules fringed. **Flowers** mostly perfect, mostly 6-parted, in terminal clusters, the clusters depressed, the involucre of 2 unequal pairs of bracts. Sepals lanceolate or ovate-lanceolate, 1-1.5 mm long, about as long as the hypanthium, stiff-hairy. Corolla white, 4-6 mm long, tube funnelform, usually 6-lobed, triangular or triangular-lanceolate, more or less hairy on the margins, much shorter than the tube. Carpels oblong, 3-3.5 mm long. **Fruit** of 3-4 carpels, crowned with the sepals. **Seed** 2-grooved on the lower face.

FLOWERING May-September.

WHERE FOUND Sandy soils.

ORIGIN Introduced from tropical America.

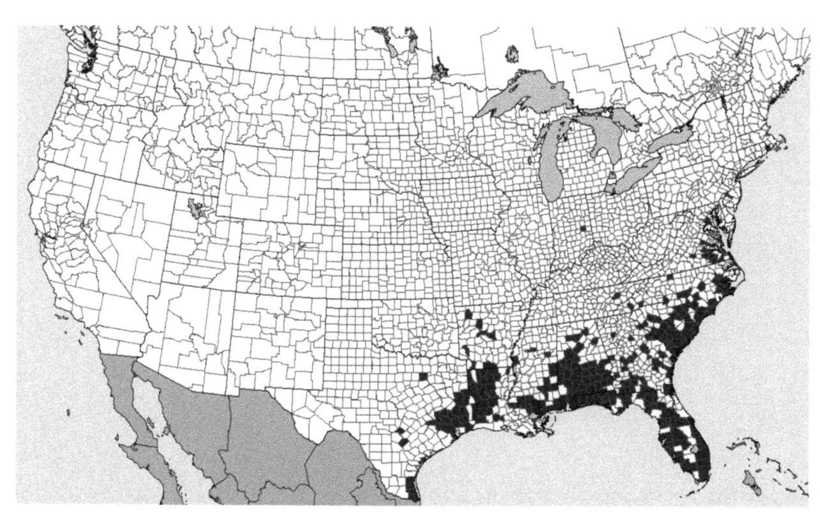

Distribution of **Richardia scabra,** ROUGH MEXICAN-CLOVER

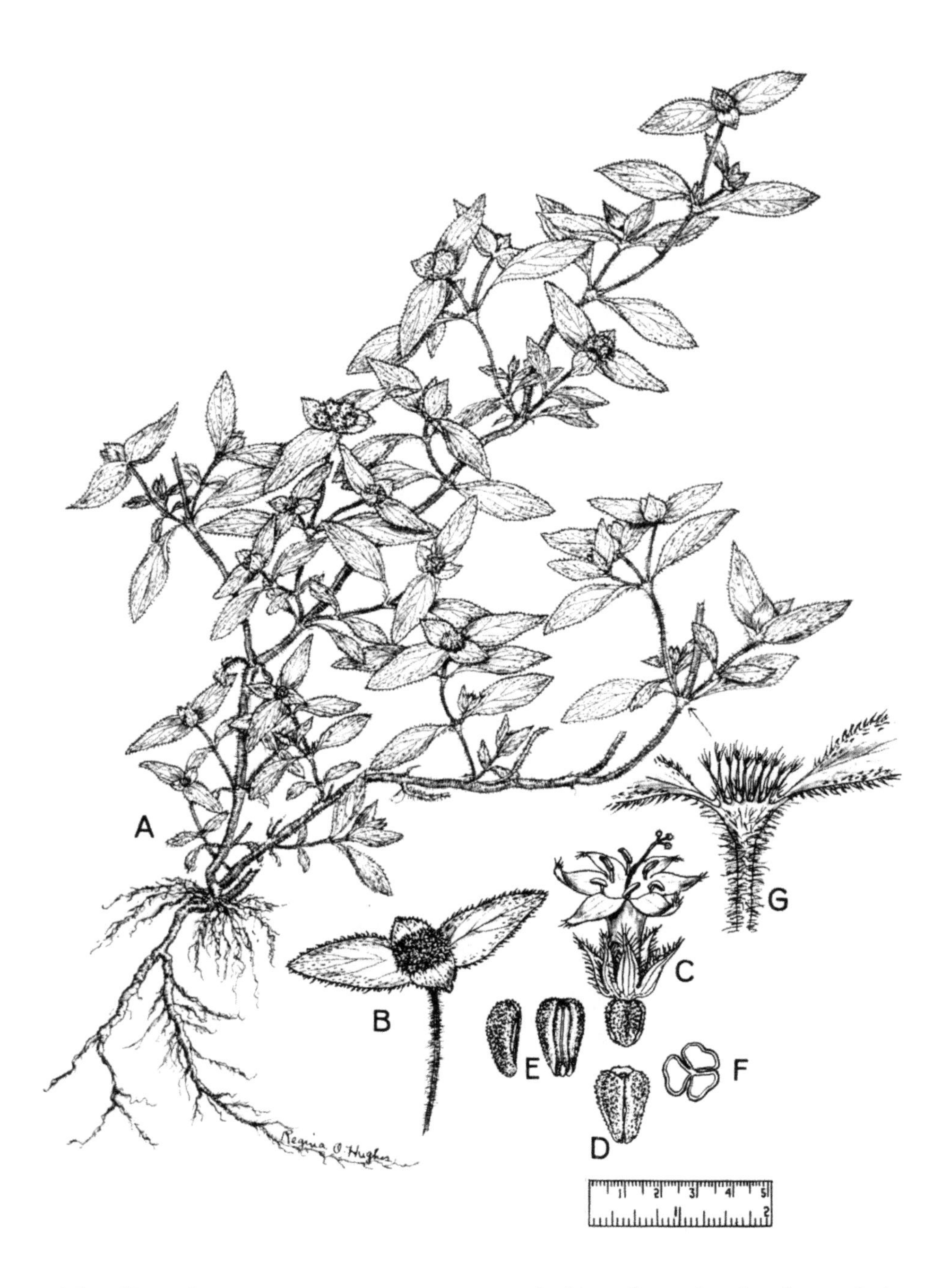

Richardia scabra, ROUGH MEXICAN-CLOVER. A, habit. B, flower head. C, flower, fruit. E, cocci (lobes of the fruit). F, cross section of fruit, showing 3 cocci. G, enlarged node, showing hairs.

Verbascum blattaria L. MOTH MULLEIN

Biennial herb, reproducing by seeds.

DESCRIPTION Stem erect, slender, glabrous or with glandular hairs near the top, 6-15 dm tall, simple or sometimes branched. **Leaves** on lower part of the stem in a rosette, 2-6 dm in diameter, dark-green, smooth or only slightly hairy, oblong, toothed or pinnately lobed, the upper leaves alternate, simple, sharp-pointed, sessile or partly clasping. **Flowers** perfect, yellow or pinkish-white, in long loose terminal racemes, on slender pedicels in the axils of bracts. Calyx 5-parted. Corolla about 2 cm in diameter, circular, 5-lobed. **Capsule** 2-celled, many-seeded, globose, 5-10 mm in diameter. **Seed** dark-brown, blunt, 0.8 mm long, the apex slightly rounded, with vertical rows of deep pits, those in rows next to each other alternating.

FLOWERING June-September.

WHERE FOUND Meadows, old fields, pastures, and waste places; common on dry gravelly soils.

ORIGIN Naturalized from Eurasia.

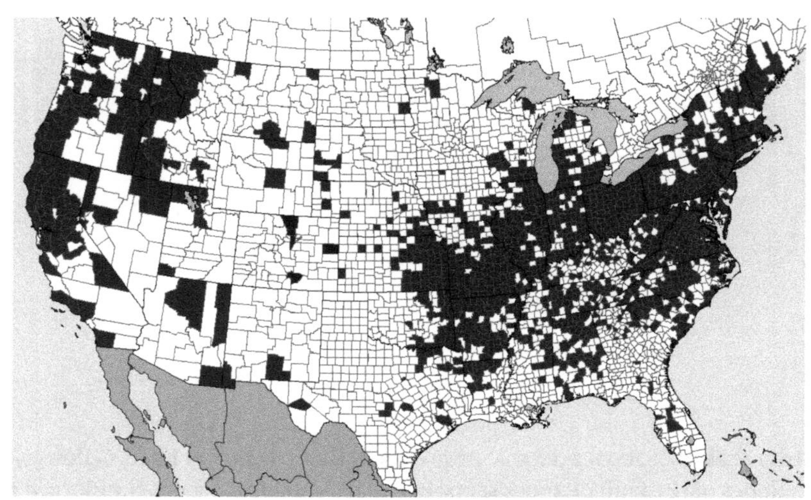

Distribution of **Verbascum blattaria**, MOTH MULLEIN

Verbascum blattaria, MOTH MULLEIN. A, habit. B, flower. C, capsule. D, seeds.

Verbascum thapsus L. COMMON MULLEIN

Biennial herb, reproducing by seeds.

DESCRIPTION Stems, 9–25 dm tall, very stout, usually unbranched, or with a few upright branches near the top, covered with woolly hairs. **Leaves** at the base of the plant in a rosette, 2–6 dm in diameter, the leaves 15–45 cm long, oblong, densely woolly, the upper leaves smaller and more pointed, the bases attached to the stem and extending down it to the next leaf. **Flowers** perfect, nearly sessile in long dense terminal cylindrical spikes. Calyx 5-lobed and very woolly. Corolla 5-lobed, circular, nearly regular, sulfur-yellow, 2–3 cm in diameter. **Capsule** globular, 2-celled, many-seeded, about 6 mm in diameter, downy. **Seed** numerous, brown, about 0.8 mm in diameter, wavy ridges alternating with deep grooves.

FLOWERING June–September.

WHERE FOUND Pastures, fence rows, roadsides, old fields, and waste places; common in dry gravelly and stony soils and railroad yards.

ORIGIN Naturalized from Eurasia; throughout most of the United States.

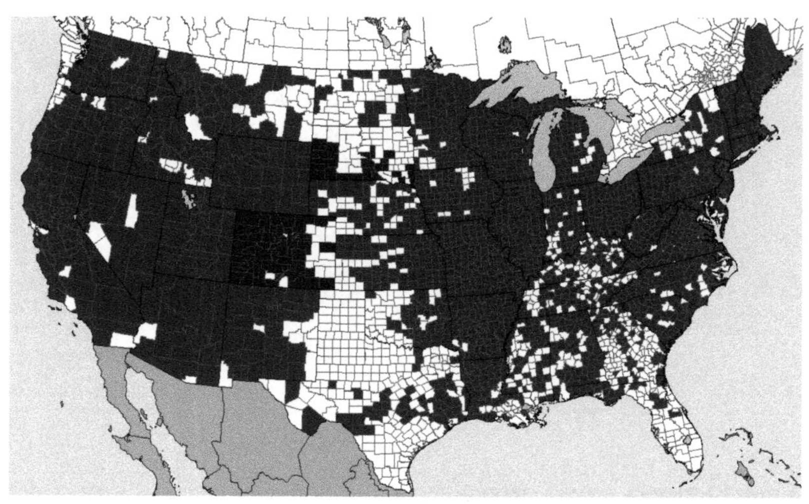

Distribution of **Verbascum thapsus**, COMMON MULLEIN

Verbascum thapsus, COMMON MULLEIN. A, habit. B, flowers. C, capsules. D, seeds.

Datura stramonium L. JIMSONWEED

Annual herb, reproducing by seeds. Root thick, shallow, extensively branched.

DESCRIPTION Stems erect, stout, 3–15 dm tall, with spreading branches above, glabrous, green or purple. **Leaves** alternate, simple, ovate, unevenly toothed, glabrous, dark-green above, 7–20 cm long on stout petioles, strong-scented. **Flowers** borne singly on short stalks in the axils of the branches. Corolla funnel-shaped, the border with 5 teeth, white to pinkish, 5–12 cm long. **Capsule** elongate-globular, about 2.5 cm in diameter, erect, 4-valved, 4-locular except near the 2-locular top, covered with short, sharp spines, but sometimes unarmed. **Seed** dark-brown to black, kidney-shaped, flattened, surface irregular and pitted.

FLOWERING July–October.

WHERE FOUND In cultivated fields, on rich soils, old feed lots, hog pens, and waste places; a troublesome weed toward the south.

ORIGIN Introduced and naturalized from Eurasia and Africa.

NOTE Plant narcotic-poisonous.

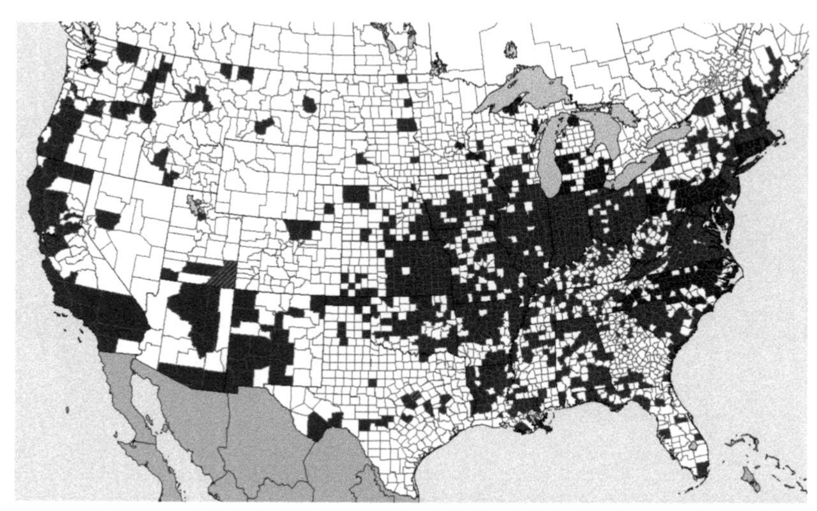

Distribution of **Datura stramonium**, JIMSONWEED

Datura stramonium, JIMSONWEED. A, habit, upper part of plant. B, cauline leaf. C, ripe capsule. D, seeds.

Physalis heterophylla Nees CLAMMY GROUNDCHERRY

Perennial herb, reproducing by seeds and rootstocks.

DESCRIPTION Stems erect, branched or widely spreading, 3-8 dm tall, downy-pubescent, usually sticky or glandular, ridged. **Leaves** alternate, simple, broadly ovate, with a rounded or cordate base, sticky-pubescent, wavy or bluntly toothed, 5-7.5 cm long. **Flowers** perfect, solitary in the axils of the leaves or branches. Calyx 5-cleft, reticulate and enlarging after flowering into a much-inflated balloon, inclosing the berry. Corolla bell-shaped to wheel-shaped, 1.5-2.2 cm in diameter, 5-angled, 5-toothed. Stamens 5, inserted on the corolla. **Fruit** a 2-celled, many-seeded, globose, yellow berry, surrounded by the inflated papery calyx. **Seed** numerous, about 2 mm in diameter, flattened, obovate to semicircular, dull, granular, light-orange or straw-colored.

FLOWERING June–August.

WHERE FOUND In cultivated fields, pastures, roadsides, meadows, and gardens; on gravelly and stony soils.

ORIGIN Native.

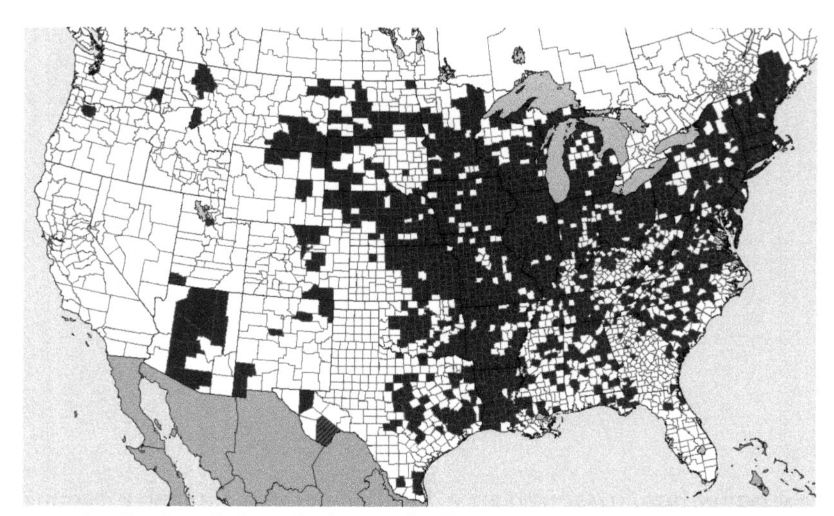

Distribution of **Physalis hetrophylla**, CLAMMY GROUNDCHERRY

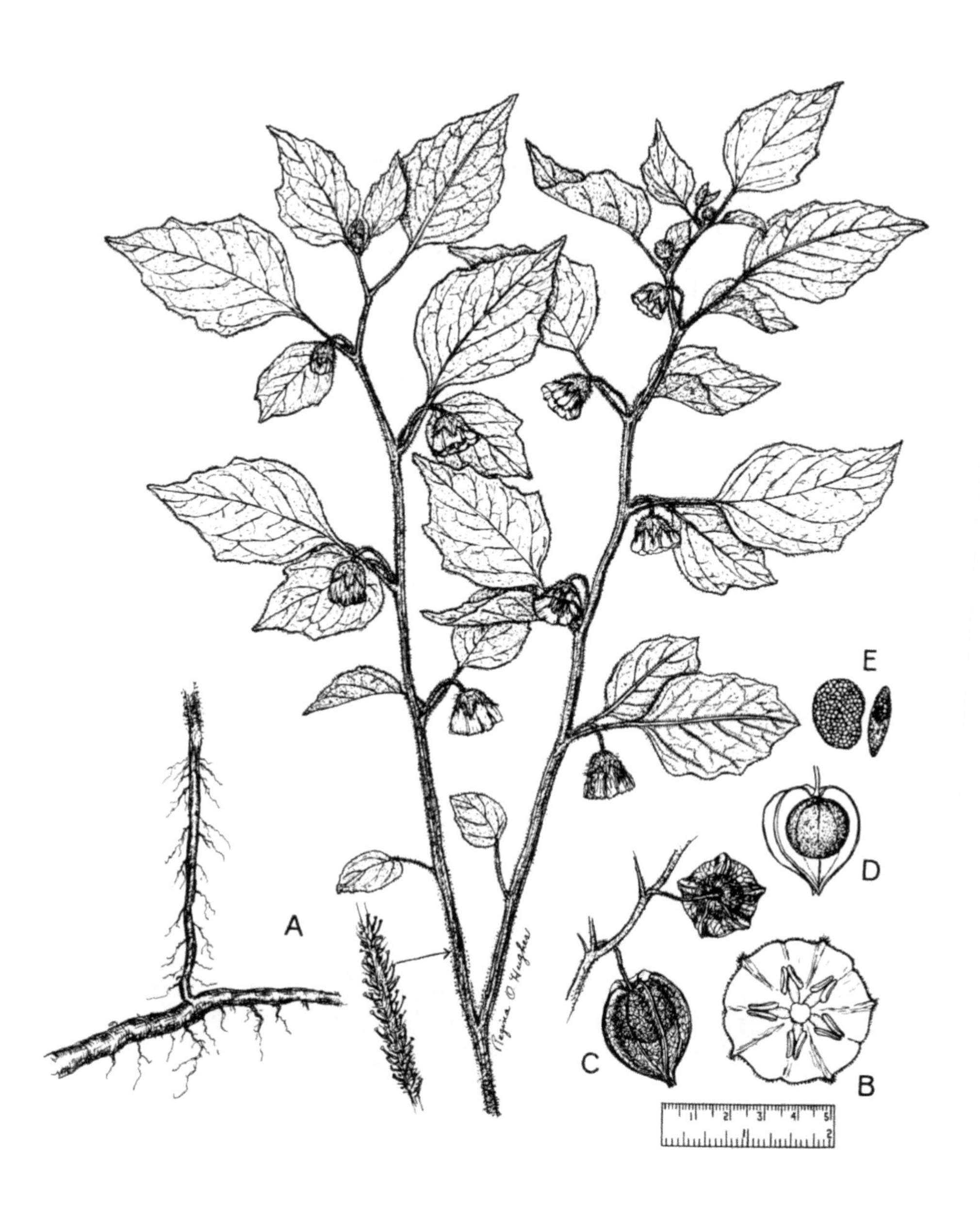

Physalis hetrophylla, CLAMMY GROUNDCHERRY. A, habit. B, flower, open to show wide filaments. C, berries in transparent calyx. D, section of fruit showing berry. E, seeds.

Solanum carolinense L. HORSE-NETTLE, CAROLINA NETTLE

Perennial herb, prickly, reproducing by seeds and by creeping underground rhizomes.

DESCRIPTION Stem simple or branched, 30–120 cm tall, hairy with star-shaped hairs (4- to 8-rayed), the slender prickles straw-shaped. **Leaves** alternate, green, mostly 7–12 cm long and about half as wide, elliptic-oblong to oval, margin divided into lobes and coarsely toothed (2–5 teeth), rough, yellow prickles on petioles, midribs, and veins. Racemes several-flowered, soon becoming 1-sided. **Flowers** about 2.5 cm across. Calyx lobes lance-acuminate. Corolla violet, bluish, or white, about 2 cm wide. Anthers equal, tapering to a tip. **Fruit** a yellow, juicy berry, smooth at first, becoming wrinkled late in the season, globose, 1–1.5 cm in diameter. **Seed** numerous, about 1.5 mm in diameter, round, flattened, yellowish.

FLOWERING May–October.

WHERE FOUND Fields, waste places, and gardens; often a troublesome weed in sandy areas.

ORIGIN Native, but considered adventive in western United States.

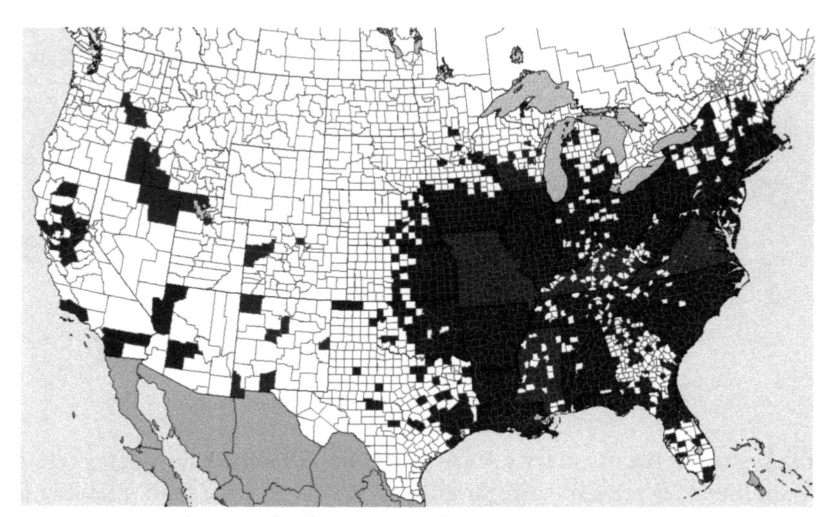

Distribution of **Solanum carolinense**, HORSE-NETTLE

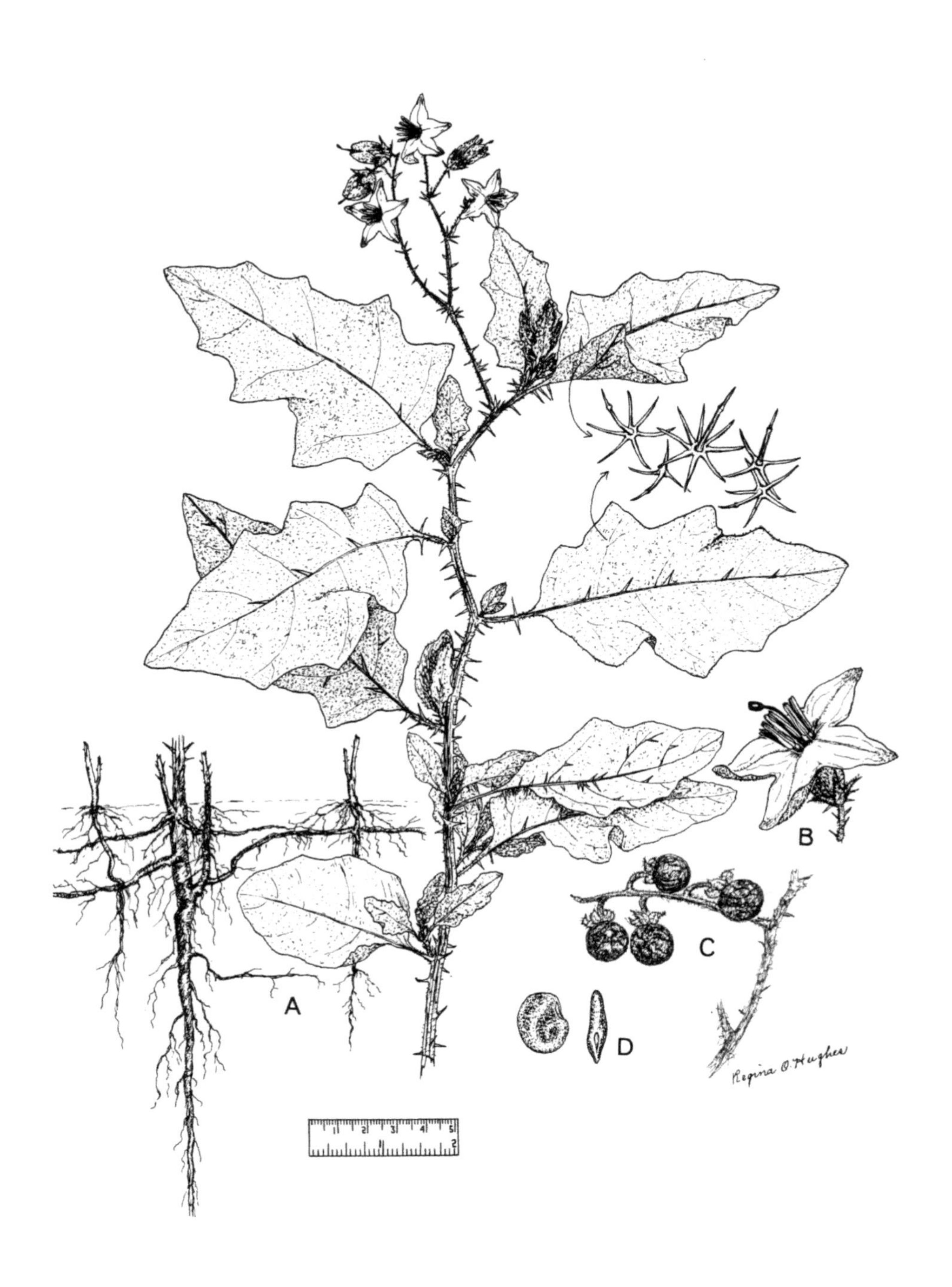

Solanum carolinense, HORSE-NETTLE. A, habit. B, flower. C, berries. D, seeds.

Solanum nigrum L. BLACK NIGHTSHADE

Annual herb, without tubers, reproducing by seeds.

DESCRIPTION **Stems** glabrous or only sparsely or remotely pubescent, erect or spreading, 30–60 cm tall. **Leaves** simple, bluntly lobed at the base, ovate, 2.5–7.5 cm long, thin, membranaceous and translucent to thickish, dense or opaque to transmitted light, narrowed, the edges wavy. **Inflorescences** umbellate, or often corymbose or nearly racemose, with 5–10 flowers on pedicels. Calyx lobes with rounded tips. Corolla lobes 5–7.5 mm long, white. Anthers 1.3–2.6 mm long. **Fruit** a berry, green turning black at maturity, 5–13 mm in diameter, lustrous to dull. **Seed** numerous, 1.2–2.3 mm in diameter, usually without concretions, dull, pitted, yellow to dark-brown.

FLOWERING May–November.

WHERE FOUND An occasional weed in waste places, roadsides, disturbed and cultivated fields, and sea-beaches.

ORIGIN Naturalized from Europe.

NOTE The unripe berries of some races are poisonous to sheep and other grazing animals. However, the ripe berries are eaten raw or cooked for preserves or pies by humans. *Solanum nigrum* and *Solanum americanum* are difficult to distinguish, except as races of a very polymorphic species. *S. americanum* is native to south-central United States (adventive elsewhere), in rocky or dry open woods, thickets, shores, and openings, often spreading to cultivated or waste grounds. *Solanum nigrum* mostly found along West Coast and in northeastern United States.

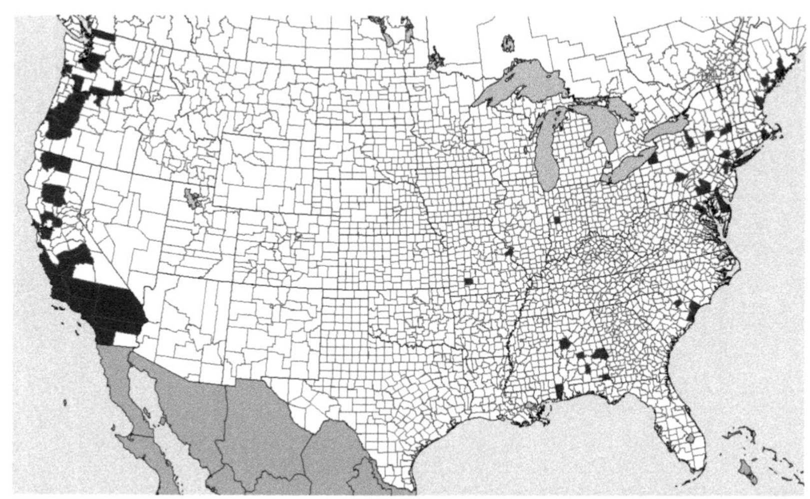

Distribution of **Solanum nigrum**, BLACK NIGHTSHADE

Solanum nigrum, BLACK NIGHTSHADE. A, habit. B, inflorescence. C, seeds.

Typha angustifolia L. NARROW-LEAF CAT-TAIL

Perennial herb, from a creeping rhizome, plants up to 15 dm tall.

DESCRIPTION Leaves usually less than 10, somewhat convex on the back, full green, herbaceous, 3-8 mm wide; **pistillate** (below) and **staminate** (above) portions of the spike usually separated by a short interval; the pistillate portion reddish-brown, in fruit 6-15 mm in diameter 0.3-1.5 dm long, its surface minutely bristly with persistent linear stigmas; the staminate half of spike 0.7-2 dm long, the **pistillate flowers** with a linear fleshy stigma and usually with a hairlike bractlet with dilated blunt tips among the bristles; denuded old axis covered with stout compound papillate pedicels 0.5-0.7 mm long. **Fruit** 5-8 mm long, subtended by copious white hairs arising distinctly above the middle.

FLOWERING Late May-July.

WHERE FOUND Chiefly in basic or alkaline waters; waste wet areas, especially along the seaboard eastward.

ORIGIN Introduced.

Typha latifolia L. COMMON CAT-TAIL

Perennial herb, from a creeping rhizome, stout; plants to 2.5 m tall.

DESCRIPTION Leaves flat, sheathing, pale or grayish-green, 6-23 mm wide. **Staminate** (7-13 cm long) and dark-brown **pistillate** (2.5-20 cm long) parts of the spike usually contiguous, in fruit 1.2-3.5 cm thick, its surface appearing minutely pebbled with crowned persistent stigmas and scarcely bristly, **pistillate flowers** without branchlets among the bristles; stigmas lance-ovate, fleshy, persistent; denuded axis of old spike retaining slender pedicels 1-2 mm long. **Fruit** about 1 cm long, with copious white hairs arising near the base.

FLOWERING Late May-July.

WHERE FOUND Marshes and shallow water, ditches, and wet wastes along rivers; the common inland species.

ORIGIN Native.

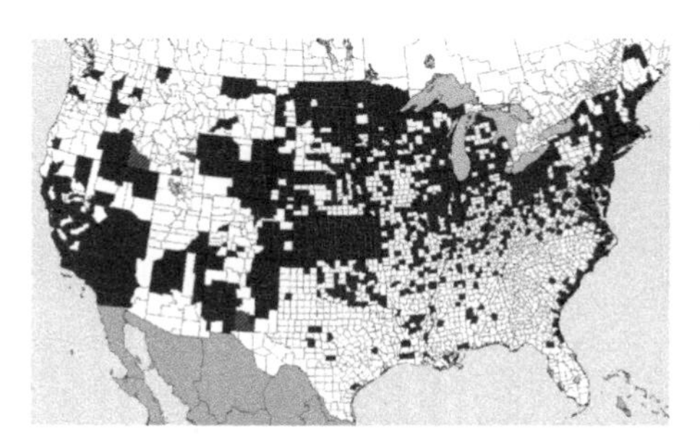

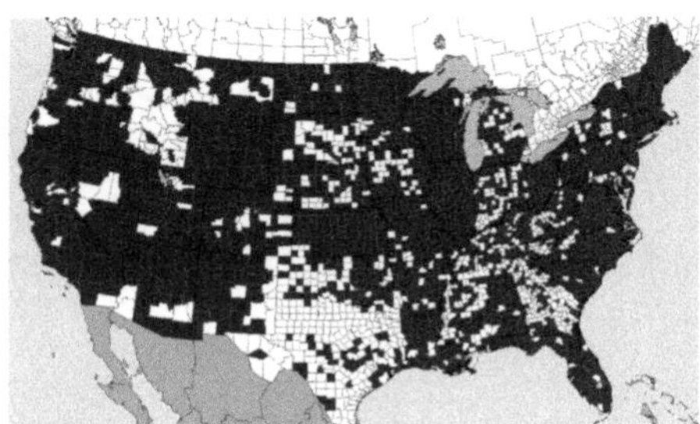

Distribution of **Typha angustifolia**, left, **Typha latifolia**, right

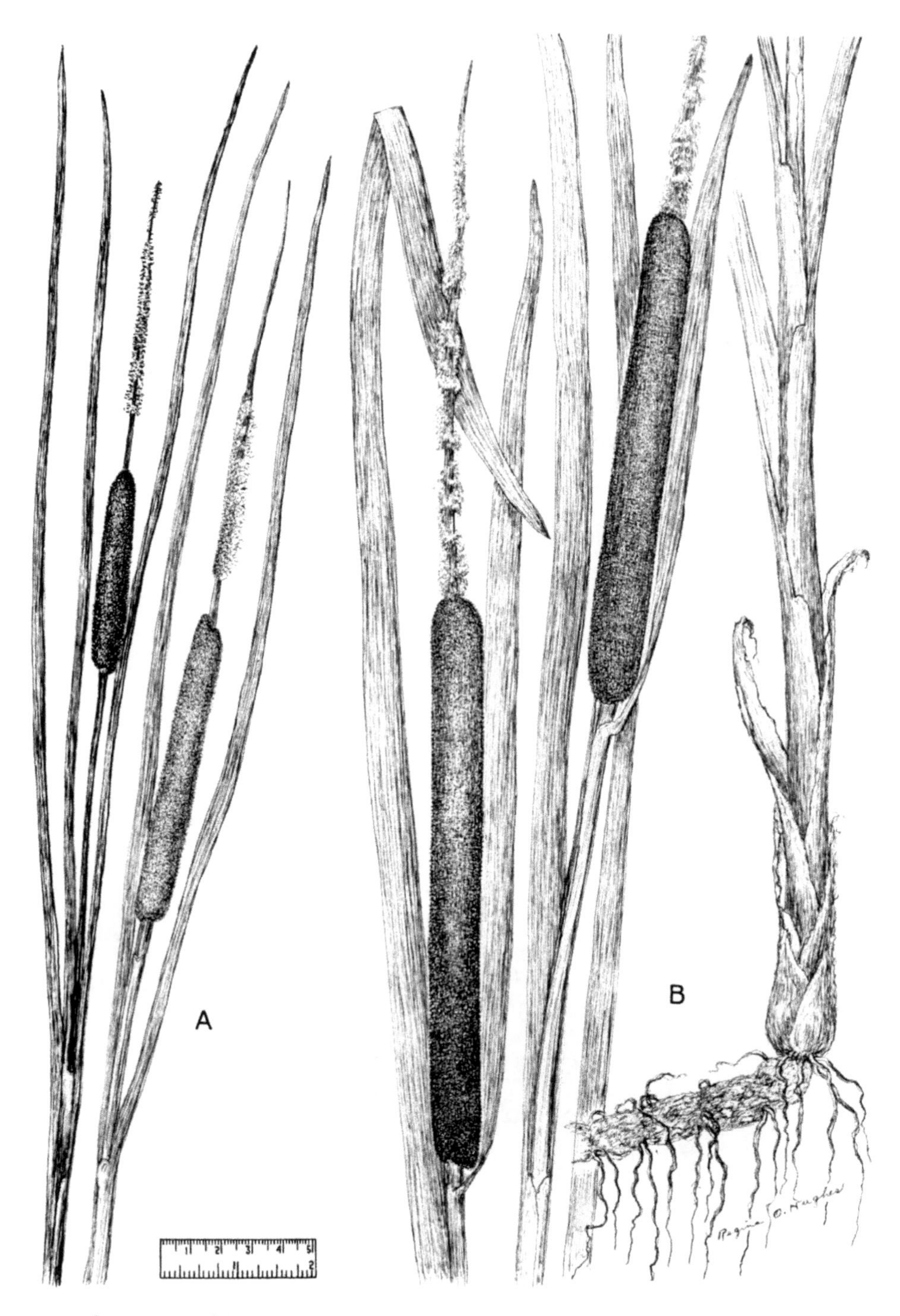

A, **Typha angustifolia**, NARROWLEAF CAT-TAIL. Habit. B, **Typha latifolia**, COMMON CAT-TAIL. Habit.

Urtica dioica L. STINGING NETTLE

Perennial, reproducing by seeds and creeping rootstocks.

DESCRIPTION Stems erect, 1-2 m tall, ridged, bristly-hairy with stinging bristles (1-2 mm long). **Leaves** opposite, simple, egg-shaped to heart-shaped, coarsely serrate, hairy or glabrous, with or without stinging bristles, usually twice as wide as the length of the petiole, the stipules linear-lanceolate, 5-15 mm long, green to pale-brown, minutely pubescent. **Inflorescence** branched, many-flowered, loose to dense, panicled spikes. **Flowers** mostly dioecious, small, greenish. Staminate flowers with 4 perianth segments and 4 stamens. Pistillate flowers with 4 perianth parts and a 1-celled ovary. **Fruit** (achene) 1-1.5 mm long, flattened, egg-shaped, minutely glandular, yellow to grayish-tan, the calyx and remnant of the style often persistent.

FLOWERING June-September.

WHERE FOUND Waste places, roadsides, vacant lots, rich soil, and edge of damp woods.

ORIGIN Native.

NOTE Sting hairs are a skin irritant.

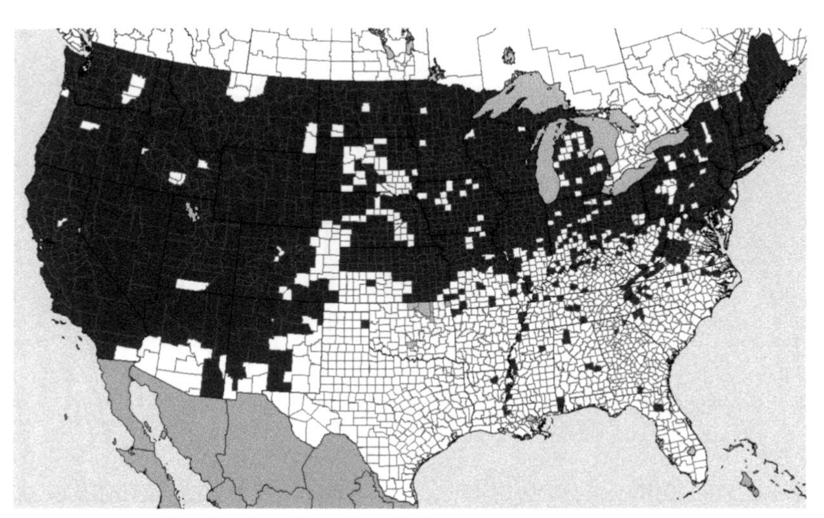

Distribution of **Urtica dioica**, STINGING NETTLE

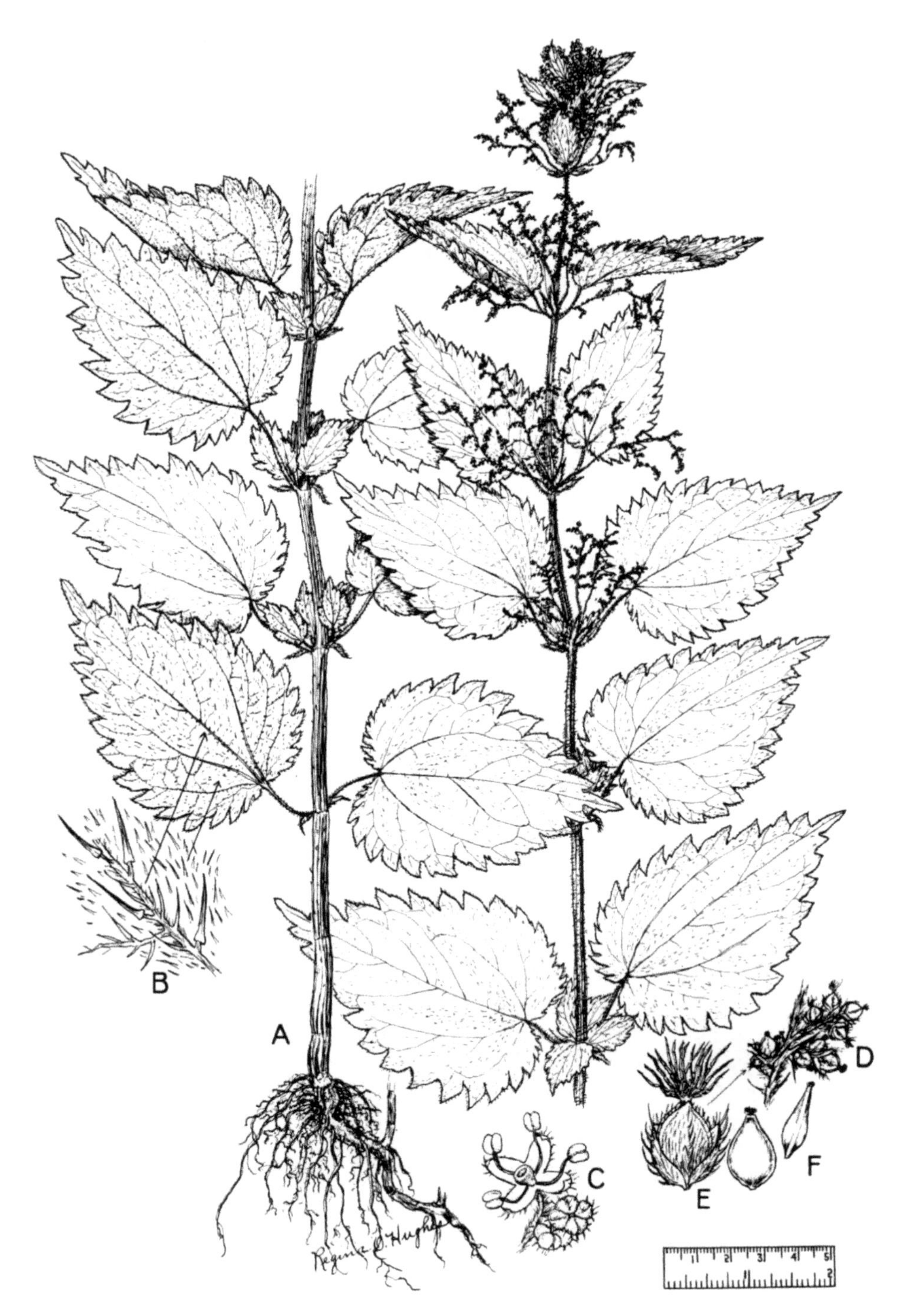

Urtica dioica, STINGING NETTLE. A, habit. B, stinging hairs, enlarged. C, flower, closed and open. D, fruiting spike. E, fruit. F, achenes, face and edge views.

Verbena bracteata Cav. ex Lag. & Rodr. PROSTRATE VERVAIN

Annual herb, reproducing by seeds.

DESCRIPTION Stems diffusely branching at the base, with some spreading and some ascending branches, 1–4 dm long, rough-hairy, somewhat 4-angled. **Leaves** opposite, simple, wedge-shaped in outline, pinnatifid and much-toothed, 1–8 cm long, rough-hairy, short-petioled. **Flowers** perfect, sessile, in dense terminal spikes, almost hidden by the conspicuous bracts. Bracts stiff, hairy, longer than the calyx, 1–2 cm long. Calyx short, 3–4 mm long, hairy, tubular, 5-toothed. Corolla purplish-blue, irregular, tubular, 5-cleft, its limbs 2.5–3 mm broad. Stamens 2 long and 2 short, included. Ovary 4-celled, not 4-lobed, but in fruit (schizocarp) splitting into 4, 1-seeded, indehiscent nutlets. **Nutlet** linear-oblong, 2 mm long, margin slightly winged, with 2 flattened sides and 1 rounded side, the convex surface netted above, with a white scar near the base, dark-brown.

FLOWERING June–August.

SYNONYMS *Verbena bracteosa* Michx.

WHERE FOUND In meadows, pastures, barnyards, sandy prairies, lawns, and waste places; rarely in cultivated fields.

ORIGIN Native.

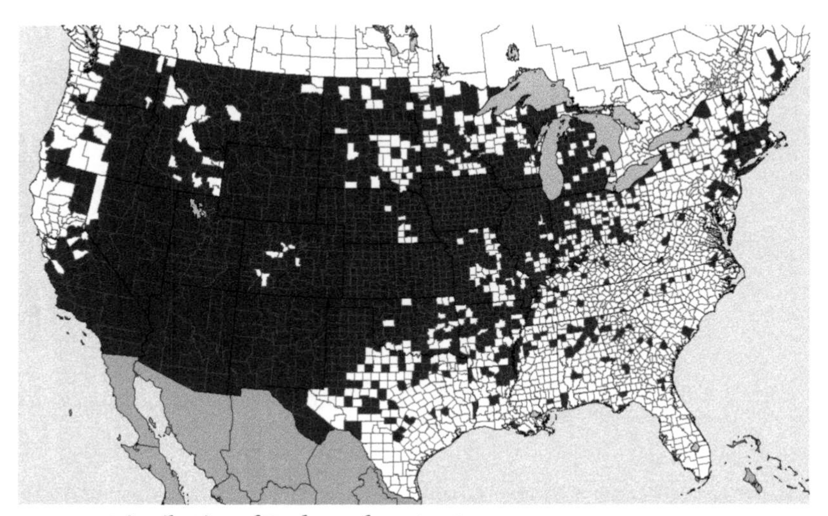

Distribution of **Verbena bracteata,** PROSTRATE VERVAIN

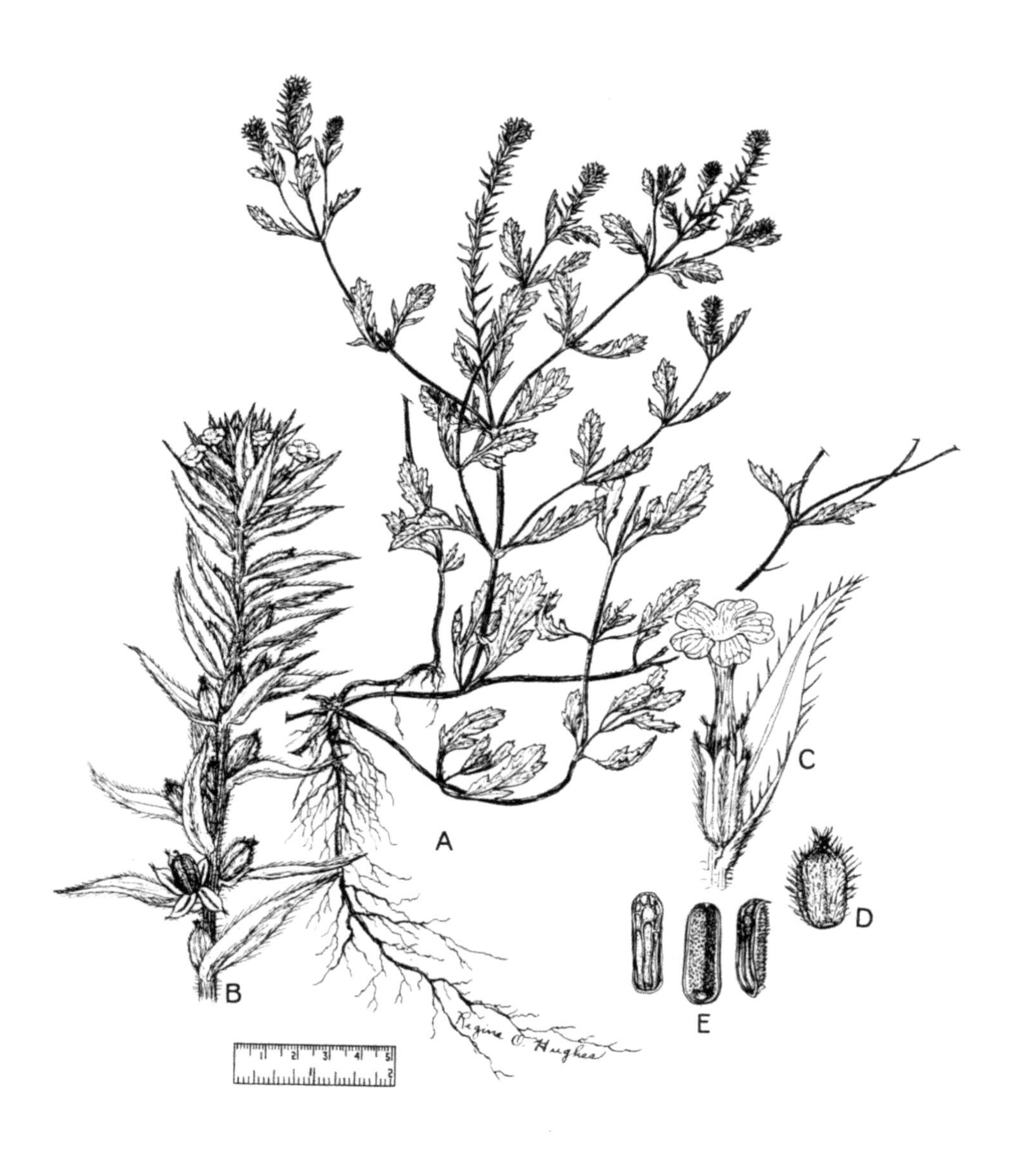

Verbena bracteata, PROSTRATE VERVAIN. A, habit. B, flowering spike, showing the long bracts. C, flower. D, schizocarp. E, mericarps.

Verbena stricta Vent. HOARY VERVAIN

Perennial herb, reproducing by seeds.

DESCRIPTION Stems erect, stoutish, nearly round, simple or with a few branches above, 3–12 dm tall, velvety with soft whitish hairs. **Leaves** opposite, simple, oblong to broadly obovate or rounded-elliptic, 2–10 cm long, 3–6 cm broad, doubly toothed, sessile, downy with whitish hairs, hairy beneath, pinnately veined. **Flowers** in dense spikes. Bracts hairy, shorter than the calyx. Calyx white-pilose, overlapping. Fruiting calyces 2.5–5 mm long. Corolla purple to rosy-pink, or white, its limbs 8–9 mm broad. **Nutlet** 2.5–3 mm long, netted above, dark or grayish-brown.

FLOWERING June–September.

WHERE FOUND Prairies and barrenlands, pastures, old fields, and waste places.

ORIGIN Native.

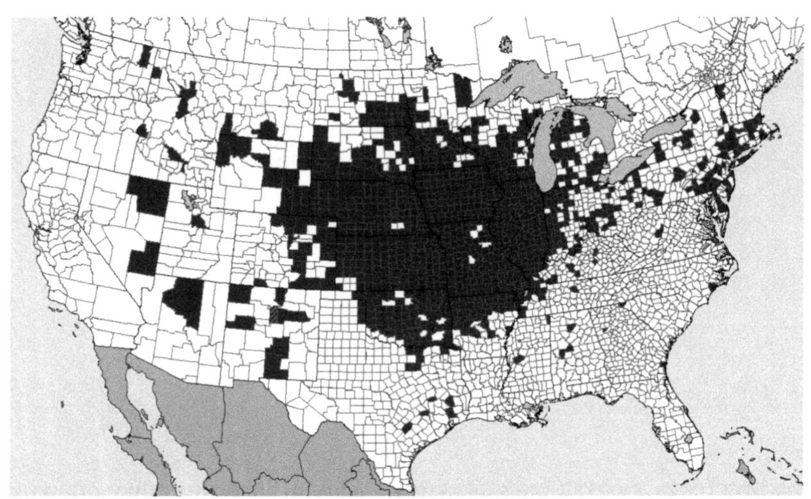

Distribution of **Verbena stricta**, HOARY VERVAIN

Verbena stricta, HOARY VERVAIN. A, habit. B, flowers. C, schizocarps. D, mericarp.

Tribulus terrestris L. PUNCTUREVINE, CALTROP

Annual herb, reproducing by seeds. Root system simple, taprooted.

DESCRIPTION Stems prostrate, branching from base to form dense mats of slender branches, to 2.4 m long, pubescent to stiff hairy. **Leaves** opposite, short-petioled, 2-6 cm long, oblong, pubescent, divided into pinnate leaflets, the leaflets in 5-8 pairs, oblong, 5-15 mm long. **Flowers** small, yellow, 5-petaled, 5-10 mm wide, axillary, the peduncles 5-10 mm long. **Seed pods** about 1 cm thick, containing 5 burs each with 2-4 sharp, stout prickles (strong enough to puncture bicycle tires or penetrate shoe soles).

FLOWERING June-September.

WHERE FOUND Pastures, roadsides, wastes, along railroads, ballast lots, sometimes in cultivated fields; dry waste areas and open sandy ground.

ORIGIN Naturalized from Europe.

NOTE A serious weed and considered a noxious weed in some states; cattle will not graze in infested areas; most common in the Southwest.

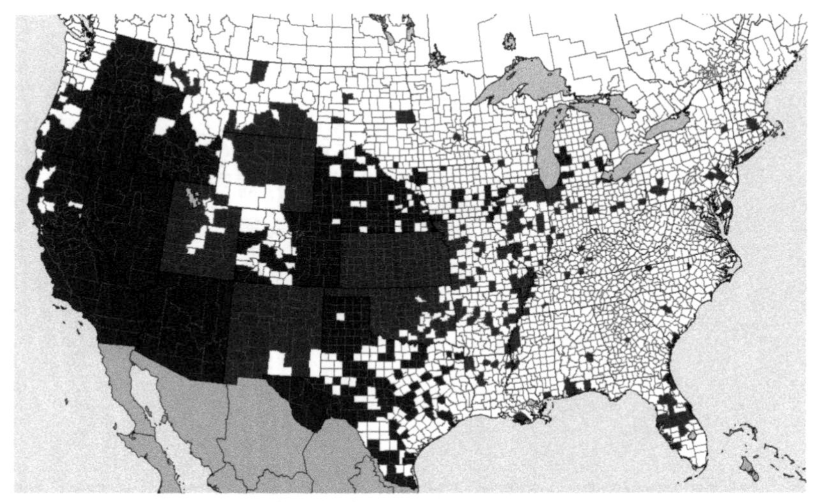

Distribution of **Tribulus terrestris**, PUNCTUREVINE

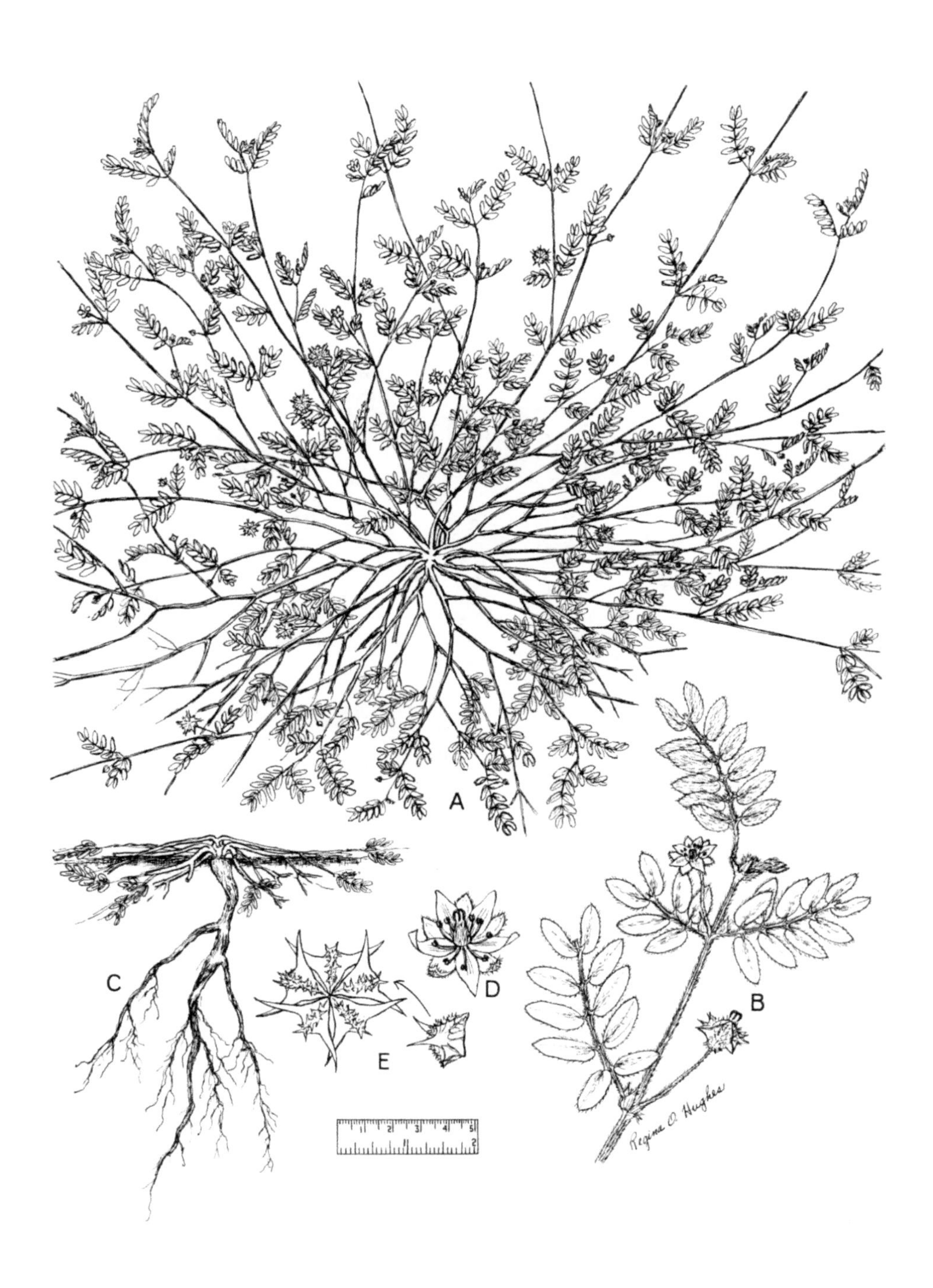

Tribulus terrestris, PUNCTUREVINE. A, Habit, seen from above. B, flowering branchlet. C, rootstock, side view. D. flower. E, fruits.

Abutilon theophrasti, VELVETLEAF. A member of the Mallow Family and a common weed in cultivated fields, especially in the Midwest (see page 262).

GLOSSARY

Acaulescent — Stemless, or nearly so.
Achene — A dry indehiscent, one-seeded fruit, formed from a single carpel, the seed distinct from the fruit, as in Asteraceae.
Acuminate — Tapering gradually to the apex or tip.
Acute — Sharp-pointed.
Adnate — Attached to.
Alternate — With a single leaf or other structure at each node.
Amplexicaul — Said of a sessile leaf with its base clasping the stem horizontally.
Annual — Living one growing season.
Anthesis — The opening of the flower bud.
Areola (pl. -ae) — Small pits, usually surrounded by a ridge.
Aristate — Having awns, or awn-shaped.
Articulate — Jointed; breaking into distinct pieces without tearing at maturity.
Attenuate — Narrowed, tapered; elongated.
Auricle — Ear-shaped appendages, usually on leaves, petals, or stamens.
Auriculate — Having ear-shaped appendages.
Awn — A bristle on the flowering glumes of grasses.
Axil — The angle between the leaf or branch and the main axis.
Biennial — Living two growing seasons.
Bipinnate — Twice pinnately compound.
Blade — The expanded part of a leaf or floral part.
Bract — A small, rudimentary or imperfectly developed leaf.
Bulb — A bud with fleshy bracts or scales, usually subterranean.
Caducous — Nonpersistent; falling away readily.
Calyx — The outer set of sterile, floral leaveshe sepals considered collectively.
Campanulate — Bell-shaped.
Canescent — Becoming gray or grayish.
Capitate — Arranged in a head, as the flowers in Asteraceae.
Capitellate — Small-headed, as the flower heads in Asteraceae.
Capsule — A dry fruit of two or more carpels, usually dehiscent by valves.
Carpel — A portion of the ovary or female portion of the flower.
Caruncle — An outgrowth near the micropyle and hilum of the seed, as in Euphorbiaceae.
Caryopsis — "Seed" of grasses, a fruit, a grainn achene in which the pericarp layers are fused.
Caudex — A trunk or stock.
Caulescent — Having a stem.
Cauline — On the stem, as stem leaves, as opposed to rosette or basal.
Cilia — Fine hairs or projections.
Ciliate — Having fine hairs or projections, usually as marginal hairs.
Ciliolate — Minutely ciliate.
Circumscissile — Opening all around by a transverse split.
Cleistogamy, Cleistogamous — Pollination and fertilization before the flower has opened.
Cochleiform — Shell-shaped.
Coma — A tuft of hairs attached to the testa of a seed.
Compound — Composed of several parts or divisions.
Contiguous — Touching but not united.
Cordate — Heart-shaped.
Corolla — The inner set of sterile, usually colored, floral leaves; the petals considered collectively.
Corymb — A raceme with the lower flower stalks longer than those above, so that all the flowers are at the same level.
Crenate — With roundish teeth or lobes.
Crenulate — Minutely crenate.
Culm — The stem of a grass or sedge.
Cuneate — Wedge-shaped.
Cuspidate — Having a rigid point.
Cyathium (pl. -ia) — An inflorescence reduced to look like a single flower.
Cyme — An inflorescence; a convex or flat

flower cluster, the central flowers unfolding first.
Cymos— Having cymes, or cyme-like.
Deciduous — Dying back; seasonal shedding of leaves or other structures; falling off.
Decompound — Several times compounded or divided, as in leaves of carrot.
Decumbent — Lying flat, or being prostrate, but with the tip growing upwards.
Decurrent — Applied to an organ extending along the sides of another, as the leaf blade extending as wings down the petiole on the stem.
Deltoid — Triangular.
Dentate — Toothed, with outwardly projecting teeth.
Denticulate — Finely toothed.
Diffuse — Loosely spreading.
Digitate — Diverging like the spread fingers.
Dioecious — Only one sex in a plant; with male or female flowers only.
Disk (disc) — A flattened enlargement of the receptable of a flower of inflorescence; the head of tubular flowers, as in sunflower.
Dissected — Divided into many segments.
Divaricate — Widely spreading; forked.
Drupe — A fruit with a fleshy or pulpy outer part and a bone-like inner part; a single-seeded fleshy fruit.
Drupelet — A small drupe, as one section of a blackberry.
Eciliate — Without cilia or hairs.
Elliptic — Oval.
Emarginate — Lacking a distinct margin; with a notched apex.
Emersed — Amphibious; protruding upwards out of the water.
Endocarp — The inner layer of a fruit wall (pericarp); it is usually woody.
Entire — Without teeth, serrations, or lobes, as leaf margins.
Evanescent — Disappearing early; not permanent; or as veins fading away toward the margins of a leaf.
Falcate — Sickle-shaped.
Fascicled — A tuft of leaves or other structures crowded on a short stem, as the flowers in Cuscuta.
Fibrous — A mass of adventitious fine roots.
Filiform — Threadlike.
Flaccid — Limp or flabby.
Floccose — Cottony.
Follicle — A many-seeded dry fruit, derived from a single carpel, and splitting longitudinally down one side.
Fruit — The ripened ovary or ovaries with the attached parts.
Fugacious — Lasting for a short time; soon falling away.
Fuscous — Dingy-brown.
Fusiform — Elongated and tapering towards each end.
Glabrate — Nearly without hairs.
Glabrescent — Very thinly covered with hairs, becoming hairless as it matures.
Glabrous — Smooth or hairless.
Glaucous — Covered with a bluish or white bloom.
Glomerules — Cluster of short-stalked flowers, as in Cuscuta.
Glume — One of a pair of dry bracts, at the base of and enclosing the spikelet of grasses.
Gynophore — Said of a style that arises near the base of the carpels or ovary lobes, as in Euphorbiaceae.
Hastate — Arrow-shaped with the basal lobes spreading.
Head — A dense inflorescence of sessile or nearly sessile flowers, as in Asteraceae.
Hilum — Scar on a seed where it is attached to the fruit.
Hirsute — Having rather coarse, stiff hairs.
Hispid — With rough, bristly hairs.
Hyaline — Clear and translucent.
Hypanthium — An enlargement of the axis of a flower under the calyx, as in Euphorbiaceae.
Imbricate — Overlapping.
Incised — Cut into sharp lobes.
Indehiscent — Not opening at maturity.
Indusium (pl. -ia) — Covering of the sorus in a fern.
Inflorescence — The arrangement of the several flowers on the flowering shoot, as a spike, panicle, head, cyme, umbel, raceme.
Internode — The stem between two

successive nodes.
Involucre — Any leaflike structure protecting the reproducing structure, as in flower heads of Asteraceae and Euphorbiaceae.
Keel — A projecting ridge, as in the flowers of Fabaceae.
Laciniate — Cut into narrow segments or lobes; fringed.
Lacuna (pl. -ae) — A large multicellular cavity, or a cavity anywhere in a plant.
Lanceolate — Flattened, two or three times as long as broad, widest in the middle and tapering to a pointed apex; lance-shaped.
Leaf sheath — The lower part of a leaf, which envelops the stem, as in grasses.
Leaflet — One of the divisions of a compound leaf.
Legume — A dry fruit consisting of one carpel, splitting by two longitudinal sutures with a row of seeds on the inner side of the central suture; pod, as in Fabaceae.
Lemma — The outer bract of a grass flower.
Lenticular — Bean-shaped; shaped like a double convex lens.
Ligule — A membrane at the junction of the leaf sheath and leaf base of many grasses.
Linear — A long and narrow organ with the sides nearly parallel.
Lobed — Divided to about the middle or less.
Locule — A compartment divided by a septa.
Lyrate — Pinnatified with the terminal lobe the largest.
Membranaceous — Thin and rather soft, often papery.
Mericarp — A one-seeded portion of a fruit that splits up at maturity.
Midrib — The central rib of a leaf or other organ; midvein.
Monadelphous — Said of stamens whose filaments are united in bundles or to form a tube.
Moniliform — Like a string of beads.
Monoecious — Both sexes in the same flower or on the same plant.
Mucronate — With a sharp, abrupt point.
Muricate — Having a surface roughened by short, sharp points.
Nerved — Veined or veiny.
Node — The part of a stem where the leaf, leaves, or secondary branches emerge.
Nutlet — A one-seeded portion of a fruit that fragments at maturity.
Obcordate — Inversely heart-shaped.
Oblanceolate — Inversely lanceolate.
Oblique — With part not opposite, but slightly uneven.
Oblong — Elliptical, blunt at each end, having nearly parallel sides, two to four times as long as broad.
Obovate — Inversely ovate.
Obtuse — Blunt or rounded.
Ocrea (pl. -ae) — A thin, sheathing stipule or a united pair of stipules.
Ocreolae — Stipular sheaths on the secondary stems, as in Polygonaceae.
Orbicular — Nearly circular in outline.
Ovate — Egg-shaped.
Palea or Palet — The inner bract of a grass floret.
Palmate — Diverging like the fingers of a hand.
Panicle — An inflorescence, a branched raceme, with each branch bearing a raceme of flowers, usually of pyramidal form.
Paniculate — Borne in panicles; resembling a panicle.
Pappus — A ring of fine hairs developed from the calyx, covering the fruit, as in Asteraceae; acting as a parachute for wind-dispersal, as in dandelion.
Pedicel or Peduncle — A short stalk.
Pedicelled — Having a short stalk, as a flower or fruit.
Peltate — More or less flattened, attached at the center on the underside.
Perennial — Growing many years or seasons.
Perfect — A flower having both stemens and carpels.
Perfoliate — Leaves clasping the stem, forming cups, as in Dipsacus.
Perianth — The calyx and corolla together; a floral envelop.
Pericarp — The body of a fruit developed from the ovary wall and enclosing the

seeds.
Perigynium (pl. -ia) — The perianth in Cyperaceae, consisting of scales, bristles, or an inflated sac, as in *Carex.*
Persistent — Remaining attached after the growing season.
Petal — One of the modified leaves of the corolla; usually the colorful part of a flower.
Petiole — The unexpanded portion of a leaf; the stalk of a leaf.
Phyllary — An involucral bract in Asteraceae.
Pilose — Having scattered, simple, moderately stiff hairs.
Pinnate — Leaves divided into leaflects or segments along a common axis; a compound leaf.
Pinnatifid — Pinnately cleft to the middle or beyond.
Pistillate — Female-flowered, with pistils only.
Polymorphic — With many forms or variations, as in leaves; variable.
Prickle — A stiff, sharp-pointed outgrowth from the epidermis, as in *Solanum.*
Procumbent — Lying on the ground.
Puberulent — With very short hairs; woolly.
Pubescent — Covered with fine, soft hairs.
Punctate — With translucent dots or glands.
Raceme — An inflorescence, with the main axis bearing stalked flowers, these opening from the base upward.
Racemose — Like a raceme or in a raceme.
Rachilla — The axis in the center of a grass spikelet.
Rachis — The axis of a pinnately compound leaf; the axis of inflorescence; the portion of a fern frond to which the pinnae are attached.
Ray — A marginal flower with a strap-shaped corolla, as in Asteraceae.
Receptacle — The end of the flower stalk, bearing the parts of the flower.
Reniform — Kidney-shaped.
Reticulate — Netted, as veins in leaves; with a network of fine upstanding ridges, as on the surface of spores.
Retrorse — Pointing backwards and downward.
Retuse — Having a bluntly rounded apex with a central notch.
Rhizome — An elongated underground stem, as in ferns.
Rootstock — An elongated underground stem, usually in higher plants.
Rosette — A cluster of leaves, usually basal, as in dandelion.
Sagittate — Arrowhead-shaped.
Samara — A single-seeded, indehiscent fruit, having a winglike extension of the pericarp.
Scaberulous — Slightly scabrous.
Scabrous — Having a surface covered with small wartlike projections; scurfy or rough.
Scale — A highly modified, dry leaf, usually for protection.
Scape — A leafless or nearly leafless stem, coming from an underground part and bearing a flower or flower cluster, as in *Allium.*
Scarious — Dry, thin, and with a dried-up appearance, usually at the tips and edges.
Schizocarp — A fruit that splits up at maturity into mericarps.
Scurfy — Covered with minute, membranous scales, as in *Chenopodium.*
Segment — A division of a compound leaf or of a perianth.
Sepal — One of the members of the calyx.
Serrate — With teeth projecting forward.
Serrulate — Finely serrate.
Sessile — Lacking a petiole or stalk.
Sigmoid — S-shaped.
Silicle — Similar to a silique, but short and broad, never more than four times as long as broad.
Silique — A dry elongated fruit divided by a partition between the two carpels into two sections.
Sinuate — With long, wavy margins.
Sinus — A depression or notch in a margin between two lobes.
Sorus — The fruiting structure in ferns, usually on the underside of the frond.
Spatulate — Widened at the top like a spatula.
Spike — An elongated inflorescence with sessile or nearly sessile flowers.

Spikelet — A small spike; the ultimate flower cluster of the inflorescence of grasses and sedges.
Spine — A short, thornlike organ.
Spinose — With spines, or spinelike.
Spinule — Small spine, giving a prickly effect.
Spinulose — With small, sharp spines.
Spreading — Diverging from the root and nearly prostrate, as the growth habit of Mollugo.
Stamen — The organ in the flower which produces pollen grains.
Staminate — Male-flowered, with stamens only.
Standard — The large petal that stands up at the back of the flower in Fabaceae.
Stellate — Star-shaped.
Stipule — Appendage at the base of a leaf, petal, or other plant part.
Stolon — A basal branch rooting at the nodes.
Stramineous — Straw-colored.
Striate — Marked with fine, longitudinal, parallel lines, as grooves or ridges.
Strigose — Bearing stiff, appressed, or ascending hairs.
Subcaulescent — Nearly tree-like, or having a short stump-like stem.
Subinvolute — Parts rolled inward and under, as leaves, sepals, or petals.
Submersed or Submerged — Growing under water.
Subulate — Awl-shaped.
Taproot — A root system with a prominent main root, bearing smaller lateral roots.
Tendril — A slender, coiling structure.
Terete — Circular in cross section.
Ternate — Three-parted.
Thyrse — A densely branched inflorescence, with the main branching racemose but the lateral branching cymose; a compound panicle.
Tomentose — Covered with dense, wool-like hairs.
Toothed — Dentate.
Trifoliate — A compound leaf with three leaflets, as in legumes.
Trigonous or Trigonal — Three-angled, as the stems of sedges.
Truncate — Terminating bluntly.
Tuber — A modified branch, usually underground and for storage of food.
Tuberous — Thickening and forming tubers.
Ubiquitous — Everywhere; in all types of habitats.
Umbel — A raceme in which the axis has not elongated, so the flower stalks arise at the same point, as in Apiaceae.
Undulate — Wavy, as the margins of leaves.
Veins — The vascular portions of leaves, as in *Galium*, or flowers, as in Potamogeton.
Villosulous — Nearly villous.
Villous — Covered with short, soft hairs; nappy.
Viscid — Sticky.
Whorled — Three or more structures at a node, as leaves, branches, or floral parts.

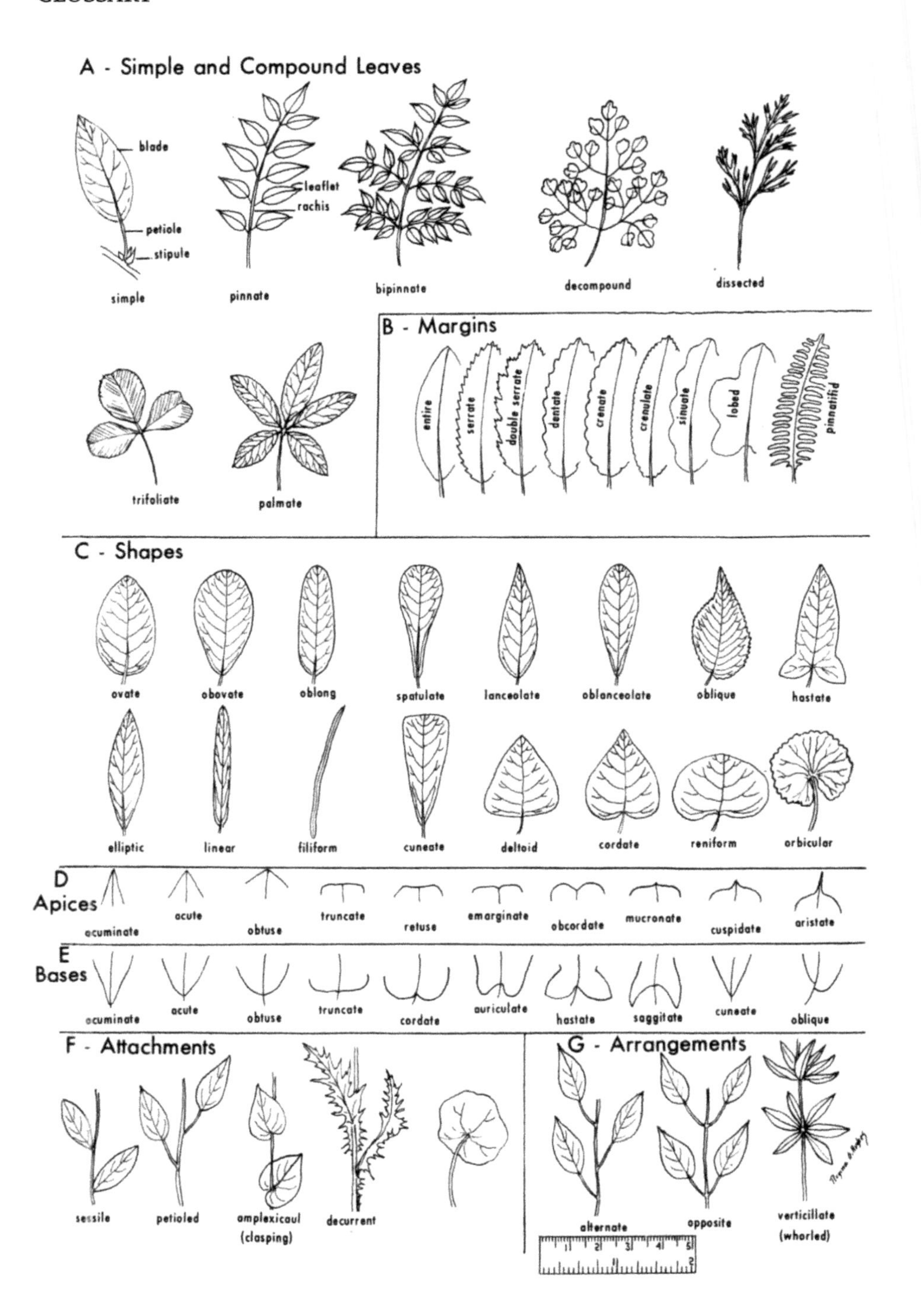

Leaf characters.

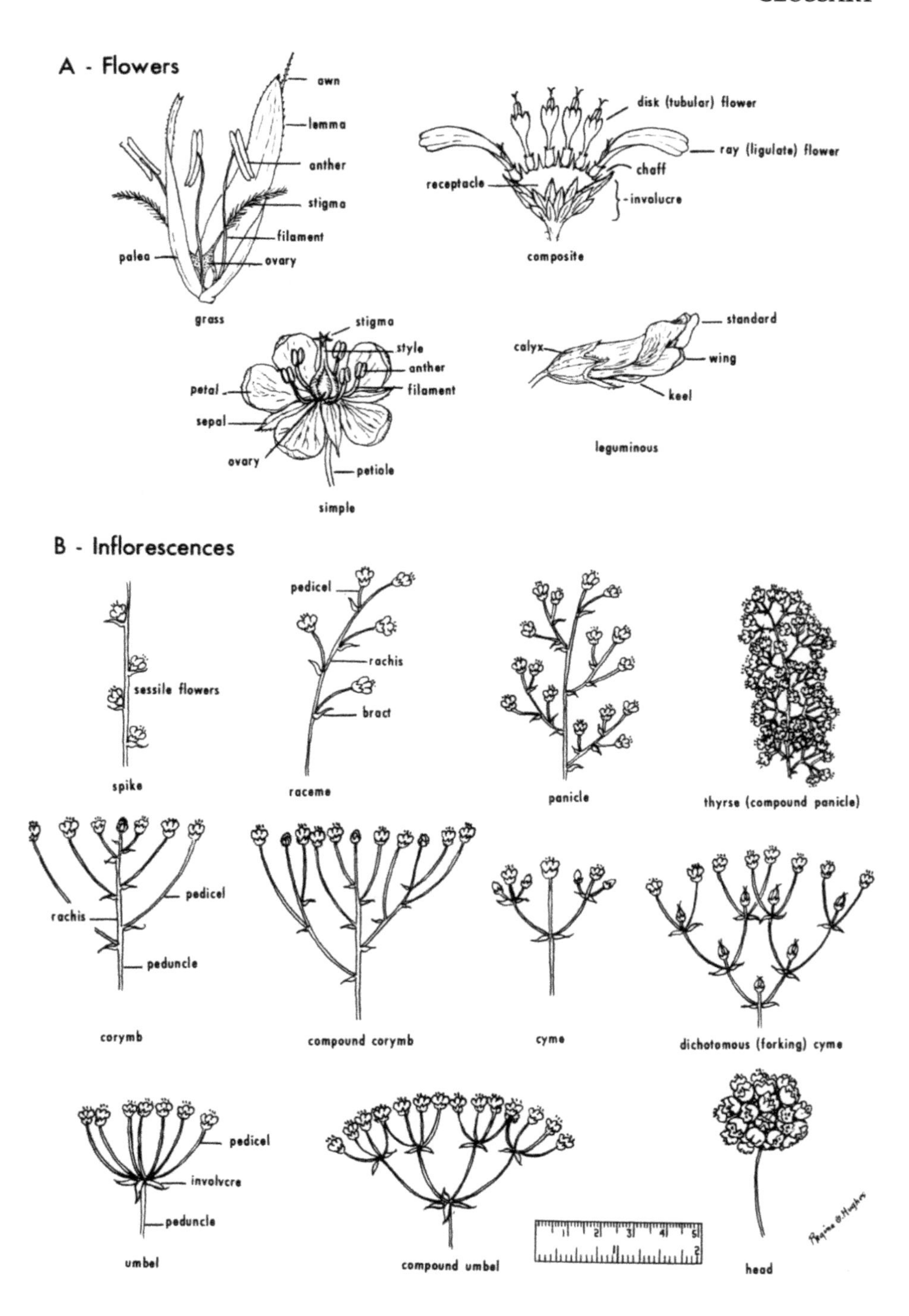

Types of flowers and inflorescences.

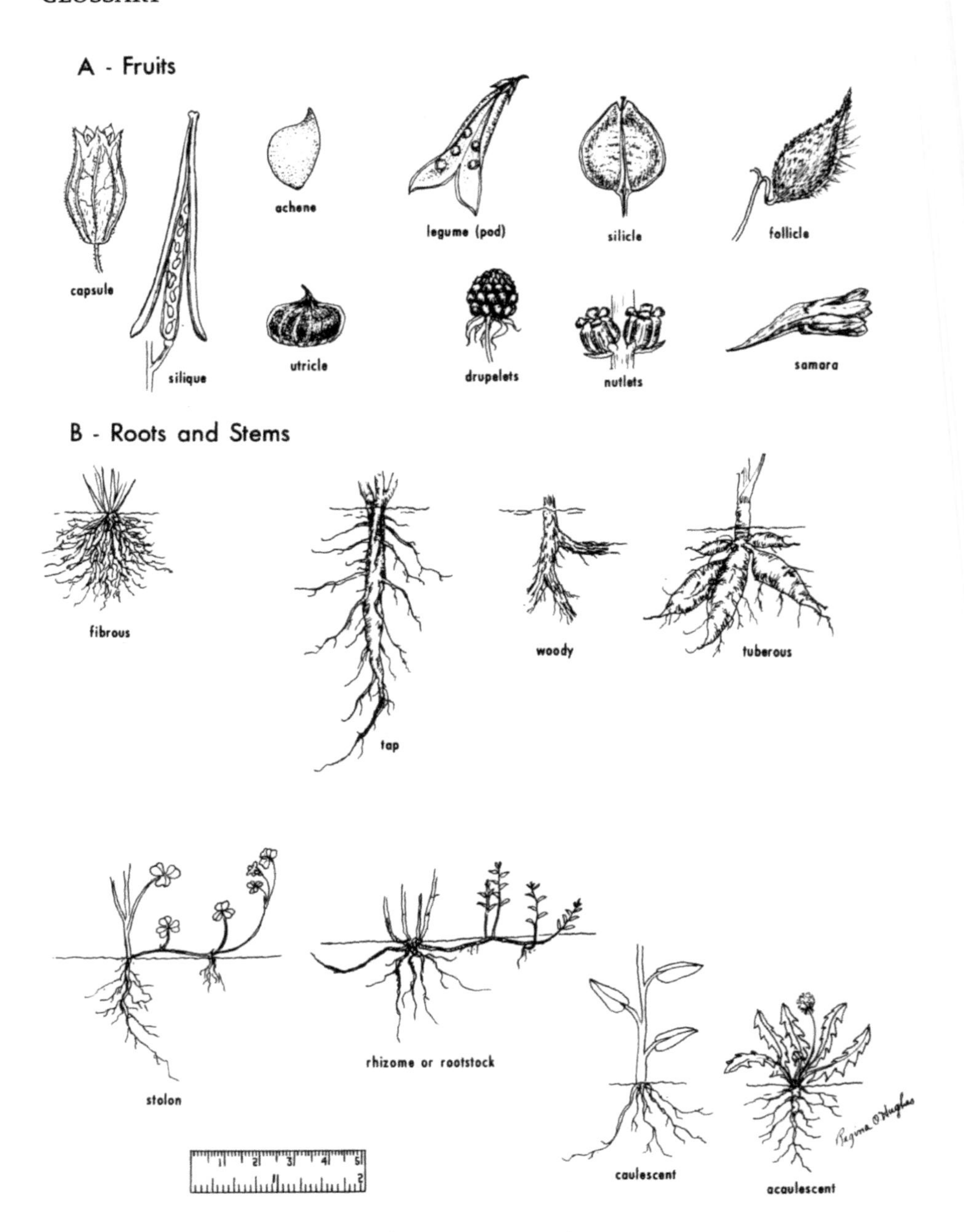

Types of fruits, roots, and stems.

INDEX

NOTE Synonyms are listed in *italics.*

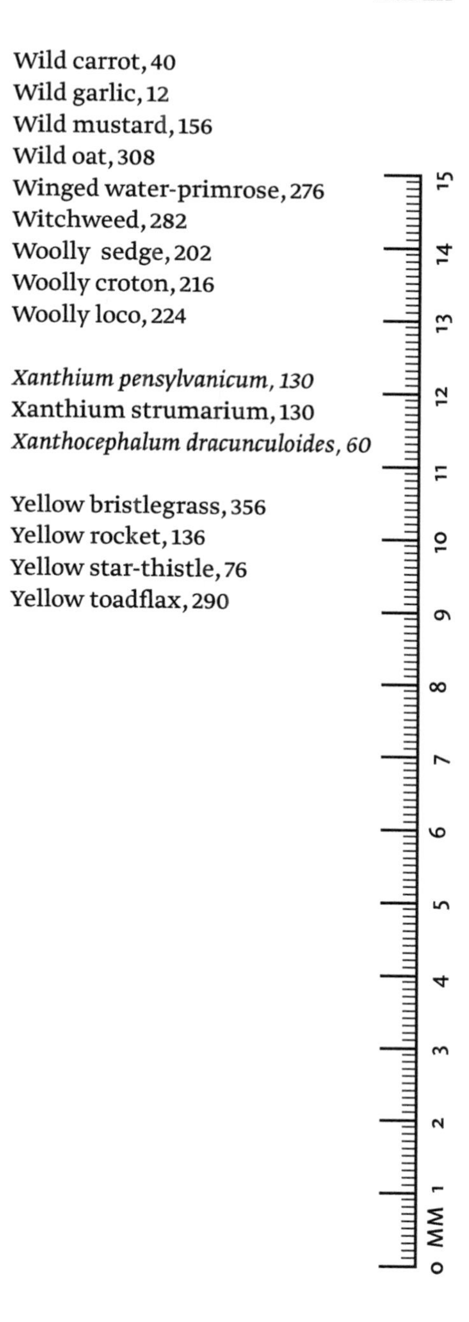
0 MM 1 2 3 4 5 6 7 8 9 10 11 12 13 14 15

CPSIA information can be obtained
at www.ICGtesting.com
Printed in the USA
LVHW050833181221
706563LV00028B/1588

9 781951 682149